Rewiring the "Nation"

Rewiring the "Nation"

The Place of Technology in American Studies

Edited by

Carolyn de la Peña

& Siva Vaidhyanathan

The Johns Hopkins University Press
Baltimore

 Published 2007
Printed in the United States of America on acid-free paper
9 8 7 6 5 4 3 2 1

The Johns Hopkins University Press
2715 North Charles Street
Baltimore, Maryland 21218-4363
www.press.jhu.edu

ISBN 10: 0-8018-8651-1 (pbk.: alk. paper)
ISBN 13: 978-0-8018-8651-5

Library of Congress Control Number: 2006936638

A catalog record for this book is available from the British Library.

For more information about *American Quarterly*, please see:
www.press.jhu.edu/journals/american_quarterly/

Book design by William Longhauser
Front Cover: Wyatt Gallery, *Cadillac,* New Orleans, Louisiana. October 2005. Digital C-Print Edition #1/5 30x38. Courtesy of the artist, Watermark Fine Arts, and www.wyattgallery.com.
Back cover: Group of tourists in front of Panama Canal, Gatun, Panama, ca. 1910.
© Underwood & Underwood / Corbis. Background image: AT&T advertisement, ca. 1920. Courtesy AT&T Archives and History Center, San Antonio.

Contents

Preface

This special issue on the role of technology in American culture aims, as have the previous *American Quarterly* special issues, to examine the intersection of the field of American studies with other fields, in this case technology studies, communication, and the history of technology. The United States has been, throughout its history, a society deeply invested in a belief in technology. As a nation defined by an uncritical embrace of technologies and technological systems, from the conviction that American technological superiority and know-how will reign globally to the often myopic belief in technological solutions as a form of problem solving, the United States is a remarkably techno-centric nation. This special issue presents a diverse array of essays that exemplify the vibrant work being produced on questions of technology by scholars working in and around American studies employing a range of interdisciplinary approaches. These essays demonstrate the crucial importance of historical and cultural analyses of the complex relationship of technologies to individual identities, social relations, and national imaginings.

This issue has been edited with great skill and thoughtfulness by guest editors Carolyn de la Peña and Siva Vaidhyanathan, who have brought together a truly exemplary group of scholars and essays to make clear how significant these questions of technology in and of American culture can be. We are grateful to the *American Quarterly* managing board for shepherding this issue through the review process, and in particular to board members Sarah Banet-Weiser and Josh Kun, and associate editors Katherine Kinney, Barry Shank, and Raúl Villa, who all read and commented extensively on manuscripts at various stages of their development. This kind of labor is necessary to the realization of such a focused work, and the authors of these essays join us in thanking these scholars for this strenuous and careful reading. We would also like to thank the additional members of the managing board, Eric Avila, Mary Dudziak, Judith Jackson Fossett, Greg Hise, Laura Pulido, John Carlos Rowe, Shelly Streeby, Helena Wall, and Henry Yu, and the *American Quarterly* advisory board, who provided helpful advice for this project. The *American Quarterly* staff has worked tirelessly to bring this issue to realization, in particular *AQ* managing editor Michelle Commander, who impressively kept it on track. *AQ* editorial assistant Cynthia Willis and *AQ* interns Jill Somers and Joe Barrett provided invaluable support as well. Stacey Lynn, our copy editor, patiently helped bring the issue to a final, clean state, and Bill Longhauser designed the wonderful cover.

We are grateful to the Program in American Studies and Ethnicity and the College of Arts and Sciences at the University of Southern California, and the Department of Culture and Communication and Dean Mary Brabeck of the Steinhardt School of Education at New York University, whose support helped this project come to fruition.

Curtis Marez
Marita Sturken
Editors
American Quarterly

Rewiring the "Nation"

Introduction
Rewiring the "Nation": The Place of Technology in American Studies

Siva Vaidhyanathan

In early July 2006 the government of North Korea tested a long-range ballistic missile that was supposed to be able to deliver a nuclear warhead as far as Alaska. The missile test did not go well for North Korea. The long-range missile fell into the sea just minutes after launch.[1] President George W. Bush's reaction to the news of the test was to boast that the freshly deployed (albeit limited) U.S. missile defense system would most likely have been able to protect the western continental United States from such a missile. "Yes, I think we had a reasonable chance of shooting [the North Korean missile] down," Bush said at a news conference in Chicago two days after the failed Korean test. "At least that's what the military commanders told me."[2] Just the day before, Bush had reinforced his commitment to a missile defense system. "Because I think it's in—I know it's in—our interests to make sure that we're never in a position where somebody can blackmail us," Bush said at a news conference in Washington, D.C., after meeting with Canada's Prime Minister Stephen Harper. "And so we'll continue to invest and spend. And since this issue first came up, we've made a lot of progress on how to—toward having an effective system. And it's in our interest that we continue to work along these lines."[3]

In these statements, Bush expressed a dangerous level of faith in an unproven technology. Since a missile-defense system first emerged as a vision of President Ronald Reagan in the 1980s, the U.S. government has spent from $2 billion to $10 billion dollars per year on various systems that would track intercontinental ballistic missiles through the stratosphere and send small intercepting vehicles up to disable or destroy the incoming warhead. This plan gathered enough enthusiasm among defense contractors to justify development and experimentation for more than a decade, despite the ease with which any potential attacker could simply evade even the best system (overwhelming the defense with "dummy" or multiple warheads, shifting warheads to low-flying cruise missiles, relying instead on human carriers to deliver warheads in luggage, etc.). Every test of every part of every prototype of missile defense has failed.[4] After repeated embarrassing failures and news accounts of them, the United States merely opted in 2002 to cease testing the system. Despite having no evidence to suggest it might work, the Bush administration activated elements of a system over Alaska and California in response to tensions with North Korea in June of 2006.[5]

Does it matter that the technology is neither empirically viable nor theoretically effective? Such faith in technology in the absence of critical analysis or empirical support is an example of "techno-fundamentalism," the belief that we can, should, and will invent a machine that will fix the problems the last machine caused. It's an extreme form of technological optimism or Whiggishness. Techno-fundamentalism assumes not only the means and will to triumph over adversity through gadgets and schemes, but the sense that invention is the best of all possible methods of confronting problems. Techno-fundamentalism is not the exclusive property of any one political ideology or agenda. Both Thorstein Veblen and Friedrich Hayek expressed unhealthy faith in technologies to solve complex social problems. Veblen, an anticapitalist iconoclast, believed that putting important decisions in the hands of engineers was the surest path to human fulfillment. Hayek, a free-market economist and inspiration for modern conservatism, believed that distributed knowledge and unfettered competition would unleash technological creative forces that would mold human society justly and democratically.[6]

In the United States at the beginning of the twenty-first century we pay a heavy price for techno-fundamentalism. We build new and wider highways under the mistaken belief that they will ease congestion and speed traffic.[7] We rush to ingest pharmaceuticals that might alleviate our ills with no more effectiveness than placebos.[8] We invest based on self-fulfilling phenomena such as "Moore's Law," which falsely predicts that computer processing power will double every eighteen months, as if computer speed were a force of nature above and beyond specific decisions by firms and engineers.[9] Perhaps most dangerously, we maliciously neglect real problems with the structures and devices we depend on to preserve our lives, as we did for decades with the levees that failed to protect the poorest residents of New Orleans.[10] In lieu of deploying deliberation and recognizing complexity at the roots of social and political problems, we operate, it seems, in a techno-fundamentalist cloud, waiting for someone to invent the next great things that can clean up the air, reverse obesity, and magically stop missiles from landing in our cities. We need not depend on messy diplomacy or credible military force to curb the activities of hostile states. We have "Star Wars."[11]

Beyond Techno-fundamentalism

Those of us trained in the history of ideologies, ideas, and cultural processes can and must step up to challenge techno-fundamentalism. It's imperative that we employ our critical faculties to unravel the rhetorical tangles and habits that get us into so much technologically fueled trouble. American studies scholars, with their traditions of public engagement, interdisciplinary thought, and ecological predispositions, are ideally positioned to confront misguided faith in technology and progress.

In his 1959 manifesto, *The Sociological Imagination*, C. Wright Mills instructed social scientists to situate their work between the poles of grand theory and numbing empiricism. "The sociological imagination," Mills wrote, "enables its possessor to understand the larger historical scene in terms of its meaning for the inner life and the external career of a variety of individuals." Thus the "imagineer" (as Mills would never have put it) can "take into account how individuals, in the welter of their daily experience, often become falsely conscious of their social positions." Mills posits three questions, or lenses, that enable scholars to generate interdisciplinary, influential, and—most of all—interesting work: What are the essential components of a society and how are they related to each other? What historical changes are affecting a particular part or function of society? Who are the winners and losers in a society and how did they get to be this way?[12]

Revising and riffing on Mills, this volume offers a variety of examples of the ways scholars of culture are using the study of technology to examine the flows, conflicts, tensions, and hazards of American culture. The articles in this special issue are, in a sense, employing what one might call a "techno-cultural imagination."[13] If a scholar relies on a techno-cultural imagination, she asks these sorts of questions: Which members of a society get to decide which technologies are developed, bought, sold, and used? What sort of historical factors are at work that influence why one technology "succeeds" and another "fails"? What are the cultural and economic assumptions that influence the ways a particular technology works in the world, and what unintended consequences can arise from such assumptions? Technology studies in general tend to address several core questions about technology and its effects on society (and vice versa): To what extent do technologies guide, influence, or determine history? To what extent do social conditions and phenomena mold technologies? Do technologies spark "revolutions," or do concepts like "revolution" necessarily raise expectations and levels of effects of technologies? The essays in *Rewiring the "Nation"* each address such vital questions, which can be answered only by considering the complex interconnections among history, culture, politics, and power. This is not an easy task. But it is one for which American studies scholars are uniquely trained. If we want different terms of debate or lines of reasoning in policies of "progress"—at home and abroad—we need to "rewire" the conversation. The essays in this issue represent a critical mass of current scholarship on the place of technology in American culture.

Technologies of Transcendence

In *A Connecticut Yankee in King Arthur's Court,* Mark Twain has his protagonist, Hank Morgan, assume the position of "boss" in medieval England by exploiting his scientific knowledge and technical prowess. After telling the story of how he took control, Morgan declares, "That reminds me to remark, in passing, that the very first official thing I did, in my administration—and it was on the very first day, too—was to start a patent office; for I knew that a country without a patent office and good patent laws was just a crab, and couldn't travel any way but sideways or backwards."[14] For Morgan, as for James Madison, patents enable progress. Without an administered incentive system, invention would halt. This notion of necessary "progress," in contrast to the "sideways or backwards" vectors of an invention-free society, is key to understanding the relationship among law, technology, commerce, and ideology in America. For that reason, Morgan was not the only new boss who decided to install a patent system as a first move. Article I, Sec. 8, of the U.S. Constitution instructs Congress to create a copyright and patent system "to promote the progress of the sciences and useful arts."[15] So from the beginning of the republic, Americans have built their "imagined community" around faith in the idea of progress.[16]

In the opening essay of this collection, "Technology and Its Discontents: On the Verge of the Posthuman," Joel Dinerstein shows us that uninterrogated faith in the progressive promise of new technologies tends to reinforce two crucial and dangerous myths: that of Whiggish "progress," and that of inevitable (some might say "natural") white and Western superiority. "The real questions we need to confront," Dinerstein writes, "are these: What is progress for? What is technology for?" Dinerstein issues a stern call for engagement by American studies scholars, particularly those who consider the privileges of whiteness to be central to an understanding of the dynamics of power within American culture, with the rhetorics and realities of technology. Dinerstein posits that the "posthuman" denies the "panhuman." In other words, technologically driven concepts of human progress via genetic engineering or mechanical intervention is an attempt by technologists to solidify a future "man" according to arbitrary standards of whiteness and maleness to combat the increasing multiplicity of human types and the multicultural face of human power.

There is much at stake in who designs technologies and what particular vision of "progress" they achieve (and for whom). David Nye's essay "Technology and the Production of Difference" argues that technologies can, simultaneously, restrict and enable individual freedoms. Nye shows us that those who have argued since the rise of mass media that communicative technologies would necessarily generate cultural and political conformity overstated their case. But just as important, Nye paints as naive those who see the newest forms of communicative technologies—digital content and global networks—as necessarily enriching and diversifying the human experience. In general, Nye argues, cultural groups use new technologies (communicative and other) to shape identity, construct and maintain distinction and diversity, and customize their life experiences. But the power to customize, Nye reminds us, is not the power to shape one's conditions. "The challenge for American studies," Nye writes, "is not only to examine how technologies have been incorporated into cultures of difference, but also to prepare students to take part in the social construction of emerging technologies. Too often these are left to the private sector, as if the market alone can adjudicate the best uses of new machines."

While Nye dissolves any grand claims we might make about the progressive and transformative nature of communicative technologies, Susan Douglas explodes them. In "The Turn Within: The Irony of Technology in a Globalized World," Douglas introduces us to "the irony of technology." No matter how potentially transformative or liberating a particular technology may seem, "the economic and political system in which the device is embedded almost always trumps technological possibilities and imperatives." Douglas helps us understand how video technology, once touted as a way to put a camera in the hands of individuals and diversify our views of the world has instead enabled a diverse group of individuals to collectively mistake their navels for the world. Technological devices can indeed enable individuals to challenge and transform power structures, but only if they "work" within systems that will allow it. Our challenge is to see the system and the material when we look through technological lenses: we need to look both to the television and to the constant negotiations and synergies among technologies, economics, politics, and ideologies.

The winners in these negotiations have had a disproportionate say over which technologies matter in our lives. As Dinerstein, Nye, and Douglas point out, American studies scholars can help us more fully understand the cost when particular ideologies of technological "progress" win out over others. They can also help us look outside the techno-fundamentalism box and discover alternative systems of technological meaning making. And certainly, there are more important technological divides in our present condition than the digital divide. In "Say It Loud, I'm Black and I'm Proud: African Americans, American Artifactual Culture, and Black Vernacular Technological Creativity," Rayvon Fouché proposes a "black vernacular technological creativity" that moves beyond standard and reductive notions of African American expressive culture limited to dance, music, literature, the visual arts, and athletic mastery. When one surveys the "great works" of technological history, theory, and sociology, "it would appear as if African Americans, throughout American history, did not have the ability to make technological decisions of their own and have led lives in which technology was foisted upon them." Fouché offers us the example of white plantation owner Oscar J. E. Stuart's attempt to patent the double cotton scraper that his slave Ned invented. The U.S. Patent Office made it abundantly clear that inventions by slaves were not worthy of patent protection. Thus Africans were systematically omitted from the grand narrative of American progress that defined success and citizenship. Fouché guides us through a set of problems and questions that would "produce a more textured understanding of the roles that black people have played as producers, shapers, users, and consumers of technology within American society and culture."

The Cultural Work of Technological Systems

As my coeditor, Carolyn de la Peña, explains in her overview of the literature in the field in this volume, the historian Thomas Hughes first articulated the "systemic" approach to studying technologies. Studying particular devices or artifacts as distinct phenomena disconnected from the economic, political, and cultural ecosystems that join and motivate us necessarily fails to explain the impact of technology on everyday life, Hughes argued.[17] Ricardo D. Salvatore reveals the importance of this approach in his article, "Imperial Mechanics: South America's Hemispheric Integration in the Machine Age." The Panama Canal was a technological marvel, but considering it outside its systems of economic, labor, and commodity flows both between the Atlantic and Pacific Oceans (the obvious role of the canal) and between North and South America (the implicit vector of such flows) misses the big issues. Salvatore introduces two new and powerful theoretical constructs, that of "spectacular machines," such as the locks and gates of the Panama Canal, and of "transportation Utopias," such as the road system that facilitates commerce and migration up and down the two continents. By envisioning the hemispheric engineering of "connectivity," via such projects as canals, highways, and airports, Salvatore shows how U.S. policy within the Americas was built on, by, and for machines and the infrastructures that supported them. Systems, of course, need not be grand structures of concrete and asphalt.

In "Precision Targets: GPS and the Militarization of U.S. Consumer Identity," Caren Kaplan also suggests that we can best understand the influence of technology by looking to systems of information flow and management rather than specific objects. Her argument focuses on geographic information systems (GIS) and global positioning systems (GPS). With its use in activities from military targeting to bass fishing, GPS has become ubiquitous in modern America, just as precise and powerful geodemographic data about U.S. residents has allowed for both commercial and political exploitation of our presumed desires and habits. Kaplan raises the concern that "discourses of precision" promote the rather anti-Kantian public ethic of viewing citizens as means rather than ends. "Yet even as these modes of identification promise greater flexibility and pleasure through proliferation of 'choices' among myriad specificities," Kaplan writes, "they also militarize and thus habituate citizen/consumers to a continual state of war understood as virtual engagement." So technologies that we purchase as tools of access, choice, opportunity, and freedom, Kaplan asserts, actually acculturate us to an invisible rigidity by keeping us always logged on, always present and accounted for.

In contrast to Kaplan's exploration of invisible structures tethered to particular technologies, Robert MacDougall's article, "The Wire Devils: Pulp Thrillers, the Telephone, and Action at a Distance in the Wiring of a Nation," connects explicit expressions to visible technologies of mobility and commerce. MacDougall examines a set of pulp novels from the late nineteenth century that reveal the state of American anxiety over "nightmares of 'reach.'" Americans, MacDougall writes, were keenly aware of being enveloped by a network of connection unprecedented in human history: railroads, petroleum pipelines, telephone and telegraph wires, and invisible corporate organs. These "Wire Devil" novels illustrate that "the pace of technological and economic change was indeed violent and wrenching to many Americans," MacDougall writes. "Each advance in the technology of communication and transportation gave new powers to its users, yet also compounded the ability of distant people and events to affect those users' lives." This was a new way of being in the world. The terms of such new powers (and weaknesses) were worked out through literature and conversations about literature.

Technology and Knowledge Systems

Knowledge systems often depend intimately on the technologies that facilitate them. Knowledge systems might include universities, professions, libraries, or entire industries. They are nexuses of functions such as commercial practices and markets, cultural judgments and norms, legal regimes, and social capital. And since the rise of global digital networks, most of the major knowledge systems in our lives have been in flux.

Technologically framed battles over copyright have risen in prominence and importance in recent years. They play a large part in both public and scholarly conversations about the future of knowledge and communication.[18] Yet as Andrew Ross explains in his essay, "Technology and Below-the-Line Labor in the Copyfight over Intellectual Property," it's time to move beyond the liberal (or in many cases libertarian) frame through which too many scholars (myself included) have presented the conflicts and tensions over copyright, technology, and the regulation of global information flows. Lost in the debate over the culture and commerce of copyright is the status of workers. "The crusade against the [intellectual property] monopolists continues to be dominated by strains of techno-libertarianism that lie at the doctrinal core of the 'information society,'" Ross writes, "obscuring the labor that built and maintains its foundations, highways, and routine production." The consequence of this narrow conception of what is at stake in these policy debates, Ross declares, consists of "voices proclaiming freedom in every direction, but justice in none."

What happens when a knowledge system works, but bureaucracies fail? In "Failing Narratives, Initiating Technologies: Hurricane Katrina and the Production of a Weather Media Event," Nicole R. Fleetwood reads the U.S. Congress's report on governmental failures when Hurricane Katrina hit the U.S. Gulf Coast in August 2005 as an exercise in techno-cultural bad faith: the tendency to look at technological failures and solutions instead of examining human catalysts and responses. The horrors of Katrina, the congressional report concludes, were a result of a series of technological failures both large and small. This conclusion, Fleetwood asserts, shifts blame from the human failures that underlay the collapses of communication and transportation that left so many people stranded and endangered. Instead of examining the complete and systematic breakdown in the social and political fabric, enabled by the malign neglect of many generations of political leaders at every level who have consistently shown themselves willing to sacrifice thousands of African Americans at times of stress, the congressional report focuses on the shortage of cars and buses in metropolitan New Orleans. "Through the Katrina event," Fleetwood writes, "we see both the vulnerability and recalcitrance of the nation-state's investment in a deterministic narrative of progress, one in which technology is at its core and the marginalization and disposability of certain populations are essential." Examining Katrina ecologically—with attention paid to inputs, outputs, flows, dynamics, and long-term problems—would yield a better understanding of complex phenomena and ways to deal with their repercussions. The analytical spirit of environmentalism is a healthy response to technological bad faith.

Fields of knowledge can intersect when examining complex phenomena that do not hold to the boundaries we have drawn in our perceptions. In "Boundaries and Border Wars: DES, Technology, and Environmental Justice," Julie Sze connects the history of technology to environmental justice by examining a particular environment that has been the subject of much reckless technological experimentation, women's bodies. Her account of the effects of diethylstilbestrol (DES) on women and fetuses reveals the extent to which human bodies are "technologically polluted." Sze concludes, "beyond knowledge and activism, we need a cultural and political analysis and a vocabulary that makes sense of DES's roots and impacts, which American studies can provide." Because all the standard categories of environmental analysis are porous (mother and fetus, human and animal, production and consumption, and environmental and technological), a properly ecological examination of a phenomenon such as DES demands an interdisciplinary account that goes beyond cultural analysis, medical analysis, or technological analysis. It requires a combination of all these fields.

Offices can be ecosystems as well—complex systems of machines, bodies, information, inputs, and outputs. And thus they demand a full ecological examination when trying to make sense of the rapid changes within commercial and financial institutions. Caitlin Zaloom documents some radical changes in financial markets in her article, "Markets and Machines: Work in the Technological Sensoryscapes of Finance." Zaloom takes us to the trading floors of commodity exchanges in Chicago and London to show how firms are adopting remarkable technologies to create new market environments. By paying particular attention to the ergonomic, environmental, and sensory conditions of their employees, trading firms are standardizing the experience of commodity trading while opening opportunities to classes and ethnicities previously excluded. This is not necessarily an innocent or progressive story, however. Zaloom demonstrates that the sensory regulation these firms employ serve the expansion of a U.S-centric neoliberal ideology. There are costs as well as benefits to the shift from the traditional cacophonous trading floor to the serene cubicle.

Technology, Mobility, and the Body

American studies scholars are at the forefront of connecting the corporeal to the mechanical. In "Educating the Eye: Body Mechanics and Streamlining in the United States, 1925–1950," Carma R. Gorman revises the history of streamlining and industrial design and demonstrates that humans were made by revealing the influence of the educational and public health movement to improve American posture. Both producers and consumers of the new streamlined products were well versed, Gorman shows, in the "body mechanics" school, which emphasized "good form." Through this study, Gorman shows that it is insufficient to examine a technology or a design phenomenon as an artifact alone. One must study the cultural and ideological system in which the artifact operates. "Further," Gorman writes, "paying attention to the body mechanics literature explains more directly than many previous theories do why technological products that looked like human bodies (i.e., streamlined, formerly artless goods) would have resonated with consumers: they had been taught, through body mechanics instruction, to understand both bodies and formerly artless goods as 'mechanical,' and as products of similar Taylorist technologies of improvement."

Vision and the sensory experience of transport are essential corporeal experiences. Technologies that move our bodies can impact how we interpret our environments and the time lines we impose on them. In "Farewell to the El: Nostalgic Urban Visuality on the Third Avenue Elevated Train," Sunny Stalter delves into the almost instant nostalgia that arose in the 1950s once New York City abandoned and destroyed its elevated train system in favor of buses and subways. Stalter argues that the elevated train "structured New Yorkers' knowledge of urban space in a way that was no longer possible on other modes of transportation." At a time of rapid technological and urban change, Stalter writes, "the El showed postwar artists a city dissociated from progress, haunted by machines of the past and visions of modernity from seventy-five years ago." The defining aspect of the El, Stalter declares, "was its openness to city space." Thus in its waning days, the El's importance to New York City was less about mobility than about visuality.

Transporting and manipulating one's body become part of the same process when people seek partners, opportunities, and new lives through online dating services that connect Latin American women with North American men. In "Flexible Technologies of Subjectivity and Mobility across the Americas," Felicity Schaeffer-Grabiel connects women's discourses of self-improvement to technologies and the marketing methods that spread the "promise of modernity and mobility" by investigating why they see the need to surgically alter their bodies while engaging in online match-making. For the Latin American women Schaeffer-Grabiel has interviewed, there seems to be a unified sense of ascendancy connected to such technologies as Internet connections with potential partners, life in the North, and plastic surgery. Using ethnography as a means to push beyond the liberatory analyses of flexible subjectivity that have marked early studies of identity and the Internet, Schaeffer-Grabiel argues that "the concept of transcending the body or recombining one's identity in Cyberspace is a privileged position that elides the labor of the body and asserts neoliberal values of choice and democratic nations of upward mobility." At a time when U.S. borders appear less permeable than ever, Schaeffer-Grabiel shows that there is nothing playful about malleable identities and neoliberal fictions of both upward and northward mobility.

American Studies and the Techno-cultural Imagination

The techno-cultural imagination need not be anti-technology. Far from it. A healthy and effective attitude toward technology demands an appreciation of the pragmatic value of inventions, devices, and conveniences tempered by a critical stance toward the ways they are promoted, marketed, and adopted. Debra DeRuyver and Jennifer Evans deploy the techno-cultural critical stance in their review essay of electronic resources for American studies research. While walking through the rich garden of primary sources for cultural scholarship, they warn against relying on these resources without subtle human guidance. "Regardless of format, it takes time to locate primary sources," DeRuyver and Evans write, "and while search engines, like card catalogs, are useful, they cannot totally replace a well-informed librarian or the human thought behind a portal site." In that spirit, they provide a service to all American studies scholars by guiding us through the most valuable sites and the best ways to use them.

As Carolyn de la Peña writes, with this issue of *American Quarterly* we are attempting to "revitalize a 'technology studies' core" within American studies. Technology was for a brief time at the center of a particular—almost quaint—notion of American exceptionalism that orbited the "myth and symbol" school of cultural history.[19] But for most of the past thirty years the most exciting moves within American studies have included fresh analyses of ethnicity and diversity within the United States and flows of people, cultural processes, and ideologies across borders and within borderlands. As de la Peña asserts in her essay, "American studies has largely left questions of technology to others, in spite of our early leadership in innovative methods of technological analysis and cultural critique." But, as she rightly points out, because of the emphases within American studies on questions of identity, power, and ideology, American studies is the ideal scholarly community to ask powerful and poignant questions about technology and its effects on our bodies, eyes, ears, minds, families, economies, armies, and academies. Our current political and economic conditions are overdetermined by techno-fundamentalism (as well as too many other fundamentalisms). The essays in this collection are examples of healthy ways to think through technological, social, and cultural problems. The unhealthy ways are too dominant in our daily deliberations. Technology is an entry point for American studies scholars to engage in public conversations over issues ranging from race to labor to war. Such contributions are imperative.

*****

Along with many other scholars of American studies, communication, and cultural studies, Carolyn de la Peña and I were saddened by the news of the passing of James Carey in May 2006. Jim was a brilliant and influential scholar. His work had a profound effect on both of us when we first encountered it in graduate school. He influenced many of the questions we asked in our own work and guided how we would grow to view the relationships between technology and culture. Jim will be best remembered, however, as a passionate and influential teacher who has guided the intellectual pursuits of dozens of important American scholars. He remains a role model and an influence to many. We dedicate this collection to his legacy and his memory.

Notes

 I wish to thank Marita Sturken and Carolyn de la Peña for their helpful comments on this article. Working with both of them over the months leading up to the completion of the issue taught me much about their strengths and my limitations. Marita Sturken deserves particular praise for her patience and wisdom, and for the leadership and vision she has demonstrated as editor of *American Quarterly* these past three years. For many discussions leading up to the conception of this issue and the critical sensibility that made it possible, I must thank Joel Dinerstein, Shelley Fisher Fishkin, and Jeffrey Meikle.

1. Dana Priest, "North Korea Tests Long-Range Missile: Controversial Rocket Fails as Other Types Are Fired; U.N. Session Set after U.S., Japan Condemn Action," *Washington Post*, July 5, 2006.

2. George W. Bush, "President Bush Holds a News Conference in Chicago," July 7, 2006, *Washington Post*, at http://www.washingtonpost.com/wp-dyn/content/article/2006/07/07/AR2006070700727.html (accessed July 8, 2006).
3. George W. Bush, "President Bush Meets with the Canadian Prime Minister," *Washington Post*, July 6, 2006, at http://www.washingtonpost.com/wp-dyn/content/article/2006/07/06/AR2006070601022.html (accessed July 8, 2006).
4. BBC, "Missile Defence Shield Test Fails," December 15, 2004, at http://news.bbc.co.uk/2/hi/americas/4097267.stm (accessed July 8, 2006). For a comprehensive analysis of the problems with missile defense in general, including the ease with which an aggressor might evade or fool even an effective system, see Steven Weinberg, "Can Missile Defense Work?" *New York Review of Books*, February 14, 2002. For an analysis of the steady degradation of the standards of testing elements of the missile defense systems, see Andrew M. Sessler, "Countermeasures: A Technical Evaluation of the Operational Effectiveness of the Planned U.S. National Missile Defense System" (Cambridge, Mass.: Union of Concerned Scientists/Massachusetts Institute of Technology, 2000).
5. Bill Gertz, "N. Korean Threat Activates Shield," *Washington Times*, June 20, 2006.
6. Thorstein Veblen, *The Instinct of Workmanship, and the State of the Industrial Arts* (New York: Macmillan, 1914), and *The Engineers and the Price System* (New York: B. W. Huebsch, 1921); Friedrich A. von Hayek, *The Counter-Revolution of Science* (Glencoe, Ill.: Free Press, 1952), and *The Road to Serfdom* (Chicago: University of Chicago Press, 1944).
7. Martin Wachs, *Curbing Gridlock: Peak-Period Fees to Relieve Traffic Congestion* (Washington, D.C.: National Academies Press, 1994).
8. Bo Carlberg, Ola Samuelsson, and Lars Hjalmar Lindholm, "Atenolol in Hypertension: Is It a Wise Choice?" *The Lancet* 364.9446 (November 6, 2004).
9. Gordon E. Moore, "Cramming More Components onto Integrated Circuits," *Electronics* 38.8 (1965). For a critical analysis of Moore's law, see Ilkka Tuomi, "The Lives and Death of Moore's Law," *First Monday* 7.11 (November 2002).
10. Timothy H. Dixon et al., "Space Geodesy: Subsidence and Flooding in New Orleans," *Nature* 441.7093 (2006). Also see Ivor van Heerden, *The Storm: What Went Wrong and Why during Hurricane Katrina—the Inside Story from One Louisiana Scientist*, ed. Mike Bryan (New York: Viking, 2006).
11. For an excellent historical account of the follies of missile defense and the ideologies and corruptions that have kept the dream alive through two decades and billions of dollars, see Frances FitzGerald, *Way out There in the Blue: Reagan, Star Wars, and the End of the Cold War* (New York: Simon & Schuster, 2000).
12. C. Wright Mills and Todd Gitlin, *The Sociological Imagination* (New York: Oxford University Press, 2000).
13. In another context I have used "techno-cultural imagination" to describe the conditions and habits that contemporary artists have enjoyed since the dissemination of digital technologies and networks. See Siva Vaidhyanathan, "The Technocultural Imagination," in *2006 Whitney Biennial: Day for Night*, ed. Chrissie Iles et al. (New York: H. N. Abrams/Whitney Museum of American Art, 2006).
14. Mark Twain and Bernard L. Stein, *A Connecticut Yankee in King Arthur's Court* (Berkeley: University of California Press in cooperation with the University of Iowa, 1983).
15. For a valuable historical account of the patent clause and its importance to the early republic, see Doron S. Ben-Atar, *Trade Secrets: Intellectual Piracy and the Origins of American Industrial Power* (New Haven, Conn.: Yale University Press, 2004).
16. Benedict R. Anderson, *Imagined Communities: Reflections on the Origin and Spread of Nationalism*, rev. ed. (London: Verso, 1991).
17. Thomas Parke Hughes, *American Genesis: A Century of Invention and Technological Enthusiasm, 1870–1970* (Chicago: University of Chicago Press, 2004), and *Human-Built World: How to Think about Technology and Culture, Science, Culture* (Chicago: University of Chicago Press, 2004).
18. Siva Vaidhyanathan, "Critical Information Studies," *Cultural Studies* 20.2–3 (March–May 2006).
19. Leo Marx, *The Machine in the Garden: Technology and the Pastoral Ideal in America*, 2d ed. (New York: Oxford University Press, 2000). To put Marx and his work in the context of both the history of technology studies and American studies, see Jeffrey L. Meikle, "Leo Marx's *The Machine in the Garden*," *Technology and Culture* 44.1 (2003), and "Reassessing Technology and Culture," *American Quarterly* 38.1 (1986).

Technology and Its Discontents: On the Verge of the Posthuman

Joel Dinerstein

Immediately after 9/11, a Middle East correspondent for *The Nation* summarized the coming war on terrorism as "[their] theology versus [our] technology, the suicide bomber against the nuclear power."[1] His statement missed the point: *technology is the American theology*. For Americans, it is not the Christian God but technology that structures the American sense of power and revenge, the nation's abstract sense of well-being, its arrogant sense of superiority, and its righteous justification for global dominance. In the introduction to *Technological Visions*, Marita Sturken and Douglas Thomas declare that "in the popular imagination, technology is often synonymous with the future," but it is more accurate to say that technology is synonymous with *faith in* the future—both in the future as a better world and as one in which the United States bestrides the globe as a colossus.[2]

Technology has long been the unacknowledged source of European and Euro-American superiority within modernity, and its underlying mythos always traffics in what James W. Carey once called "secular religiosity."[3] Lewis Mumford called the American belief system "mechano-idolatry" as early as 1934; a few years later he deemed it our "mechano-centric religion." David F. Noble calls this ideology "the religion of technology" in a work of the same name that traces its European roots to a doctrine that combines millenarianism, rationalism, and Christian redemption in the writings of monks, explorers, inventors, and NASA scientists. If we take into account the functions of religion and not its rituals, it is not a deity who insures the American future but new technologies: smart bombs in the Gulf War, Viagra and Prozac in the pharmacy, satellite TV at home. It is not social justice or equitable economic distribution that will reduce hunger, greed, and poverty, but fables of abundance and the rhetoric of technological utopianism. The United States is in thrall to "techno-fundamentalism," in Siva Vaidhyanathan's apt phrase; to Thomas P. Hughes, "a god named technology has possessed Americans." Or, as public policy scholar Edward Wenk Jr. sums it up, "we are . . . inclined to equate technology with civilization [itself]."[4]

Technology as an abstract concept functions as a *white mythology*. Yet scholars of whiteness rarely engage technology as a site of dominant white cultural practices (except in popular culture), and scholars of technology often sidestep the subtext of whiteness within this mythos. The underlying ideology and cultural practices of technology were central to American studies scholarship in its second and third generations, but the field has marginalized this critical framework; it is as if these works of (mostly) white men are now irrelevant to the field's central concerns of race, class, gender, sexuality, and ethnic identity on the one hand, and power, empire, and nation on the other. In this essay I will integrate some older works into the field's current concerns to situate the current posthuman discourse within an unmarked white tradition of technological utopianism that also functions as a form of social evasion. By the conclusion, I hope to have shown that the *post*human is an escape from the *pan*human.

This is an important moment to grapple with the relationship of technology and whiteness since many scientists, inventors, and cognitive philosophers currently hail the arrival of the "posthuman." This emergent term represents the imminent transformation of the human body through GNR technologies—G for genetic engineering or biotechnology, N for nanotechnology, and R for robotics. "The posthuman," as N. Katherine Hayles defined it in *How We Became Posthuman* (2000), "implies not only a coupling with intelligent machines but a coupling so intense and multifaceted that it is no longer possible to distinguish meaningfully between the biological organism and the informational circuits in which the organism is enmeshed." To be reductive, the posthuman envisions the near future as one in which humans are cyborgs—in which the human organism is, for all practical purposes, a *networked being* composed of multiple human-machine interfaces. Underlying cultural beliefs in technological determinism matched with the inalienable right of consumer desire will soon produce what even cautious critics call "a social transformation" at the level of the individual body, as consumers purchase genetic enhancements (to take one example). In other words, steroids, cloning, gene mapping, and surgical implants are just the tip of an iceberg that, when it melts, will rebaptize human beings as cyborgs.[5]

William J. Mitchell calls this new self-concept "Me++"—a pun on the computer language C++—and claims this future is already present. When Mitchell claims to "routinely exist in the condition . . . [of] 'man-computer symbiosis,'" or that he "now interact[s] with sensate, intelligent, interconnected devices scattered throughout my environment," who can argue with him? An eminent design theorist and urban planner at MIT, Mitchell breezily describes a near future of "high-tech 'wearables'" with implanted computers (e.g., clothes, eyeglasses, shoes) that extend our sense of self over an increasingly permeable body surface. If each person is "jacked in" to dozens of computers within a "few millimeters" of the human shell, will that transform human nature (as many GNR enthusiasts claim)? As Mitchell declares, "increasingly I just don't think of this as computer interaction," but as something like an expansive self. "Me++" is a consumer gold rush: the evolution of the fragile human body into a silicon-based cyborg with superhuman capacities. Here's a complementary—and unexceptional—claim from Rodney A. Brooks, the chair of the Artificial Intelligence Lab at MIT: "We are about to become our machines . . . [we] will morph into machines." Brooks admits this process may bring short-term metaphysical confusion, but he assures readers in *Flesh and Machines: How Robots Will Change Us* that GNR technologies will bring long-term progress.[6]

What do claims for "man-computer symbiosis" have to do with whiteness and religion? Brooks and Mitchell are technological determinists for whom the blithe morphing of the human organism into cyborgs recapitulates the Western tendency to universalize its own perspective. Their works consider the coming of GNR technologies as inevitable, progressive, and beneficial, and their rhetoric assumes universal, equitable distribution of such changes. Moreover, their disregard of social realities perpetuates an unspoken racialized (white) narrative of exclusion that treats technology as an "autonomous" aspect of cultural production illuminating the road to a utopian future that will not require social or political change.[7]

Technological progress has long structured Euro-American identity, and it functions as a prop for a muted form of social Darwinism—either "might makes right," or "survival of the fittest." Here is the techno-cultural matrix: progress, religion, whiteness, modernity, masculinity, the future. This matrix reproduces an assumed superiority over societies perceived as static, primitive, passive, Communist, terrorist, or fundamentalist (depending on the era). The historian of technology Carroll Pursell points out that "the most significant engine and marker" of modernity is "technology ([which is] almost always seen as masculine in our society)," and that only the West invokes modernity as "a signal characteristic of its self-definition."[8] In *Machines as the Measure of Man: Science, Technology, and Ideologies of Western Dominance*, Michael Adas traced the rhetoric of technology as it became the primary measure of intelligence, rationality, and the good society, supplanting Christianity for nineteenth-century colonial powers. Weapons, mass production, and communication networks became the fetishes of colonial dominance and racial superiority, which were disseminated (for example) in numerous British best sellers through binary opposites of dominance/passivity: "machine versus human or animal power; science versus superstition and myth; synthetic versus organic; progressive versus stagnant."[9] Such oppositions still inform contemporary theories of Western superiority (e.g., "the clash of civilizations," "the end of history"). Casting preindustrial (or premodern) peoples as risk-averse and enslaved to obsolescent ideologies—that is, as not progressing—sentences them to second-class status with regard to the future.

Sturken and Thomas ask two crucial questions about the role of technology in the American cultural imagination: "Why are emergent and new technologies the screens onto which our culture projects such a broad array of social concerns and desires?," and consequently, "Why is technology the object of such unrealistic expectations?" I extrapolate the following two answers from the field's critical framework, by way of Leo Marx, Kasson, Nye, Carey, and Noble (among many others). *New technologies help maintain two crucial Euro-American myths: (1) the myth of progress and (2) the myth of white, Western superiority.*[10]

In a given society, a myth functions as "a play of past paradigm and future possibility," according to Laurence Coupe's study, an act of "remembering and re-creating the sacred narratives of the past." Progress secularized the idea of Christian redemption by inventing (and instantiating) a near-sacred temporal zone—the future—to contain its man-made utopian dreams. A myth cannot be declared in rational terms; it "resist[s] completion" in order to keep up its "dialectic . . . of memory and desire, of ideology and utopia." For a myth to have cultural force, it must be unarticulated; it works as "a disclosure rather than . . . a dogma," an opening into unspoken systems of belief.[11]

Technological progress is the telos of American culture, the herald of the future, the mythic proof in the nation's self-righteous pudding. "Nowhere . . . can we find a master narrative so deeply entrenched in popular imagination and popular language as the mythic idea of progress," notes the historian of technology John Staudenmeier, "particularly technological progress." Yet at the intellectual level, historians Carl Becker and J. B. Bury deconstructed the myth nearly a century ago. Becker even identified progress as a covalent religion at the 1935 Stanford lectures: "the word Progress, like the Cross or the Crescent, is a symbol that stands for a social doctrine, a philosophy of human destiny." For both Bury and Becker, the myth of *social* progress emerged from the Enlightenment idea of the perfectibility of man through the application of reason. That man-made future would be "a more just, more peaceful, and less hierarchical republican society based on the consent of the governed." Instead, over two centuries, technology has piggybacked onto social progress by creating the *rush* of change without social improvement.[12]

"We have confused rapidity of change with advance," John Dewey wrote in a 1916 essay titled "Progress," and four years later noted that "these four facts[—]. . . natural science, experimentation, control, and progress[—]have been inextricably bound up together." For Dewey, the "attitude . . . toward change" itself had changed during the Enlightenment due to scientific advance. What had once been "the Christian idea of the millennium of good and bliss" had been reworked into a man-made ideal spoken of "under the names of indefinite perfectibility, progress, and evolution." The result was that "the Golden Age for the first time in history was placed in the future instead of at the beginning." Once the future replaced heaven as the zone of perfectibility—as powered by technology—"progress" began to function as a religious myth that substituted a sacralized temporal zone (the future) for a sacred spatial one (Heaven).[13]

The sacraments of this belief system are new technological products. The presumption of a continuous flow of new technologies has been inscribed in the cultural imagination and has become a teleological signifier of social progress that helps to *structure* the nation's self-congratulatory can-do optimism in a better future. As historian Michael L. Smith summed it up, "the artifacts of technological innovation—[from] electric lights, automobiles, airplanes, [to] personal computers—have come to signify progress . . . [and] Americans have been asked to visualize the future as a succession of unimaginable new machines and products."[14] More than a century ago, Edward Bellamy, himself a lay pastor preaching a vision of technologically driven utopia in *Looking Backward*, caught the paradox:

> This craze for more and more and ever greater and wider inventions for economic purposes, coupled with apparent complete indifference as to whether mankind derived any ultimate benefit from them or not can only be understood . . . [as] one of those strange epidemics of insane excitement which have been known to affect whole populations at certain periods. . . . Rational explanation it has none.

Yet the hunger for "greater and wider invention" has not ebbed; it's not the tulip craze. It is instead the sign of a cultural dis-ease, of an ongoing gold rush of the American mind. At sites such as the EPCOT center, Americans pay for the privilege of being "indoctrinated" into a progressive history of technology and "faith in a sanitized, inexorably beneficial, technological future."[15]

For the GNR enthusiasts (as I will call them), the *agency* of progress has shifted—from society (social planning, good government, virtuous leadership) to the individual (quality-of-life, obtainable through constant consumption). In a sense, the Enlightenment utopia of the mind—as the rational host of self-control, self-mastery, and perfectibility—has shifted to the body. As self-actualization now seems possible through technological advance, the body has become the locus of consumer desire and the (literal) base for layers of technological prosthetics. As Vivian Sobchack notes, "the desire for transformation through technology has . . . detached itself from visions of rationality and [social] progress and attached itself (with some anxiety) to more subjective states of technological being."[16] In other words, social relations will not improve through moral elevation or a more equitable distribution of resources, but through self-mastery available over the counter. *Social* progress was vested in a faith in political institutions, centralized planning, and democratic participation. As late as the 1970s, the utopian ideals of technological transformation tended more toward the national (transportation networks, nuclear energy, NASA) and even the domestic (consumer appliances, television) than the personal. The desire for technological transformation of "subjective states"—via the body—can be traced to the simultaneous emergence of miniaturized electronics (e.g., the Walkman, video games, cell phones), psychotropic drugs, and the Internet.

The 800-pound gorilla in the discursive room is the need for a new definition of "the human"—without which the term "posthuman" is meaningless. In various ways, this is at the heart of Sherry Turkle's and Donna Haraway's work, of cognitive scientists and philosophers from Daniel Dennett to Francisco Varela, and in the utopian claims of futurists such as Hans Moravec and Ray Kurzweil.[17] A few years ago, veteran *Washington Post* editor Joel Garreau wondered why the past generation's technological changes transformed work and home, but left social and political life unchanged. "*Where is the social impact of this change? Where is the Reformation? Who are the new Marxists?*" he wondered. Turns out they're GNR enthusiasts, and they're predicting "the [imminent] transcendence of human nature." At first blinded by cultural lag, Garreau has produced a balanced presentation of posthuman utopian claims in *Radical Evolution: The Promise and Peril of Enhancing Our Minds, Our Bodies—and What It Means to Be Human.* Of course, many non-Western peoples are (by and large) locked out of the discussion since the symbiosis of "human" and "technology" excludes them (almost by definition). So if, as Sturken and Thomas claim, new technologies are always "a Rorschach test for the collective concerns of a particular age," what does the enthusiasm for the "posthuman" say about postmodern consumer society?[18]

Such a scenario seems tailor-made for the field, since American studies produced the critical framework necessary to confront this question directly: *Are our new, improved cyborg bodies waiting for us just around the bend, or is this just another cycle of technological utopianism promoted through Leo Marx's "rhetoric of the technological sublime*"?[19] That is the guiding question of this article. Building on this long introduction, I will first sketch the roots of technological worship in the European past, and then map the posthuman discourse onto the resultant myths of progress and the Adamic. In the conclusion, I will address the two most important questions: *What are the consequences of the myth of technological progress as it informs white, Western superiority? What possibilities open up if the myth can be delegitimated at the level of national identity?* This is an exploratory essay into a more conscious future.

Progress and the White Adamic

If technology is equivalent to dominion over Nature, then "the religion of technology" (according to Noble) emerged from a few early medieval monks who resurrected the symbolic ideal of the original Adam. They believed the pre-Fallen Adam, immortal and created in the divine likeness, was *recoverable* through individual piety and work in the "mechanic arts," such that men could be co-workers with God in making over the planet to prepare for the second coming. The reach of this concept is long (as I will show), but an American strain took shape in nineteenth-century New England. In his classic work *The American Adam* (1955), R. W. B. Lewis showed how American writers secularized the Puritan ideal of a new Jerusalem by sending male loners out to the frontier, where each could work for "a restoration of Adamic perfection, knowledge, and dominion, [and] a return to Eden." For Oliver Wendell Holmes, only science could bring the "new man," and such "restoration" would owe much to technological transformation. Lewis illuminates a pattern in the texts of Cooper, Whitman, and Thoreau, wherein male bodies mark territory in new (and potentially redemptive) landscapes:

> The hero of the new adventure [was] an individual emancipated from history . . . self-reliant and self-propelling, ready to confront whatever awaited him with the aid of his own unique and inherent resources. It was not surprising, in a Bible-reading generation, that the new hero . . . was more easily identified with Adam before the Fall.[20]

The concept of *the Adamic* is invested in recuperating an Edenic purity earned through virtuous work: it informs the Euro-American myth of Columbus's discovery, Euro-American dreams of space, and the posthuman. A quick sketch is in order.

The first intellectual figure to valorize the "mechanic arts" (i.e., technology) as a means to access the divine was an influential ninth-century Irish monk, John Scotus Erigena. Calling the mechanic arts "divinely inspired," Erigena elevated practical activity to works of grace and helped masculinize carpentry and crafts. His writings provided an ideological foundation to the "medieval industrial revolution" of the twelfth-century homosocial monastic world. As Mumford showed in *Technics and Civilization*, the creation of watermills, windmills, the spring wheel, and the mechanical clock, along with innovative mechanisms for metal forging and ore crushing, created early systems of mass production that valorized order, rationality, and system; the creation of steady mechanical power created new methods of milling, tanning, and blacksmithing. What Noble calls "the monastic mechanization of the crafts" found its sublime dynamic agent in waterpower and its sublime artistic form in the cathedrals that formed a sacred geography in Europe for five centuries.[21]

The next conceptual element in the Adamic was provided by a twelfth-century Italian monk, Joachim of Fiore, who called for an avant-garde of "spiritual men" to act as agents of the second coming and "recover mankind's original perfection." His influential ideas were later taken up by Francis Bacon, whose widely read utopian work, *The New Atlantis*, published in 1627, imagined a society in which humankind became purified through rational order as applied to social organization; in other words, he imagined a monastic society on a national level. The engineering school is the center of learning in Bacon's work, and he called it "The College of the Six Days' Work"; the spiritual men were in charge of a second, rational, *man-made* creation ("Six Days") meant to improve and redeem the first. The influence of *The New Atlantis* on the history of science, technology, exploration, and globalization cannot be overestimated: it served as a literary blueprint for the Royal Society of London and anticipated the modern industrial laboratory.[22]

In an American context, Columbus arrives as one of those "spiritualized men"—the embodiment of cutting-edge maritime technology. Columbus's closest friends were monks and Franciscan friars, and he spent a great deal of time in monasteries. Columbus's hunger for finding a passage to the Indies was always couched in the language (and vision) of God's purpose and the practice of the sailor's "art [that] predisposes one . . . to know the secrets of the world." Columbus called his voyages to the New World an "enterprise to Jerusalem," and wrote that "God made me the messenger of the new heaven and the new earth of which he spoke in the Apocalypse of St. John . . . and he showed me the spot where to find it." After his second voyage, Columbus walked the streets of Cadiz and Seville in sackcloth; he was dressed on his deathbed in a Franciscan habit and buried in a Carthusian monastery. As many scholars have noted, "after Columbus, paradise became . . . a place," and I will not here rehearse the vision of the New World as a site to redeem an exhausted

Europe through domestication, improvement, and progress, from the parable of *The Tempest* to Europe's "second discovery of America" through technology—Fordism, Taylorism, aviation, speed, and skyscrapers—as rendered in Hughes's *American Genesis.*[23]

The rise of a scientific perspective and the waning of religion during the Enlightenment created fertile ground for transforming the sacred image of the human organism into a mechanical one. For Mumford, the "worship" of machines by white elites was a fait accompli by the late seventeenth century—"the world [had] a new Messiah: the machine"—and this faith was manifested in the "compulsive urge toward mechanical development without regard for . . . the development in human relations." My touchstone would be Julian Offray de la Mettrie's *Man, a Machine* (1748). To de la Mettrie, human beings were "a collection of springs which wind each other up," the human body "a large watch" powered by wheels, and the soul itself nothing but "an enlightened machine." At first reviled, the idea of the "human machine" became a commonplace in the nineteenth century. This fusion of the religious and the mechanical helped usher in an astonishingly fertile period of domestic invention between 1830 and 1860. Leo Marx found in the technological discourse of the period a palpable sense "that inventors [believe they] are uncovering the ultimate structural principles of the universe."[24]

More important, in the mid-nineteenth century technology and the Adamic come together at the level of national myth. In his brilliant synthetic work, *America as Second Creation: Technology and Narratives of New Beginnings*, David E. Nye found that nearly all Euro-American "foundational narratives" of nineteenth-century frontier settlement understood their right to the land, not as the New Jerusalem of the Puritans, but "as the technological transformation of an untouched space." Whether by the axe, the mill, the canal, the steamboat, the railroad, the dam, or the steel plow, the technology "caused" a chain of events, allowing the settlers to participate in what they called a "second creation." The white settlers legitimated their presence from New England to California by putting forth the technology as the agent that conquered the wilderness, thus eviscerating Native American (i.e., "first creation") claims to the land, and giving the United States nothing less than "a national myth of origin." Here's the boilerplate: "A group enters an undeveloped region," and using "one or more new technologies . . . transforms a part of the region." The region becomes prosperous, attracts new settlers, and "the original landscape disappears and is replaced by a second creation largely shaped by the new technology." The narratives are often written or told "in the passive voice and emphasize the technology" as the agent "of a developmental process." It is a minor-key version of manifest destiny, one town at a time—an "exemplary tale of progress"—often told less for the purpose of establishing national borders than to justify "the assimilation of nature" for industrial society. To use Heidegger's well-known terms, Americans began to "enframe" nature—natural resources—as a set of raw materials, as "standing-reserve" for human consumption.[25]

All founding myths partake in religious concepts: they posit a story of origin, explain a people's right to a given geography, and grant a transcendent reason for that existence (in this case, progress and improvement). Nye's analysis reveals how the concept of the Adamic evolved here from an individual, male ideal to a group (and then national) identity as a result of frontier experiences. Instead of a transhistorical symbolic ideal of a materialist, autonomous male (Adam), a group of bodies laid claim to a new land through their participation in its (technological) improvement. In other words, Euro-American bodies affirmed their right to new geographical space without recourse to conquest narratives; instead, this "technological creation story" became the formula for justifying the "improv[ement] of any Eden whose inhabitants were few or ignorant or lacked a railroad." One of Nye's myriad examples is an 1859 lecture by Abraham Lincoln on inventions in which he proclaimed that "old fogie, father Adam . . . [was] a very perfect physical man," but he lacked sophisticated communication networks (the telegraph, the railroad), did not enjoy food "brought from the other side of the world," and was thus "no equal of Young America." The New Adam *was* Young America.[26]

The fusion of progress, technology, and religion into a white mythology is then continually reinscribed. The massive social transformations brought on by telegraph, railroad, and electricity created a sense that technology was "white magic" (to use Franco Moretti's term), and "the awe and reverence once reserved for the Deity . . . [became] directed toward technology." History as "a record of . . . progress" became doctrine during the Enlightenment, but with "rapid industrialization . . . the notion of progress became palpable; 'improvements' were visible to anyone." As new machines continually altered the workplace, as communication and transportation networks collapsed time and space, modernity became a social fact: one's life did not resemble one's parents' or grandparents' lives. As Marx summed up the transformative moment, "To look at a steamboat [or a locomotive] . . . is to see the sublime progress of the race."[27] The markers of the difference were machines, technological products, and the effects of technological networks.

Carey and Quirk revealed the role of "the future" in this ideological mix, and showed how Americans colonize this temporal zone with the utopian repercussions of new technological networks. Their examples were the telegraph, railroad, telephone, electricity, automobiles, and finally, computers; for each, advocates predicted a new day and "a radical discontinuity from history." This quasi-sacred nature of technology has marked it, for Americans, as "a force *outside* history and politics." Technology thus becomes the prime mover of an ongoing "millenarian impulse," and futurists (public relations workers, scientists, writers, and businessmen) "cast themselves . . . [as] secular theologians composing theodicies for . . . [their] technological progeny." For example, utopian claims for computers in the 1970s were seen less as marketing reports than as dispatches from the future's frontlines by "self-abnegating servants" plugged in "to the truth and the future as determined by the inexorable advance of science and technology." Grafting the rhetoric of technological revolutions onto the millennial impulse creates the necessary conditions for the mythic system of progress; over and over again in "contemporary futurology," the emergent technological network reboots the national faith. "In modern futurism," Carey wrote as if anticipating posthuman rhetoric, "*it is the machines that possess teleological insight.*"[28]

Rather than cast any doubt on the technological fix, Americans have instead witnessed the rise of the contemporary Adamic: first, in the white, homosocial world of NASA, which has functioned as a monastic guild for two generations, and second, in popular culture. From the 1950s through the 1970s, nearly all of NASA's key positions were filled by evangelical Christians. NASA's director, Werner von Braun—ex-Nazi rocket scientist, father of the U.S. space program, and born-again Christian—declared that the purpose of sending men into space was "to send his Son to the other worlds to bring the gospel to them" and to create a "new beginning" for mankind. In the 1950s, scientists and physicists believed new planets and space colonies might become a safety valve for a planet poisoned by nuclear winter. Physicist Freeman Dyson wrote the "Space Traveler's Manifesto" in 1958, and he supported the development of nuclear energy to secure a power source for a starship that was mankind's best chance to survive apocalypse. The claim was seconded by Rod Hyde, NASA's group leader for nuclear development: "What I want more than anything is to get the human race into space . . . It's the future. If you stay down here some disaster is going to strike and you're going to be wiped out." Directed by the "spiritual men" of NASA, humanity would restart on another world so that human beings could still be headed for a redemptive future even as they left behind the mess of the impure.[29]

Ninety percent of American astronauts have been "devout Protestants"; many carried Bibles and Christian flags in their spacesuits. "I saw evidence that God lives," Frank Borman reflected of his experience as Apollo 8's commander;

Apollo 11's Buzz Aldrin received radio silence from NASA to read the first fourteen lines of Genesis while walking on the moon. Aldrin took communion with a kit packed by his pastor containing "a vial of wine, some wafers, and a chalice," as well as "[a] reading from John 15:5"; he later reflected with joy that "the very first liquid ever poured on the moon and the first food eaten there were communion elements." In 1969, technology and religion fused with national myth and political power: President Richard Nixon pronounced the week of Apollo 11's flight and landing on the moon "the greatest week since the beginning of the world, [since] the Creation." Nixon was immediately reprimanded by Reverend Billy Graham—his personal religious leader—who declared there had indeed been three greater events: Jesus Christ's birth, crucifixion, and resurrection.[30]

If landing on the moon was the fourth greatest week-long event since the Creation, it is significant that it was accomplished by the first cyborgs: astronauts. Continually attached to technological networks through spacesuit (synthetic second skin) and spaceship (nurturant, home, environment), astronauts were the first human-machine interface. Norman Mailer captured the posthuman shift at the Apollo 11 launch:

> [Neil] Armstrong . . . space suit on, helmet on, plugged into electrical and environmental umbilicals, *is a man who is not only a machine himself in the links of these networks*, but is . . . in fact a veritable high priest of the forces of society and scientific history concentrated in that mini-cathedral, a general of the church of the forces of technology.

In *Me++*, Mitchell named all of the linkages necessary for space travel, from "extravehicular mobility units (EMUs) . . . [with] internal and external plumbing systems," to "backpack primary life support systems (PLSSs), with supply and removal systems . . . [as] controlled from chest-mounted consoles." Each astronaut wore a "maximum absorption garment (MAG) to collect urine, a liquid cooling and ventilation garment (LCVIG) to remove excess body heat, an EMU electrical harness (EEH) to provide communication and bioinstrument connections, a communications carrier assembly (CCA) for microphones and earphones . . . and a polycarbonate helmet with oxygen supply and carbon dioxide purge valve." To Mitchell, as an astronaut, "you got to sleep with your extrabiological body double."[31]

That *Columbia* was the name of the command ship of Apollo 11 and the third space shuttle connects the so-called Age of Exploration with the age of space travel: visionary quests for new worlds rationalized by the search for new economic markets (domestic or foreign), historicized as the pursuit of "pure" knowledge, and informed by a search for God (and directed *by* Him).

Whether the Adamic of the seas or the skies, this mythic concept is still concerned with the same dialectic of individual and national accomplishment: it's still about being the first body on some new frontier, planting seed and flag, and rationalizing national fantasy in the quest to redeem Fallen Man on a land claimed for Edenic purification. Think of *Star Trek*'s opening as a spell or a chant that reproduces the Adamic in space, *the final frontier*. "These are the [spiritualized] men of the *Starship Enterprise*. Their ten-year mission . . . [is] to boldly go where no man has ever gone before." Again, the first body (or bodies) in (a new) space. Again, the muted martial trumpet cry to the future and imperialist dominion over new spaces. As Sobchack points out, "one need only remember that the *first* American space shuttle—*The Enterprise*—was named after (and perhaps carried the same ideological baggage as) the flagship of . . . *Star Trek*."[32]

It may seem like a long jump from Erigena to Columbus to NASA, but it is an identifiable (and ongoing) tradition. The posthuman Adamic reproduces the tradition's three elements: (1) the valorization of the "mechanized arts" through the thrill of scientific discovery and exploration; (2) the shadow Christian tradition of redeeming Fallen man (or an exhausted geography); and (3) the competitive challenge of being the first body in a new environment—whether physically on a new continent or a new world, or now, mentally, in cyberspace. This is not an essentialist genealogy, but a record of cultural practices vetted by ideology. This white mythology has already produced a posthuman Adamic discourse—and myth—that promises nothing less than the technological transcendence of the individual human organism. It bears repeating: *technology is the American theology*. As Rosalind Williams states succinctly, "to affirm that technology drives history is to deny what [or that] God does."[33]

The Posthuman Adamic

Here's a provocative statistic: a Harris Poll showed that 42 percent of American parents would "use genetic engineering on their children to make them smarter . . . [and] 43 percent, to upgrade them physically." A separate poll found that more than a third would even be willing to "tweak their children genetically to make sure they had an appropriate sexual orientation." As Lori B. Andrews has argued, therein lies a market as large "for prebirth genetic enhancement" as motivated the sales of Prozac or Viagra. Thus, genetic manipulation will probably be at the cutting edge of transcending the limitations of the human body. In fact, when the film *Gattaca* was released—a dystopic vision of a society based on genetic hierarchy—a toll-free number used in one of its trailers (1-

888-4-BEST-DNA) was swamped with calls for genetic upgrades; the American Society for Reproductive Medicine had to issue an official denial of its participation. According to Princeton biologist Lee Silver, "the use of reprogenetics is inevitable. It will not be controlled by governments . . . or even the scientists who create it. . . . the global marketplace will reign supreme."[34]

At the moment, Kevin Warwick and the Australian performance artist StelArc represent the posthuman Adamic from the robotics side—a man-made "second creation" of the human body for an imminent future in which (to quote StelArc's website), "THE BODY IS OBSOLETE." StelArc has had a third electronic arm surgically implanted in his stomach and, in performances, creates art and performs mathematical equations coordinating his three arms.[35] In 2002, robotics professor Kevin Warwick had an electrode array surgically implanted in his forearm that allowed him to transmit his thoughts over the Internet and to access all electronic interfaces in his office building. Having an electronic connection means that "your physical capabilities extend as far as the internet will take you, but so too your powers of absorbing information. . . . You are not limited . . . to taking in information from your local vicinity via your eyes and ears." Being directly jacked in to the computer transcends "the simple human-body perimeter" through direct electrical connection of human and machine; Warwick has successfully sent his "brain signals" all over the world. "A cyborg body is truly a global one."[36]

In *I, Cyborg*, Warwick gave three passionate reasons for his experiments: curiosity of human-computer symbiosis; the feeling of reentry when "returned to the ranks of humanity"; the thrill of discovery, "my desire to be the first cyborg . . . the pioneer." He thought himself "in a similar position" to Charles Lindbergh or the explorers: "I could go where no one had ever gone before." For example, Warwick and his wife, Irena, have communicated over the Web in near-telepathic connection through electrode arrays in both their hands. Their first successful "direct nervous system to nervous system link-up" was thrilling; Warwick's first thought was to run over and give his wife a hug, but he realized "we both still had our nervous systems wired up. . . . We didn't have to hug, we didn't have to say anything to each other. We had a new way—our own way—of communicating now." Here is the Adamic reproduced in all its glory: being first in a new landscape, transcending the organic body, redemption and progress powered by the mechanized arts.[37]

Various kinds of permanent silicon implants are already working in hundreds of human bodies: cochlea implants in the inner ear, implants that control the tremors of Parkinson's disease, direct neuroelectronic interfaces that allow fluid, integrated motion with prosthetic legs. In each case, "there is direct electrical connection between the electronics of the silicon device and the nervous system of the patient."[38] These technologies are emergent, not speculative, so GNR enthusiasts cannot be dismissed as crackpots promising us a Jetson-like future and individual jetpacks. And, often as not, they are the academic cream of MIT, Stanford, and Cal Tech.

And isn't moving the machines inside the body the logical next step in technological evolution? Since the early 1980s, the consumer fruits of the electronic and computer revolutions have stirred every younger generation to a faith in a peaceful, plentiful, groovy future made possible by technological fables of abundance. In addition, ads now help young consumers visualize every individual as a mobile network, with iPods and Blackberrys enabling 24/7 mobile connectivity. GNR enthusiasts have simply reenvisioned the human body as a computer array. Brooks imagines implanting a screen in the part of the brain responsible for visual processing with an internal on/off system that jacks the mind into the Internet. Like many others, he is exalted by these ideas: "What if we could make all these external devices internal, what if they were all just part of our minds, just as our ability to see and hear is just a part of our mind?" As Garreau reflects, "for all previous millennia, our technologies have been aimed outward, to control our environment. . . . [Now] we have started a wholesale process of aiming our technologies inward."[39]

This is perhaps most true for nanotechnology (N), the science of manipulating subatomic particles to take advantages of properties (such as conductivity) present only at that level. Medicine is its primary locus of application, and its researchers foresee robot surgeons dispatching tiny programmed "nanobots" into infected areas to cure the body without surgery or chemotherapy. Ray Kurzweil predicts that smart nanobots will easily "reverse the environmental destruction left by the first industrial revolution," provide safe, cheap, clean energy, and "destroy . . . cancer cells, repair DNA, and reverse the ravages of aging." Nanotech pioneer Robert A. Freitas, whose research concerns the potential for nanobots to perform brain scans, goes even further. Here's an FAQ from his website: "What would be the biggest benefit to be gained for human society from nanomedicine?" His response: "Nanomedicine will eliminate virtually all common diseases of the twentieth century, virtually all medical pain and suffering, and allow the extension of human capabilities, most especially our mental abilities." Humans will soon have the storage capacities of computers, according to Freitas, since a "nanostructured data storage device" is no bigger than a neuron and can hold "an amount of information equivalent to the entire Library of Congress." Once all human beings have "extremely rapid access to this [volume of] information"—and, I suppose, regular upgrades—Freitas foresees immense social changes. What will be nanomedicine's biggest success? "The most important long-term benefit to human society" will be "the dawning of a new era of peace." Why will that occur? "People who are independently well-fed, well-clothed, well-housed, smart, well-educated, healthy, and happy will have little motivation to make war." Apparently, in the nanotech-powered utopia, immortality supplants the will to power and the seven deadly sins, since "human beings who have a reasonable prospect of living many 'normal' lifetimes will learn patience from experience."[40]

As noted above, posthuman utopianism differs from all past visions—from Plato's to Thomas More's to Charlotte Perkins Gilman's—in being based not on social planning but on self-actualization. As one futurist rhapsodizes over the advantages of cyborg life:

> Just a small piece of silicon under the skin is all it would take for us to enjoy the freedom of no cars, passports, or keys. Put your hand out to the car door, computer terminal, the food you wish to purchase, and you would be dealt with efficiently. Think about it: total freedom; no more plastic.[41]

This superficial notion of "total freedom" aside, the obvious context is our radically individualistic consumer society. Posthuman utopians ignore what Albert Borgmann simply calls—in response to their claims—"the social dimensions of human being." Humans are socialized by parents, siblings, teachers, and environments, and acquire consciousness (and self) in dialogue with friends, co-workers, and nature itself. As Borgmann claims about the relationship of individuality and socialization: "Each of us is a unique and inexhaustible locus of convergence and transmission through our ancestry, both evolutionary and historical, through our descendants, through the sensibility of each of our organs, through our pleasures and pains, through our wounds and our scars, through our attacks and our embraces." To assume that upgrading the "wetware" will lead to utopia misunderstands the mundane aspects of being human. As Borgmann claims for himself: "I shape my conduct in emulation, competition, or companionship with others."[42]

To be reductive, GNR enthusiasts are able to evade the "social dimensions" of human *being* because they work from the "computational model" of the brain (the dominant paradigm of cognitive science). To GNR enthusiasts—as to cognitive scientists and philosophers—the human organism is simply a special kind of machine composed of electro-chemical networks. As Hayles has shown, in the decade after World War II, cybernetic theorists shifted the idea of mental processes from "thought" and "consciousness" to "information." As they ratcheted up "analogs between machines and humans," early cybernetic theorists began to "*construct the human in terms of the machine.*"[43] In other words, they externalized the model of the mind and bracketed off subjective experience as irrelevant. Over three generations, "information" supplanted "thought" and the speed of computation became the measure of mental power. According to John Searle, most cognitive scientists do not "regard consciousness as a genuine scientific question," and most textbooks on the brain "have no chapters on consciousness." There are alternative models ("the embodied mind," "the phenomenological mind"), but cognitive science remains devoted to the computational model, which has "virtually nothing to say about what it means to be human in everyday, lived situations."[44]

GNR enthusiasts thus conceive of human organisms as nodes in technological networks rather than in social ones. To cognitive scientist Andy Clark, for example, human beings have always been "natural-born cyborgs," and the human organism is simply "tools all the way down." ("Tools-R-Us," he quips.) In reading back into history the model of the "man-machine," computationalists like Clark shift the idea of technologies (or tools) from *prosthesis*—tools used externally, by the body and as an extension of it—to something like *techno-symbiosis* (i.e., tools wired into the mind and body, neural subsystems as computer programs). As Hayles has shown, cyberneticists consider human beings as simply "body surfaces . . . through which information flows"; if so, she wonders, "who are we?" In *Posthumanity* (2004), philosopher Gerald Cooney spins it slightly differently: will a body outfitted with a dozen prosthetics—thin body armor, internalized electronics—feel "just as much *my* body and part of *myself*?" Is that kind of cyborg "as much a human being?" Either perspective "implies the deconstruction of the autonomous self," and both cognitive scientists and cultural theorists now meditate on how the cyborg "self" would work at the level of mind.[45]

It is certainly possible that the transformation from the liberal human subject to Me++ will be no more chaotic than, say, the concurrent experiences of modernism and antimodernism from 1880 to 1940. As Clark argues, at any given moment, the self is "a conception of whatever matrix of factors we experience as being under our direct control." If the thought of moving a prosthetic leg creates motion in it, the mind accepts it as part of "self." Cooney concurs:

> as long as my synthetic hip joint interacts with adjoining bone structures in a way that is equivalent to that of the natural joint, and as long as synthetic neurons interact with each other . . . in the relevant ways, there will be walking and thinking. It isn't just that we couldn't tell the difference, but there *would be no difference at the level at which these functions occur.*

The same would go for implanted silicon chips and connection with others through external networks; there is obvious precedent with regard to pacemakers and hearing aids. If so, then "truly hybrid biotechnological selves" requires only individuals who welcome such a self-image, and even posthumanists agree "social acceptability" may take two generations. Still the networked being—the cyborg—has every possibility of "feeling" like a self; *Me++* renders this upgraded self-concept in an apt and very American locution.[46]

Yet I am troubled by the technological utopianism in Clark, Brooks, Mitchell, Kurzweil, and others. For example, Clark claims his computer is now central to his self-concept and "the recent loss of my laptop had hit me like a sudden and somewhat vicious type of . . . brain damage."[47] This is metaphorical, not metaphysical; anecdotal, not representative. As an evacuee from New Orleans during Hurricane Katrina, I read dozens of testimonies that first week from people who lost all kinds of material possessions, professional equipment, art and lab work, sentimental talismans. Not once did anyone equate life and/or limb with his computer or her brain power. In fact, I read at least a dozen articles with some version of this offer: "I left my car on Dryades St. near Valence; if you can read this, smash the window and drive to safety." Once safe, people cried out for those lost and dead, not their computers. Since my return, I have spoken to people who lost their homes, their cars, their musical instruments, their computers—not a single mention of a ghost-PC nagging at the corners of sensorimotor memory. Is it only technological utopians who are trying to drag us into a posthuman future?

Equally suspect is the ebullience with which GNR enthusiasts disdain their own human bodies. Warwick thinks of cyborgs as "upgraded humans," and damns all who will not take advantage of implants to "the subspecies human race." Clark believes the brain has outgrown its "biological skinbag," and insists it will not be "bound and restricted" anymore; he thinks of himself less "as a physical presence than . . . a kind of *rational or intellectual* presence." For Brooks, we need not fear the transformation into cyborgs, since "we, the man-machines, will always be a step ahead of them, the machine-machines." Kurzweil predicts that by 2099 all humans will be "machine-based" (i.e., "they [will] no longer have neurons, flesh or blood") except for the most obstinate Luddites. In *Radical Evolution*, Garreau sketches the coming divide between "the enhanced" and "the naturals," with the latter becoming increasingly uncompetitive. Apparently, the present human organism is obsolete—and good riddance to it. Yet Sherry Turkle reports that people most often distinguish between human and computer characteristics by "dwell[ing] long and lovingly on those aspects [of life] . . . that are tied to the sensuality and physical embodiment of life."[48]

The presumptions of GNR enthusiasts in *declaring* the shift to the post-human are the arrogant proclamations of a white technological priesthood presiding over globalization. As Langdon Winner points out in "Are Humans Obsolete?," not only do

> post-humanists show little awareness of their deep cultural biases and . . . the breathtaking cultural arrogance their proposals involve . . . [but they] have not looked carefully at how their notions reflect unstated, unexamined preconceptions rooted in their own highly rarified, upper-middle-class, white, professional, American and European lifestyles.

Winner also rightly wonders whether submitting genetic enhancement to market forces will create a genetic divide that will dwarf the digital divide.[49]

The rapid pace of innovative GNR technologies has imbued scientists with a sense that "humanity [is] now conceived as godlike in its utterly free creative power and its responsibility for the future."[50] They, in turn, try to convey that sense of mastery to the future's consumers: "Having such things implanted in our brains will make us tremendously more powerful," Brooks writes, and "we will be[come] superhumans in many respects."[51] Yet when enthusiasts herald the wave of new miraculous technologies, there is never a mention of "the spectacular series of disasters" within living memory of technologies meant to give humans mastery over their environment: "Hiroshima, the nuclear arms race . . . Chernobyl, Bhopal, the Exxon oil spill, acid rain, global warming, ozone depletion."[52] Not only are there many levels of "*unintended* consequences" to technological products—as documented in Edward Tenner's *When Things Bite Back*—but in the United States, scientists can barely garner funding to confront *obvious* consequences (e.g., global warming, coastal erosion). The forecast of a GNR-powered utopia sustains itself by ignoring wasted landscapes and offering in its place the image of our individual bodies as theme parks.

This future depends on the practical reality of the computational model of the brain. According to Searle, computationalists "howl with outrage" when alternative models of the brain are presented—in a manner more like "the adherents of traditional religious doctrines of the soul" than scientists. His theory is that their defensiveness is rooted in "the conviction . . . that computers provide the basis of a new sort of civilization—a new way of giving meaning to our lives, a new way of understanding ourselves." The computational model is "profoundly antibiological" and ignores that the brain is embedded in the body and its biological processes. The model reeks of both dualism and atavistic rationalism—Aristotle's godlike (or machine-like) mind imprisoned in the impure or loathed animal body (the "meat body," "wetware").[53] But if the brain is a computer, it can be integrated with all other computers, and all the predictions of GNR enthusiasts make sense.

It is the only future GNR enthusiasts can imagine. Technology—or, more precisely, scientism—is their belief system: it provides faith in the future, continues the myth of progress, and keeps all definitions of the human in the hands of Western science. Consciousness is bracketed off as so much white noise; to cognitive scientists and GNR enthusiasts, the human being is a sophisticated network of neural subsystems whose constant upgrading becomes the purpose of posthumanity. "If we can create minds simply by designing computer programs," Brooks predicts, "[then] we will have achieved the final technological mastery of humans over nature."[54] Again, the "second creation" ideal.

Cooney asks a simple question: "Do we even want this [posthuman] future?" He first puts progress on the stand: "What is our criterion for judging that major technological developments constitute progress rather than posing unacceptable dangers and threatening [a] . . . loss of fundamental values?" Conceding that market forces will likely drive genetic enhancement, he wonders, "Can we allow the market to determine the outcome when our human nature is in play?" This leads to his crucial question: "What [about human life] . . . requires a posthuman future?"[55] I reiterate: only the myths of progress, the Adamic, and white, Western superiority *require* a posthuman future. The posthuman is the dream of bodies of pure potentiality—ones that do not decay but plug into networks of information and pleasure.

With the computational model we get a posthuman utopian future—the human being as perfected machine. If the brain is a computer, human neurocircuitry finds a nonbiological home. If it's not, then we get the same old greedy humans abusing power and rationalizing environmental devastation and global misery via technological progress. This synthesis of Western rationalism, Christian disdain for the animal body, and superiority over the Other comprises a faith-based narrative whereby scientists eliminate all subjective experience to focus on the dream of a mechanical brain from which they can then upgrade the human race. On the first day of the future, Western scientists create a cyborg and the long-sought-after second creation is achieved. In the meantime, since all technology is reflected in cultural narratives, a posthuman Adamic myth emerged in popular culture concurrently with GNR technologies.

The Terminator is the future's posthuman Adam. In each of the three Terminator films, beginning in 1984, the cyborg model T-800 (Model 101) arrives stark naked—ushered in by bolts of lightning, hurricane gusts of wind, and a breach in the time-space continuum. His buff, perfect, white Aryan male body is fetishized in each opening scene, and we watch his naked perfection acquire clothes and weapons by making "girlie-men" of punks, bikers, or slackers. He is a bolt from the future blue, the rejuvenated Adam of the Next Testament: on the first day of the future, God creates the Terminator.

The Terminator is the cyborg successor to fallen, puny man. A flesh suit zipped over a computer matrix combining GNR technologies, the Terminator must be martyred at the end of every film. Each time he returns, reprogrammed and reborn from the future to restart the story. Each reappearance challenges audiences to confront the underlying tensions of imminent posthumanity: will we still be recognizable to ourselves if we morph into cyborgs? If the "dream of robotics [is] that we will gradually replace ourselves with our robotic technology," Bill Joy repudiates that dream, asserting that "the[se] robots would in no sense be our children." Cooney raises this issue as well: will we recognize cyborgs as "a human outcome"?[56]

Cooney's and Joy's responses point up the significance of the Terminator as a floating signifier of the posthuman Adamic.[57] So long as cyborgs are imagined as superhuman male bodies—as the perfect, desired, mechanical Other, as motorcycle-driving, shades-wearing, gun-toting, Western heroes of the future, as *male* technological society's *man*-made technological saints—then the posthuman dream of evolving into cyborgs both perpetuates the mythic triumphalism of progress and constitutes a refusal to acknowledge the limits of an individual human body and an individual life. To adapt Richard Slotkin's classic formulation, call it *regeneration through technology*.[58]

Judgment Day: The End of Progress

I will conclude with Sturken and Thomas's most important question of all: "How is it possible to think about technologies outside of these frameworks?"[59]

Nearly a generation ago, Haraway recognized the need for a more "imaginative relation to technoscience that propound[ed] human limits and dislocations—the fact that we die, rather than Faustian . . . evasions." Yet as popular culture and well-funded GNR enthusiasts have more influence than academic theorists, they have commandeered cyborg iconography; the necessary corrective of conceding the "human limits" of biological processes has yet to occur. Haraway has since called for new metaphors—such as trickster figures (e.g., Coyote)—to "refigur[e] possible worlds" by thinking outside of techno-science; this hasn't happened even within the humanities. Instead, we have seen the rise of the posthuman Adamic.[60]

For Nye, the "technological creation story has long remained dominant" because questioning it required a reassessment of history, social justice, and ethics—as well as the demystification (and demythification) of every keyword in the techno-cultural matrix. Here's Nye's assessment of why the nineteenth-century "second creation" narrative remained dominant until the 1960s:

> Rejecting the foundation story . . . meant recognizing historical injustices to the first inhabitants, accepting environmental limits, and acknowledging the ideological nature of the free market. Rejecting the foundation story implied the loss of white entitlement to the continent. Discarding second-creation stories required acknowledging cultural conflicts and listening to counter-narratives.[61]

Such counternarratives of the frontier now exist: Nye points to ecofeminism and Native American accounts, the works of wilderness advocates and borderlands scholars. Perhaps there is long-term potential for the trickle-down of such counternarratives to transform the national mythos, but the techno-cultural matrix remains strong.

To become conscious of the underlying mythology guiding their utopianism, GNR enthusiasts would need to acknowledge the cyborg's white body, their ideal of white progress, and the historical conflation of technology and religion. Many scholars have traced the social construction of the white body as the normative, ideal human body (e.g., Richard Dyer's *White*), but only recently have nonwhites begun to answer back from an empowered cultural position. In *White Theology: Outing Supremacy in Modernity* (2004), religious scholar James W. Perkinson claims that if Euro-Americans aspire to maturity, "the white body must be returned somehow to its history, [and] white identity reincarnated in local community and global cosmology." To do so, Perkins claims, Euro-Americans must specifically leave *black* bodies alone: "blackness can no longer be erected as a buffer against the demands of maturity, a screen against which to play out fear and fantasy, despair and desire."[62] An interesting claim, but such a separation is impossible due to hybridity at every level: cultural, social, genetic, artistic, intellectual, philosophical.

In my *Swinging the Machine: Modernity, Technology, and African-American Culture Between the World Wars* (2003), I argue that Euro-American bodies have been specifically *colonized* by African-American music, dance, kinesthetics, and speech. In fact, white, Western bodies have appropriated the moves and grooves—and thus, the embodied philosophies—of many cultures over the past two centuries (e.g., Latin American, South Asian). Consider the acculturation involved in the practice of martial arts and yoga, or in the appropriation of various ethnic music and dance traditions (not to mention ethnic foodways).

In fact, my theory of posthuman escapism is that *it is based in the fear of understanding the human organism as a multiethnic, multicultural, multigenetic construction created through centuries of contact and acculturation.* In *Swinging the Machine*, I theorized a dialectic of technology and African American culture: I showed that in every generation, African American music (and dance) provides *survival technologies* in dialogue with that era's machine rhythms and technological systems.[63] Any new definition of the human must account for the historical *failure* of white minds and bodies to create appropriate cultural responses to oppressive technological systems. The denial of this failure gives rise to the utopian projections of GNR enthusiasm. In other words, *the posthuman is an escape from the panhuman.*

Kurzweil has mapped responses and fears of the posthuman future—like mine—onto a three-stage transition model for accepting new technologies: (1) "awe and wonderment at their potential *to overcome age-old problems*"; (2) "a sense of dread at the new set of dangers"; (3) realizing "*the only* viable and responsible path is to set a careful course . . . [to] reap the benefits while managing the dangers." Has any nation ever actually succeeded at steering such a "careful course"? Note Kurzweil's rhetorical assumptions: that new technologies "overcome age-old problems"; that autonomous technology cannot be questioned, only managed; that our dread is cowardice. Given market forces and the consumer drive for self-improvement, these technologies may be inevitable, but nothing in human history points to posthumanity living deeper, richer lives that benefit the *majority* of the earth's peoples. (As Mike Davis has recently shown, a billion people now live in slums.) Tools and technologies are not inherently good or bad, inevitable or autonomous, progressive or regressive, as scholars from Mumford to Nye have always insisted; they are merely vehicles and applications of human beings.[64]

"If progress means to go forward," Carl Becker wondered in 1936 with the Nazis in ascendance and the Stalin purges in full bloom, "[then] forward to what end, to the attainment of what object?" In 1963, theologian Paul Tillich reduced progress to the pejorative term, "forwardism." More to the point here, Bill McKibben wonders how a future of robots and genetic enhancement is "something 'higher' and 'better'" for techno-utopians when they so rarely pay even lip service to social needs such as "feed[ing] the hungry."[65] GNR enthusiasts assume technological progress will produce social progress. Yet even if GNR technologies come on-line as predicted, why would this produce a better society rather than just health, hedonism, and mobility for the upper classes? *It won't*; that's just how myth works. Technological progress is the quasi-religious myth of a desacralized industrial civilization; it is sustained through new technological products, not empirical social change.

The real questions we need to confront are these: What is progress *for*? What is technology *for*? Scientists might rightly claim such questions are not within their purview, but they surely demand attention from American studies scholars. A few suggestions: at the institutional and disciplinary level, we need to make technology indispensable to the field's mastery and our understanding of whiteness; at the rhetorical level, we need to interrogate the assumptions of "the posthuman," as it continues the white, Western tendency to universalize its own concepts; at the transnational level, we need to engage critiques of specifically Western ideals of technology and rationalism, as in the works of Ashis Nandy and Francisco Varela. Andy Clark may be right when he defines the human organism as "tools all the way down," but we are also "creolized" all the way up.[66]

In providing intellectual leadership, the first step is simple: return to the fusion of the terms "social" and "progress," and uncouple "technological progress." In the vernacular, the phrase "technological innovation" always connotes linear progress (i.e., new technologies = beneficial social change). To return to "social progress" as a criteria for judging humankind would place Western technology within a global eco-culture whose disastrous contemporary state is due at least in part to the "white" legacies of colonialism, capitalism, and technology, often in service to the Enlightenment ideal of the liberal human subject ("the human").

I reaffirm my true thesis: the *post*human is an escape from the *pan*human. I define the latter here as an emergent global identity invested in a creolized self-concept and, by extension, a creolized world history. It is admittedly a utopian concept, but I contend there can be no global (i.e., post-*Western*) definition of "the human" without it.

Notes

1. Robert Fisk, "Terror in America," *The Nation*, October 1, 2001, online edition.
2. Marita Sturken and Douglas Thomas, "Introduction: Technological Visions and the Rhetoric of the New," in *Technological Visions: The Hopes and Fears That Shape New Technologies*, ed. Marita Sturken, Douglas Thomas, and Sandra J. Ball-Rokeach (Philadelphia: Temple University Press, 2004), 6.
3. In Richard Dyer's *White* (New York: Routledge, 1997), 16–18, Dyer defines white culture within a matrix of capitalism, colonialism, and Christianity, to which I am adding technology. James W. Carey, *Communication as Culture* (Boston: Unwin Hyman, 1988), 114.
4. Lewis Mumford, *Technics and Civilization* (1934; New York: Harcourt, Brace, Jovanovich, 1963), 45, 53, 365, and *The Culture of Cities* (New York: Harcourt, Brace, 1938), 442; David F. Noble, *The Religion of Technology* (New York: Knopf, 1997); T. J. Jackson Lears, *Fables of Abundance* (New York: Basic, 1994); Siva Vaidhyanathan, *The Anarchist in the Library* (New York: Basic, 2004), xii; Thomas

P. Hughes, "Afterword," in *The Technological Fix*, ed. Lisa Rosner (New York: Routledge, 2004), 241; Edward Wenk Jr., *Tradeoffs: Imperatives of Choice in a High-Tech World* (Baltimore: John Hopkins University Press, 1986), 6.

5. This outline of the "GNR" acronym is from Ray Kurzweil, "Promise and Peril," in *Living with the Genie: Essays on Technology and the Quest for Human Mastery*, eds. Alan Lightman, Daniel Sarewitz and Christina Desser (Washington, D.C.: Island Press, 2003), 39; N. Katherine Hayles, *How We Became Posthuman: Virtual Bodies in Cybernetics, Literature, and Informatics* (Chicago: University of Chicago Press, 1999), 35; Joel Garreau uses the "GRIN" acronym in *Radical Evolution* (New York: Doubleday, 2005), 4–5, and passim.
6. William J. Mitchell, *Me++: The Cyborg Self and the Networked City* (Cambridge, Mass.: MIT Press, 2003), 7, 34, 82, and passim; Rodney A. Brooks, *Flesh and Machines: How Robots Will Change Us* (New York: Vintage, 2002), 212.
7. Langdon Winner, *Autonomous Technology* (Cambridge, Mass.: MIT Press, 1977).
8. Carroll Pursell, *White Heat: People and Technology* (Berkeley: University of California Press, 1994), 215.
9. Michael Adas, *Machines as the Measure of Men: Science, Technology, and Ideologies of Western Dominance* (Ithaca, N.Y.: Cornell University Press, 1989), 144.
10. Sturken and Thomas, "Introduction," 3; John Kasson, *Civilizing the Machine* (New York: Grossman, 1976); David E. Nye, *American Technological Sublime* (Cambridge, Mass.: MIT Press, 1994), and *Narratives and Spaces: Technology and the Construction of American Culture* (New York: Columbia University Press, 1997). Other works by these authors are discussed in depth below.
11. Laurence Coupe, *Myth* (New York: Routledge, 1997), 196–97.
12. John M. Staudenmaier, "Rationality versus Contingency in the History of Technology," in *Does Technology Drive History?: The Dilemma of Technological Determinism*, eds. Merritt Roe Smith and Leo Marx (Cambridge: MIT Press, 1994), 262–63; Carl Becker, *Progress and Power* (Stanford, Calif.: Stanford University Press, 1935), 1–4; Becker's work was preceded by historian J. B. Bury, *The Idea of Progress: An Inquiry into Its Origin and Growth* (London: Macmillan, 1920). Leo Marx, "The Idea of Technology and Postmodern Pessimism," in *Does Technology Drive History?*, 249–50.
13. John Dewey, "Progress," in *Characters and Events*, ed. Joseph Ratner (New York: Henry Holt, 1929), 820–21; Dewey, *Reconstruction in Philosophy* (New York: Henry Holt, 1920), 42; Dewey, "Time and Individuality," in *The Essential Dewey*, eds. Larry A. Hickman and Thomas M. Alexander (Bloomington: Indiana University Press, 1988), 218.
14. Michael L. Smith, "Recourse of Empire: Landscapes of Progress in Technological America," in *Does Technology Drive History?*, 38.
15. Bellamy is quoted in Noble, *The Religion of Technology*, 100; Staudenmaier, "Rationality versus Contingency," 265–66.
16. Vivian Sobchack, "Science Fiction Film and the Technological Imagination," in *Technological Visions*, 157.
17. See, for example, Sherry Turkle, *The Second Self: Computers and the Human Spirit* (New York: Simon and Schuster, 1984), and *Life on the Screen: Identity in the Age of the Internet* (New York: Simon and Schuster 1995); Donna Haraway, *Primate Visions* (New York: Routledge, 1989), and *Feminism and Technoscience* (New York: Routledge, 1997); Daniel Dennett, *Brainchildren: Essays on Designing Minds* (Cambridge, Mass: MIT Press, 1998), and *Consciousness Explained* (Boston: Little, Brown, 1991); Francisco J. Varela, Evan Thompson, and Eleanor Rosch, *The Embodied Mind: Cognitive Science and Human Experience* (Cambridge, Mass.: MIT Press, 1993); Hans P. Moravec, *Mind Children: The Future of Robot and Human Intelligence* (Cambridge, Mass.: Harvard University Press, 1988), and *Robot: Mere Machine to Transcendent Mind* (New York: Oxford, 1999); Ray Kurzweil, *The Age of Spiritual Machines: When Computers Exceed Human Intelligence* (New York: Viking, 1999), and *The Singularity Is Near: When Humans Transcend Biology* (New York: Viking, 2005).
18. Garreau, *Radical Evolution*, 9, 11; Sturken et al., *Technological Visions*, 1.
19. Leo Marx, *The Machine in the Garden* (New York: Oxford University Press, 1964), 195–207; see also Carey, *Communication as Culture*, 120–23.
20. Noble, *The Religion of Technology*, 6, 10–11, 40, 45–46; R. W. B. Lewis, *The American Adam: Innocence, Tragedy, and Tradition in the Nineteenth Century* (Chicago: University of Chicago Press, 1955), 5, 32–41, 45.

21. Noble, *The Religion of Technology*, 14–20; Lewis Mumford, *Technics and Human Development* (New York: Harcourt, Brace, Jovanovich, 1967), 263–72.
22. Noble, *The Religion of Technology*, 24; Lewis Mumford, *The Pentagon of Power* (New York: Harcourt, Brace, Jovanovich, 1970), 105–29.
23. Quotes are in Noble, *The Religion of Technology*, 31–32, 38; on Columbus's voyages as religious missions, see also Kirkpatrick Sale, *The Conquest of Paradise* (New York: Knopf, 1990), 29–30, 93–97.
24. Mumford, *Technics and Civilization*, 45, 53, 365; a good translation of *Man, a Machine* can be found online, along with these quotes, at http://cscs.umich.edu/~crshalizi/LaMettrie/Machine (accessed June 1, 2006); Marx, *Machine in the Garden*, 202; for the body-as-machine, see Anson Rabinbach, *The Human Motor* (Berkeley: University of California Press, 1992).
25. David E. Nye, *America as Second Creation: Technology and Narratives of New Beginnings* (Cambridge, Mass: MIT Press, 2003), 1, 10, 12–13, 290–93, and passim; Martin Heidegger, *The Question of Technology and Other Essays* (New York: Harper & Row, 1977), 17–25.
26. Nye, *America as Second Creation*, 284.
27. Franco Moretti, *Modern Epic: The World-System from Goethe to Garcia Marquez* (London: Verso, 1997), 244; Leo Marx, *The Machine in the Garden* (New York: Oxford, 1964), 195–97.
28. Carey, *Communication as Culture*, 113–15, 191; emphasis in original. See also 173–200, inclusive.
29. This discussion of NASA and the quotes are taken from Noble, *Religion of Technology*, 114, 129–40.
30. Noble, *Religion of Technology*, 137–40.
31. Norman Mailer, *Of a Fire on the Moon* (Boston: Little Brown, 1969), 182; Mitchell, *Me++*, 23.
32. Sobchack, "Science Fiction Film," 148. The first space shuttle, designed in 1976 as a test vehicle only, was slated to be called *Constitution*, in honor of the bicentennial. According to the NASA Web site, "viewers of . . . *Star Trek* started a write-in campaign urging the White House to select the name *Enterprise*"; http://science.ksc.nasa.gov/shuttle/resources/orbiters/enterprise.html (accessed June 1, 2006).
33. Rosalind Williams, "The Political and Feminist Dimensions of Technological Determinism," in *Does Technology Drive History?*, 222.
34. Lori B. Andrews, "Changing Conceptions," in *Living with the Genie*, 107–8.
35. On StelArc, see Andy Clark, *Natural-Born Cyborgs* (New York: Oxford University Press, 2003), 115–19; and Kevin Warwick, *I, Cyborg* (Urbana: University of Illinois Press, 2004), 265; for StelArc's declaration of posthuman principles, see his Web site, http://www.stelarc.va.com.au/index2.html (accessed June 1, 2006).
36. Warwick, *I, Cyborg*, 175–76.
37. Ibid., 175–76, 282–83.
38. Brooks, *Flesh and Machines*, 216.
39. Ibid., 228; Garreau, *Radical Evolution*, 6.
40. Kurzweil, "Promise and Peril," 50; http://www.foresight.org/Nanomedicine/NanoMedFAQ.html#FAQ19 (accessed June 1, 2006).
41. Peter Cochrane is quoted in Warwick, *I, Cyborg*, 73.
42. Albert Borgmann, "On the Blessings of Calamity and the Burdens of Good Fortune," *The Hedgehog Review* 4.3 (Spring 2002): 12–13.
43. Hayles, *How We Became Posthuman*, 64; emphasis in original.
44. John R. Searle, *The Mystery of Consciousness* (New York: New York Review of Books, 1997), 192–93; Varela, Thompson, and Rosch, *The Embodied Mind*, xv, 52.
45. Clark, *Natural-Born Cyborgs*, 136–37; Brian Cooney, *Posthumanity: Thinking Philosophically About the Future* (Lanham, Md.: Rowman & Littlefield, 2004), xix; Hayles, *How We Became Posthuman*, 109.
46. Clark, *Natural-Born Cyborgs*, 131, 135 (emphasis in original); Cooney, *Posthumanity*, 43; for the cultural lag that will accompany "social acceptability," see, for example, Brooks, *Flesh and Machines*, 229–30.
47. Clark, *Natural-Born Cyborgs*, 4, 11.
48. Warwick, *I, Cyborg*, 89, 157, 308; Brooks, *Flesh and Machines*, 212; Clark, *Natural-Born Cyborgs*, 132; Kurzweil is quoted in Garreau, *Radical Evolution*, 105, and Garreau, 6–8; Turkle is quoted in Cooney, *Posthumanity*, xix.
49. Langdon Winner, "Are Humans Obsolete?" *The Hedgehog Review* 4.3 (Fall 2002): 41–42.

50. Gilbert Meilaender, "Genes as Resources," *The Hedgehog Review* 4.3 (Fall 2002): 73.
51. Brooks, *Flesh and Machines*, 229–30.
52. Marx, "Postmodern Pessimism," 238; see also Bill Joy, "Why the Future Doesn't Need Us," in *Society, Ethics, and Technology* (New York: Wadsworth, 2002), eds. Morton Winston and Ralph Edelbach, 231–48.
53. Searle, *The Mystery of Consciousness*, 190–92; Cooney, *Posthumanity*, xxii, 5–6, 87–88.
54. Brooks, *Flesh and Machines*, 206.
55. Cooney, *Posthumanity*, ix–xx.
56. Bill Joy is quoted in Cooney, *Posthumanity*, 155; and Cooney, *Posthumanity*, xxiv.
57. Such sci-fi myths have substantial influence on national politics. "The Terminator," now "The Governator," was elected without any political experience on the grounds of technological fantasy. Then there's the $100 billion spent over twenty years on the Strategic Defense Initiative/"Star Wars" program. Now called the Ballistic Missile Defense Organization, it has never shown the remotest potential for success.
58. Richard Slotkin, *Regeneration through Violence* (Middletown, Conn.: Wesleyan University Press, 1973), and *Gunfighter Nation* (New York: Atheneum, 1992).
59. Sturken and Thomas, "Introduction," 3.
60. Donna Haraway, "The Actors Are Cyborg, Nature Is Coyote, and the Geography Is Elsewhere: Postscript to 'Cyborgs at Large,'" in *Technoculture*, eds. Constance Penley and Andrew Ross (Minneapolis: University of Minnesota Press, 1991), 16, 21, 25.
61. Nye, *America as Second Creation*, 293.
62. James W. Perkinson, *White Theology: Outing Supremacy in Modernity* (New York: Palgrave Macmillan, 2004), 70, 189.
63. Joel Dinerstein, *Swinging the Machine: Modernity, Technology, and African-American Culture Between the World Wars* (Amherst: University of Massachusetts Press, 2003), 22–24, 312–14, and passim.
64. Kurzweil, "Promise and Peril," 62 (my emphasis); Mike Davis, *Planet of Slums* (London: Verso, 2006).
65. Becker, *Progress and Power*, 7; Paul Tillich, *The Spiritual Situation in Our Technical Society*, ed. J. Mark Thomas (Macon, Ga.: Mercer University Press, 1988), 190–91; McKibben is quoted in Cooney, *Posthumanity*, xx.
66. Ashis Nandy, *Alternative Sciences: Creativity and Authenticity in Two Indian Scientists* (Oxford: Oxford University Press, 1995); Ashis Nandy, ed., *Science, Hegemony and Violence: A Requiem for Modernity* (New York: Oxford University Press, 1990).

Technology and the Production of Difference

David E. Nye

Just as the linguistic turn once invited all Americanists to consider rhetoric and representation as crucial parts of every analysis, the field of technology is becoming unavoidable for anyone concerned with communications, material culture, labor, capitalism, globalization/Americanization, or the social construction of culture itself. In part, this is because the meanings of an artifact and its design are flexible, varying from one culture to another, and from one time period to another. Henry Petroski, one of the most widely read experts on design, argues that there is no such thing as perfect form: "Designing anything, from a fence to a factory, involves satisfying constraints, making choices, containing costs, and accepting compromises."[1] Technologies, which include all of material culture, are social constructions with political and social implications.

Once one accepts that technology is far more than steam engines, nuclear reactors, and iPods, but also includes spinning wheels, pottery, bras, hand tools, and Neolithic axes, it is difficult to think of any topic that does not have a technological dimension. Many scholars find themselves becoming historians of technology without quite realizing it. No one interested in slavery can ignore prize-winning works in the history of technology such as Angela Lakwete's *Inventing the Cotton Gin: Machine and Myth in Antebellum America* or Judith Carney's *Black Rice: The African Origins of Rice Cultivation in the Americas.* Who interested in racism can ignore Michael Adas, *Machines as the Measure of Man*?[2] No one concerned with music or public space will want to miss Emily Thompson's *The Soundscape of Modernity: Architectural Acoustics and the Culture of Listening in America, 1900–1933.*[3] Studies of technology are indispensable for understanding the social construction of the home, the landscape, the city, the workplace, transport, energy systems, and cultural reproduction.[4]

As American studies scholars discover potential synergies with this literature, they will find that few historians of technology believe machines are in the saddle and ride mankind. Instead, various forms of social construction dominate the field's theoretical agenda.[5] Outside the academy, however, technological determinism is more common than one might suppose; its adherents include optimists such as former Speaker of the House Newt Gingrich and his colleagues at the American Enterprise Institute. Inspired in part by Alvin Toffler, Gingrich has long been convinced that machines (notably computers) are improving the world.[6] In the same spirit, Nicholas Negroponte declared in a best-selling book of the 1990s: "Digital technology can be a natural force drawing people into greater world harmony."[7] This is nonsense, but widely believed nonsense. No technology is, has been, or will be a "natural force." Nor will any technology by itself break down cultural barriers and bring world peace.

Technological optimists generally believe in the virtues of the "free market," as if a beneficent determinism were the inevitable outcome of "the invisible hand" in laissez-faire economics. In declaring that television, the Internet, or some other device was "inevitable," they seem to believe that these technologies are so appealing that consumers, given the chance, "naturally" will buy them. In contrast, scholars who focus on technology generally reject this view, because they examine not only consumers but also inventors, entrepreneurs, and marketers. They see each new technology not simply as a product, but as part of a larger system of artifacts. If some popular authors (and many politicians) seem to believe that machines determine history, historians generally agree that new technologies are shaped by social conditions, competing products, prices, traditions, popular attitudes, interest groups, class differences, and government policy.[8]

Yet, many twentieth-century academics long were, just as a few contemporary authors still are, determinists. They commonly put forward two arguments, each wrong-headed in its own way: (1) advanced technologies are homogenizing the world, dissolving distinctive cultures into a global system, and (2) advanced technologies are driving society toward more choices and greater difference. The second argument emerged with particular force during the Internet boom of the 1990s, when a chorus of stockbrokers and pundits sang the praises of all things digital. Esther Dyson, George Gilder, George Keyworth, and Alvin Toffler, whom some call "cyber-libertarians," released "Cyberspace and the American Dream: A Magna Carta for the Knowledge Age." Among their many claims was this deterministic assertion: "Turning the economics of mass-production inside out, new information technologies are driving the financial costs of diversity—both product and personal—down toward zero, 'demassifying' our institutions and our culture. Accelerating demassification creates the potential for vastly increased human freedom."[9] Similar ideas permeated the promotional literature surrounding the so-called New Economy of the late 1990s.

For decades, critics of industrialization had asserted just the opposite, arguing that mass production of goods and improved communications together erased cultural differences. During much of the twentieth century, sociologists argued that industrial technologies were homogenizing people, places, and products. As the assembly line produced identical goods, it seemed to erase difference. Workers became interchangeable, and consumers with identical houses and cars seemed interchangeable as well. As technical systems became more complex and interlinked, the argument ran, human beings became dependent upon the machine and had to adjust to its demands. Technology shaped the personality and dominated mental habits. Thorstein Veblen declared: "The machine pervades the modern life and dominates it in a mechanical sense. Its dominance is seen in the enforcement of precise mechanical measurements and adjustments and the reduction of all manner of things, purposes and acts, necessities, conveniences, and amenities of life, to standard units."[10] For Veblen's generation, the advent of the assembly line and mass production became a powerful metaphor for standardization and machine domination.

The homogenizing effects of technology seemed most obvious in the United States. Veblen's contemporary, the distinguished Dutch historian Johan Huizinga noted in *Life and Thought in America*: "The progress of technology compels the economic process to move toward concentration and general uniformity at an ever faster tempo. The more human inventiveness and exact science become locked into the organization of business, the more the active man, as the embodiment of an enterprise and its master, seems to disappear." Huizinga noted the attraction of interchangeability and argued, like Alexis de Tocqueville before him that "the American *wants* to be like his neighbor." Indeed, the American "only feels spiritually safe in what has been standardized."[11] The uniformity had become so great by 1918 that a visitor, traveling a thousand miles from one city to another inside the United States, could frequently be disappointed, because the new place seemed so much like everywhere else.

In the 1930s and after, the Frankfurt School viewed modern communications as the primary means of creating uniformity and social control.[12] Its members feared that as publishing, radio, and film penetrated popular culture, they packaged, standardized, and trivialized human complexities. Max Horkheimer complained to a colleague: "You will remember those terrible scenes in the movies when some years of a hero's life are pictured in a series of shots which take about one minute or two, just to show how he grew up or old, how a war started and passed by, and so on. This trimming of existence into some futile moments which can be characterized schematically symbolizes the dissolution of humanity into elements of administration."[13] From such a perspective, cultural power had passed to the dream factories of Hollywood and the songsmiths of Tin Pan Alley.

Such pronouncements became a litany by the 1950s. The modern man in David Reisman's *The Lonely Crowd* no longer actively directed his life but was shaped by forces and movements outside himself.[14] It seemed obvious that the more technological a society became, the more uniform was its cultural life. One critic complained, "As for pluralism, differences in the technological state are able to exist only in private activities: how we eat; how we mate; how we practice ceremonies. Some like pizza; some like steaks; some like girls; some like boys; some like synagogue; some like mass. But we do it in churches, motels, [and] restaurants" indistinguishable from one another.[15]

By the early 1960s, a chorus of authors attacked industrial society. Jules Henry, in a widely discussed book, *Culture Against Man*, argued that modern people had acquired "technological drives" that were destructive. These "drives can become almost like cannibals hidden in a man's head or viscera, devouring him from inside. . . . the American then may consume others by compelling them to yield to his drivenness." These technological values were also self-destructive, in his view: "Americans get heart attacks, ulcers, and asthma from the effects of their drives, and it seems that as exotic cultures enter the industrial era and acquire drive, their members become more and more subject to these diseases."[16] Henry argued that science and technology had become the center of a "culture of death."

Likewise, the student revolt of the 1960s was in part a rejection of the values of efficiency and standardization. Mario Savio attacked the University of California for conceiving itself as a factory, in which the president was the manager, the faculty were the workers, and the students were raw materials to be manufactured into docile white-collar workers. He declared to cheering Berkeley students:

> There is a time when the operation of the machine becomes so odious, makes you so sick at heart, that you can't take part; you can't even passively take part, and you've got to put your bodies upon the gears and upon the wheels, upon the levers, upon all the apparatus, and you've got to make it stop. And you've got to indicate to the people who run it, to the people who own it, that unless you're free, the machine will be prevented from working at all![17]

Savio was hardly alone in such views. Theodore Roszak's *The Making of a Counter Culture* also attacked the attempts of government and corporate experts to manage and standardize daily life. He declared that the "prime strategy of technocracy" was "to level down to a standard of so-called living that technical expertise can cope with—and then, on that false and exclusive basis, to claim an intimidating omnicompetence."[18] A popular song from the same period by Malvina Reynolds complained of towns of "little boxes" and described how the identical people living in these boxes had identical children who all went to university where they were also put into boxes, "little boxes, all the same." The 1960s student revolt was far more than a reaction against institutionalized racism or against the Vietnam War. It was also a reaction against standardization, efficiency, and business-directed routines that young radicals felt had come to dominate their lives.

One community seemed to epitomize the world of "little boxes," Levittown, Long Island, built after World War II.[19] The same builders constructed all the houses at the same time using mass-production techniques that they had developed working under government contracts during the war. In several locations, they eventually would build more than 140,000 houses, using identical parts that were precut or prefabricated before arrival at the construction site. Levittown, Long Island, contained 17,400 houses, then the largest development ever put up by one company. Buyers literally lined up to purchase the homes, in part because there they got more floor space for their money than elsewhere. Yet, many criticized the development. Wouldn't such uniformity produce standardized, soulless people? Riesman compared the suburban world to the conformity of a small college fraternity.[20] The architectural critic Paul Goldberger declared it "an urban planning disaster." Nor was Levittown an isolated example, as similar planned communities were built in much of the Western world. Standardization seemed to run rampant.

Yet today Levittown is not a monotonous row of "little boxes." Over decades, home owners have added garages, pillars, dormers, fences, and extensions. They have painted their homes many different colors and planted quite different shrubs around them, landscaping each plot into individuality.[21] In 2006, a visitor to Levittown has to study the houses carefully to see their common elements. Three generations of home owners have used a wide range of technologies to obliterate uniformity. Starting out in the 1950s with a relatively homogenous population that was white, middle-class, and young, the mass-produced suburb of Levittown has become increasingly diverse in its appearance, demographics, and racial makeup. A house that sold for $6,700 there in 1952 was worth $300,000 a half century later, and could be sold immediately.[22]

The reshaping of Levittown finds corollaries elsewhere, notably in the automobile industry. During the first years of the assembly line, Henry Ford refused to manufacture a wide variety of cars. Instead, the Model T was available in only a few variant forms and a limited range of colors. Ford was reputed to have said that customers could get any color they wanted, so long as it was black. In fact, a few other colors were available in the 1920s. Nor was the model design completely static, as the company constantly made small improvements. But Ford eschewed the annual model change, because it was expensive to retool the assembly line to accommodate what he regarded as mere facelifts in the car's external appearance. By freezing the basic design, Ford could concentrate on improving the efficiency of his assembly line, which in turn made it possible to lower the cost of his cars. A new Model T dropped dramatically in price, from $850 in 1908 to $360 in 1916, with further reductions into the 1920s.[23] In contrast, General Motors brought out new models every year, in a changing variety of colors. To do this, GM annually retooled its assembly line and passed this cost on to the consumer. To Ford's dismay, the public embraced these changes, and gradually GM won so much market share that Ford reluctantly abandoned the Model T in 1927 (after producing more than 15 million) and began to make annual models with greater variety.[24] Today, even the least expensive cars come in many colors, and midrange automobiles can be made in literally hundreds of different ways, depending upon the upholstery, accessories, colors, and options a buyer chooses. In 2006, the Ford F-150 truck was available in more than a hundred different configurations that included variations on the cab, bed, engine, drivetrain, transmission, and trim. In addition, there were different colors in upholstery and exterior paint.[25] For this vehicle alone there were hundreds of possibilities to choose from, and once the vehicle was purchased, each owner could further customize it to the point that it literally was one of a kind.

The telephone provides another example of differentiation replacing standardization.[26] In the early twentieth century, AT&T completely dominated telephone design, and virtually all Americans had black desktop telephones. Engineers designed them to be long lasting and functional, and the consumer had no choice. Indeed, the consumer did not even own the home telephone, but rented it by the month. Engineering concerns for sound quality and national service predominated over questions of design. AT&T sought profits not by selling a product that soon became obsolete, but by renting a product that was extremely durable. It developed not a diversity of telephones but the world's most extensive system, serving a large customer base. However, just as Henry Ford could not keep selling only Model Ts, eventually "Ma Bell" had to pay more attention to consumers. For several generations, the major innovations in the telephone were technical: better sound quality, direct dialing, better long-distance connections, and so forth. In contrast, by the 1950s Europeans had telephones with more visual appeal, though the service itself almost always cost more. In the 1960s AT&T also began to offer a wider variety of telephone colors and styles. It encouraged families to acquire different designs for kitchens, living rooms, and bedrooms. This campaign was a success, but proved to be only the beginning. Once consumers could own telephones rather than rent them, literally thousands of different designs appeared. Novelty phones were sold that resembled almost any conceivable object. When AT&T was split up in the 1980s, the proliferation of phone companies diversified service as well as the phone itself.

At the same time, telephones moved from being collective to individual. In 1950, it was still common for several families to share a party line. By 1970 most households had a private line, shared by several family members. In subsequent decades, separate phones for different members of the family, especially teenagers, became more common, but the full differentiation of the telephone came with the explosive growth of mobile phones, each owned by one person who carried it all the time. Not only did many firms produce mobile phones, but they constantly upgraded the models so they seemed to have as many options as a sports car. Consumers not only had distinctive phones, but they continually changed their ring tones, which by 2004 had become a major source of revenue for the music industry. In short, as with Levittown and the Ford motorcar, the telephone metamorphosed from a uniform, mass-produced item into a highly differentiated product.

No institution better understood consumer demand for variety than the department stores. In each of the industrial nations these new emporia emerged in the second half of the nineteenth century, notably in France, Britain, and the United States, but also in Germany, Scandinavia, Canada, and elsewhere. Department stores covered an entire city block or more and had at least four floors, creating a vast commercial space that dwarfed the traditional shop. They competed not only on price but also on selection, giving consumers choices among products differentiated into the widest possible spectrum of styles and qualities. Managers found that consumer taste changed rapidly, and it became their business to shift their assortment accordingly. By the early twentieth century, manufacturers routinely relied on department store managers and buyers to find out what the public wanted.

Such "fashion intermediaries" became crucial interpreters of consumer desires. As Regina Blaszczyk concluded: "Make no mistake: supply did not create demand in home furnishings, but demand determined supply."[27] Her *Imagining Consumers* shows how firms catered to the mass market through an interactive process. They could seldom dictate taste but rather succeeded when they discovered what the consumer wanted and then delivered it quickly through flexible production. Department stores had to stock more than mass-produced goods, because their customers did not want to dress or furnish their homes all in the same way. Consumer demands were relayed to factory designers through fashion intermediaries including salesmen, retail buyers, materials suppliers, art directors, showroom managers, home economists, advertising executives, and market researchers. The "field letters of factory salesmen read like primitive market research reports."[28] A wholesale jobber in Denver listened to country storekeepers as they inspected wares and then reported their comments back to the factory. The department store did not dictate style to the consumer so much as it relayed changes in taste to the manufacturer. As a J. Walter Thompson executive put it in 1931: "The consuming public imposes its will on the business enterprise."[29]

Because consumers demanded differentiation, mass production was not the only way or necessarily the most profitable way to manufacture. While the early twentieth century was mesmerized by assembly lines and giant corporations such as Standard Oil, Ford, and General Motors, one of the leading scholars in the history of technology, Philip Scranton, has focused attention on midsized and smaller firms that made a wide variety of consumer goods through customized and small batch production. Far from being backwaters destined to be rationalized by scientific management or converted to assembly lines, such companies were just as important to an advanced economy as the large corporation. Because they were smaller than a Ford or GM and tended to remain family firms or closely held corporations, they have not attracted so much attention. But collectively, in 1909 such firms contributed a third of the value added in the economy as a whole, and employed one-third of all workers.[30] Their specialty production grew just as fast as mass production, and they added more value and employed more workers. By 1923 value added had tripled and specialty firms still accounted for a third of industrial employment. They did not produce identical goods, but responded flexibly to demands for variety. Such companies were innovative and profitable, and they made possible the endless novelty that was the hallmark of a consumer society. They had little use for standardization but explored many different productive systems. They might use scientific management or other systematic approaches to increase reliability, to reduce errors, or to reorganize, but they resisted changes that froze design or constrained innovation.[31] These companies excelled in the batch production of such items as carpets, furniture, jewelry, cutlery, hats, and ready-to-wear clothing. Then as now, middle-class consumers wanted new styles, and large profits accrued for the firms with flexible modes of production that could supply differentiated product lines to retailers.

The critics of the first half of the twentieth century overstated the degree to which mass production would be accompanied by conformity and standardization of the personality. Neo-Marxists were particularly scathing in making the case, and supposed that society passed through unavoidable stages of increasing regimentation, labeled "Taylorism and Fordism."[32] But historians of technology seldom use these terms to describe production or consumption.[33] They see a far greater complexity and variety in the organization of work, and they find that consumers relentlessly demand variety. Over time, even the companies and products that seemed to epitomize mass society dissolved into difference. Ford had to abandon his Model T, the universal black telephone from AT&T evolved into myriad sizes, colors, and designs, and Levittown's uniformity disappeared in a wave of home improvements and landscaping.

The shift from viewing technology as the harbinger of standardization to viewing it as the engine of differentiation is epitomized by attitudes toward the computer. In the 1950s, the computer represented centralized control, systematic information gathering, and the invasion of privacy. It apparently promised the ultimate negation of the individual, making possible an extension of the rationality of the assembly line and its interchangeable parts into new areas. When Jean-François Lyotard wrote *The Postmodern Condition* in the late 1970s, his first pages concerned the effects of the computer on society, and the "hegemony of computers" that brought with them "a certain logic, and therefore a set of prescriptions determining which statements are accepted as 'knowledge' statements."[34] He saw computers as an apparatus of top-down control, in the creation and maintenance of a new circulation of information that would benefit large institutions.

After the early 1980s, however, the computer, once feared as the physical embodiment of rationalization and standardization, gradually came to be seen as an engine of diversity. The key shift was the movement from giant mainframe computers that stored all the data and software at a few central locations to the personal computer, which decentralized the system and made it part of the material culture of everyday life. As individuals gained control over their own machines, they could personalize them, by choosing their favorite font to write in, downloading their favorite songs, or using art works on screen-saver programs. Granted that these were only surface changes; they signaled a shift away from the rigidity of the old mainframes. Gradually, the computer ceased to seem threatening or external and became a convenience at work and a partner in all sorts of play. During the 1990s, millions of individuals began to explore the possibilities of e-mail and the World Wide Web. Each person could, and most students did, construct a home page and broadcast it to the world.

Yet, not all groups have equal access. To what extent is the Internet limited to affluent and well-educated consumers? Is there a digital divide? In the United States in the year 2000, 56.8 percent of Asian households had Internet access, compared to 29.3 percent for African Americans and only 23.7 percent for Hispanics.[35] Income accounts for some of these differences, but the cultural values of minority groups are also important. If one looks at households with incomes below $15,000, one out of three of Asian American households had Internet access, compared to only one out of sixteen African American households or one of nineteen among Hispanic Americans. Given the same resources, these groups have different priorities. At all income levels, the gap remained much the same with Asians and whites on the one side of the digital divide and blacks and Hispanics on the other.[36] By comparison, the digital divide was not gendered, as 44 percent of both men and women had access to the Internet.[37]

Internationally, differences are greater. In 1999 a substantial percentage of the world's population had never made a telephone call, and less than one percent of India's population was on the Internet.[38] Rapid diffusion of technologies to all people everywhere can hardly be taken for granted. Rather, for most of the world, use of the Internet seems limited to the educated and wealthy in urban areas. Yet matters are not so simple. At the end of the 1990s two anthropologists, Daniel Miller and Don Slater, studied the Internet on the island of Trinidad, in order to test the widespread assumptions about its unequal distribution and social effects.[39] Their work suggested that the Internet is more widely diffused than one might expect, but that it is not erasing cultural differences. They looked to see if Internet usage was strongly marked by differences in wealth, since computers, modems, and time online are expensive. They found that Internet use in Trinidad was quite democratically dispersed through Internet cafes. They had expected that businesses might be driving forces in Internet development, notably advertising agencies and telecommunications companies. They were not. On Trinidad, the Internet was far less a project of capitalist hegemony than a grassroots movement. Network use did not conform to expectations of globalization, either. Users did not create new online identities that were more international than their off-line sense of self. Rather, they used the Internet to maintain and strengthen a web of relationships with extended families in Britain, Canada, and the United States.

Nor did the Trinidadians embrace a global culture that weakened their sense of identification with their own nation or culture. Indeed, it was difficult to find islanders who feared that the Internet might absorb them into a global mass culture. On the contrary, they used the Internet to project pride in their own nation, to broadcast its music, to educate others about their islands, and to sell its products and vacation experiences. The Internet did not seem to overwhelm but rather to strengthen the local, for example, making it easier for the music of Trinidad and Tobago to reach the rest of the world. Miller and Slater found that using the Internet did not even seem to undermine the local dialect, as the "Trini" idiom flourished on personal home pages. In short, use of the Internet strengthened and spread local cultural values rather than undermining them. The uses of the Internet on Trinidad suggest once again that people adapt technologies to express local identities. The technological specifications of the Internet in Trinidad need to be much the same as elsewhere for the system as a whole to work, but the social construction of the Internet can vary a great deal. Two decades ago, George Lipsitz noted in *Time Passages* that "the invention of magnetic recording tape made it possible to enter the record business with relatively little capital."[40] Today, the Internet opens up further possibilities for local musicians, allowing them to produce and disseminate their music worldwide, through downloading and podcasting. At the same time, the giant record companies have experienced falling sales.

Recent debate about globalization partially concerns whether technologies are being used to create a more homogeneous or a more heterogeneous world. George Ritzer has warned of "McDonaldization," arguing that fast food restaurants epitomize an impersonal standardization that Western nations aggressively export to the rest of the world.[41] Similarly, Benjamin Barber has attacked the cultural imperialism that he finds is translating a rich cultural variety into a single, bland "McWorld."[42] Francis Fukuyama disagrees with Ritzer and Barber. He does not see cultural imperialism but applauds the triumph of free markets, democracy, and Western cultural values.[43] All of these authors have little interest in technology per se.

Other critics, such as Roland Robertson, find that Western-style corporations do not obliterate local cultural constellations.[44] Instead, each local culture adopts only some of the products and practices offered in the global marketplace, adapting those selected to fit into its own routines. Rather than adjusting to a single pattern, each cultural region creates hybrid forms, which Robertson calls "glocalization." Even a corporation like McDonald's that wants to offer the same menu everywhere on the globe finds that it must give in to this process. It sells red wine to go with its hamburgers in Spain, for example, and it does not serve beef in India, where cows are sacred. Material culture registers how variety emerges out of intercultural negotiation. A process of creolization is taking place, producing such novel combinations as Cuban-Chinese cuisine, Norwegian country-western music, and "Trini" homepages.

As the leading Dutch Americanist Rob Kroes imaginatively argues, in cultural contact zones, people "are scavenging along the tide line of Western expansion, appropriating its flotsam and jetsam. They feel free to rearrange the order and meanings of what they collect. They turn things upside down, beads turn into coinage, mirrors into ornaments. Syntax, semantics, and grammar become jumbled." In this process of selective appropriation, "people at the periphery create their own environment" and this creolized cultural production often may be re-exported.[45] Just as inside the United States former slaves and immigrants created their own cultural worlds, selectively appropriating elements of different cultures, so too other cultures that come into contact with Western society engage in a creolizing process. What results is not a standardized world, but a potentially endless process of differentiation. Just as in the 1920s mass production did not obliterate batch production, which was more flexible and responded to changing consumer demands, fears of standardization were exaggerated. "Glocalization" or creolization is common. However much critics may focus on the supposed homogeneity of a machine culture, the owner of every house, car, and telephone in Levittown can express individuality through consumption.

That argument, based on consumer choice, is not the same as the argument that advanced methods of production and distribution have made it easier to manifest difference. This is an argument based on the technologies used in marketing, and can be illustrated by the shift from a small number of dominant television broadcasters to narrowcasting. In the 1950s, an oligopoly of just three national networks dominated American television, and many European countries had only state television. A half century later, the average home receives dozens of channels, including stations specializing in history, science fiction, old films, sport, nature, foreign language programs, and many other areas. The same process has emerged in marketing; instead of seeking to sell one thing to everyone, marketers have created extensive segmentation that divides consumers into smaller target groups. If the early twentieth century was the era of mass production for a mass market, advertisers in recent decades have become more discriminating. Rather than send out the same brochure to millions of homes, they compile specialized mailing lists, divide the market by postal codes, or use niche cable TV stations that address only a fraction of the national market. On the Internet, market segmentation has reached its logical culmination. The more sophisticated sites, such as Amazon.com, directly interact with customers and suggest new products to them based on their previous purchases. The proliferation of such practices, while giving the seller more chances to reach buyers, suggests top-down social control. It does not, however, suggest technological determinism.

The expansion from the neighborhood store to enormous supermarkets also registers the increasing demand for wide choice. The production and consumption of food has a central place in every culture. As the world's most energy-intensive society, the United States uses 17 percent of its power for food. Roughly a third is for food production, another third for manufacturing and processing, and a third for transportation, refrigeration, cooking, and washing dishes. Does ethnic differentiation flourish within this energy-intensive framework? By the 1980s, a typical large supermarket stocked 30,000 items. The shopping cart might contain foods to make any number of ethnic cuisines. The supermarket made possible myriad combinations, including hybrid meals that reflect the current enthusiasms and the cultural backgrounds within a household. At the supermarket, consumer individualism and advanced technological systems seem compatible. This argument suggests that consumer demands are making advanced technologies into tools for heterogeneity instead of standardization.

There is a counterargument. The price of achieving this variety was the decline of ethnic neighborhood stores and specialty shops. The ethnic variety on the supermarket shelves is based on capitalist rationalization, packaging, and distribution. The system as a whole includes not only the products and the shoppers, but investors seeking maximum profits, an increasingly rationalized agricultural sector that does not favor small farmers, the food processing industry, and the trucking industry. To see why this matters, consider the visitors to the immigration museum at New York's Ellis Island. While there, they can have lunch in a food court much like that in any large shopping mall. The food court offers choices between ethnic cuisines and seems to exemplify the idea that not a melting pot but a "salad bowl" of differentiated groups emerged from immigration.

Yet a single company, the Aramark Corporation, provides all the food served, regardless of its ethnic appearance. Nor is serving food at Ellis Island its core business. Aramark is diversified, with 240,000 employees and more than $11 billion in sales in more than twenty countries.[46] Likewise, on the level of the technological systems used to produce and deliver the food, ethnic differences evaporate. The individual serving areas all use the same kinds of freezers, steam trays, fryers, and microwaves. They prepare dishes suited to the demands of a cafeteria, an assembly line operation that functions best when food does not require much on-site preparation before it is served. They all serve food on disposable plates. These foods may not taste the same, but some of the most pungent differences have been toned down or eliminated to reach a wider customer base. The customers eat these meals in the informal American manner, using little time in the process. If one focuses on process rather than content, these meals are produced, served, and consumed in much the same way.

While every individual can select from a wide range of foods or cars or telephones, she is still enveloped in larger technological systems. It is easier to select among many telephones than it is to do without one. It is easier to make choices at the supermarket than to find an alternative source of food. And it has become almost impossible in much of Western society to live without an automobile. Such patterns have larger implications. Whatever food or automobile they prefer, Americans of all ethnic and racial backgrounds get less exercise than they once did, and they tend to eat high sugar and high fat diets that lead to obesity and the "diseases of affluence."

Furthermore, the majority of ethnic restaurants depend on customers from outside the group that the cuisine "represents." Indeed, "55 percent of America's consumer food budget is spent on restaurant meals and ready-to-eat convenience foods."[47] This is a higher percentage than in most other countries, and it is a further indication that market forces underlie the display of ethnic diversity. In addition, the food is often modified to appeal to the taste of a wider public. For example, Mexican food served abroad is often not quite so fiery as food in Mexico. Likewise, the Swedish smorgasbord in North America usually only vaguely resembles what Swedes eat at home. The surface appearance of variety should not obscure the homogenizing practices involved in marketing, advertising, and cooking cuisines so they will appeal to the general public.

The cyber-libertarians were only partially correct when they predicted a "demassification" of production that would increase differentiation and provide greater scope for the construction of new identities. Consumer liberation to a considerable degree is recontained, or limited, by the technological momentum of institutional structures. Amazon.com undoubtedly provides a massive range of titles, but the independent bookstore stocked with carefully selected titles is fast disappearing. Giant supermarkets did the same to local groceries, and a Wal-Mart can decimate main street. Consumers do get differentiation of goods in the marketplace, but like the people eating lunch at Ellis Island, they may have only one supplier. Massive institutions often deliver what appears to be "demassified difference."

This process as a whole might be seen as part of the "invention of tradition" argument put forward by Eric Hobsbawm and Terence Ranger. Much of what seems the venerable survival of ancient customs turns out to have been shaped or even created wholesale by nineteenth-century nationalists intent on establishing a pedigree for a certain cultural group, for example, through the promotion of national flags, the recovery and promotion of folk dances and traditional clothing, or the writing of national anthems. Hobsbawn wrote of "the use of ancient materials to construct invented traditions of a novel type for quite novel purposes. A large store of such materials is accumulated in the past of any society," and at times "new traditions could be readily grafted on old ones . . . by borrowing from the well-supplied warehouses of ritual, symbolism and moral exhortation—religion and princely pomp, folklore and freemasonry (itself an earlier invented tradition of great symbolic force)."[48]

Technologies are related to this process in several ways. First, technologies often have been used to disrupt the social fabric and undermine custom, creating a need for new, invented traditions as substitutes for lost routines and undermined social patterns. Indeed, Hobsbawm argues that the invention of tradition will "occur more frequently when a rapid transformation of society weakens or destroys the social patterns for which 'old' traditions had been designed."[49] This has been particularly the case for the last two centuries. Second, newly invented traditions are almost always disseminated and discussed through the media. Third, to the extent that embracing new traditions can be enhanced through tourism, the steamboat, railroad, and later the motorcar and airplane were essential to mass participation. For example, in United States certain physical structures became national symbols, notably Times Square, the Sears Tower in Chicago, and the Golden Gate bridge. The patriotic subject saw innumerable photographs of these sites, and wanted to see them in person. The national transportation network made it easier to satisfy this desire. Not incidentally, many of the railway stations, airports, and highways on this journey also became images of national greatness, and in some cases were understood to be expressions of the technological sublime.[50] Finally, the increases in transportation created opportunities to sell trinkets, souvenirs, food, and lodging to travelers and tourists.

Just as markers of Scottish nationalism, such as elaborate varieties of tartans for every clan, were invented in the nineteenth century, in the last decades of the twentieth century groups searched their histories for sources and traditions that might underpin a separate sense of identity. In cases of such corporate co-optation, as Naomi Klein notes,[51] marketers were pleased to oblige the new preferences, and advertisements of the 1980s and 1990s stressed visible differences. Yet, outside these corporate structures, in moments of creative appropriation, according to Rob Kroes, the invention of traditions also has become pervasive. Instead of creating traditions that reinforced a unifying nationalism, grassroots organizations and consumers demand products that celebrate racial, ethnic, and regional diversity.

In the invention of new traditions, however, the young are especially adept at discovering unsuspected possibilities in new technologies. Think of the myriad unanticipated uses of the World Wide Web, of the emergence of skateboarding,[52] of the creation of drag racing, of the amateurs' discovery of radio broadcasting in the 1920s, of the (re)emergence of podcasting eighty years later, of the bicycle craze of the late nineteenth century, of flagpole sitting, of Jimi Hendrix playing and destroying his electric guitar. These were all creative appropriations by the young, who continually transform the uses and meanings of technologies to suit their fancy. In contrast, older people are often slow to explore the possibilities latent in a new device. John Perry Barlow, who wrote lyrics for the Grateful Dead and later became a consultant to computer companies, argues that "it takes about thirty years for anything really new to arise from an invention, because that's how long it takes for enough of the old and wary to die."[53] A close observer of new technologies, Barlow is hardly a determinist.

In contrast, for much of the nineteenth and twentieth centuries, sociologists and historians assumed that the machine age could only lead to a crushing homogeneity. But in practice, people have often used technologies to create differences. Even otherwise conventional middle-class consumers generally prefer variety. A manufacturer bent on absolute uniformity, such as Henry Ford, eventually had to give in to the public's demand for a range of models and options. Likewise, home owners proved adept at transforming Levittown's rows of identical, mass-produced homes into variegated neighborhoods. Difference triumphed over uniformity. And were the immigrants, racial minorities, and the poor not at least as inventive?

By the end of the twentieth century, it was clear that people in highly technological societies preferred to maximize differentiation. Racial, regional, and ethnic communities developed separate identities by inventing and disseminating new traditions, and they used off-the-shelf technologies in new ways. Likewise, in contact zones between highly technological societies and the rest of the world, a creolization process took place, as many peoples selected and rearranged elements of Western culture and absorbed them into their own traditions. To be sure, corporations proved adept at co-opting or imitating these developments, creating more diverse product lines and using new sales methods. Yet, technologies were not simply being used to eradicate cultural differences and thereby create a single, global culture. In some cases, people used them to enhance the possibilities for the survival of marginal communities or the growth of minority cultures.

Consider three Los Angeles examples of creolization, each of which shows how technologies can be put to the service of creative difference. The Watts Towers stand as a representative of such reappropriations. For more than thirty years (1921–1955), the Italian immigrant Simon Rodia built the towers using ordinary hand tools. He followed no master plan, but intuitively used steel rods, concrete, and most important, a great many cast-off materials, including broken glass, fragments of pottery, seashells, and pieces of ordinary ceramic tile.[54] As the *Los Angeles Times* declared, Rodia transformed "the mundane and familiar, the mass-produced and commonplace, into something unlike anything else, something singular and surprising and strange."[55] Likewise, Mexican Americans in Los Angeles creatively appropriated the automobile, as they developed elaborately painted and restyled "lowrider" cars. Finally, black Americans have re-created a new music out of the detritus and waste of the city. As Tricia Rose put it, "hip-hop transforms stray technological parts intended for cultural and industrial trash heaps into sources of pleasure and power."[56] In these and countless other cases, people have taken control of technologies as part of their means of self-expression.

The spatial formation of Los Angeles, as analyzed by Mike Davis in *City of Quartz,* shows that control over technology, in the form of roads and the built environment, is also at the heart of local politics.[57] Technology is a social construction that can be used to build barriers as well as to build community. In the old downtown L.A., "Anglo, black, and Mexican shoppers of all ages and classes rubbed shoulders." The new downtown, built many blocks away, was cordoned off from the poor. "Ramparts and battlements, reflective glass and elevated pedways, are tropes in an architectural language warning off the underclass Other."[58] These shopping areas and wealthy, gated communities articulate a set of values embodied in their cable television, broadband connections, and new automobiles. Yet beyond their gates are the welter of ethnic communities and poor neighborhoods, which are hardly the passive recipients of whatever trickles down from the globalized economy. These highly inventive communities generate new clothing, musical forms, and other styles in material culture that often are reappropriated by "global" culture. Using automobiles and mobile phone networks, they cruise much of the city space and give it their own meanings. This dynamic between what John Fiske calls an imperializing power and the localizing powers of the margins energizes much of popular culture, moving it away from homogeneity toward differentiation.[59]

In short, every group uses technologies to express and shape identity and at times to defend it, whether it be those who customize cars into lowriders, the builders of corporate Los Angeles, or the Amish with their careful selection of which machines they will permit within their communities. Clearly, technology does not automatically empower people, in a romance of pluralism in which identity politics reigns supreme. Governments and corporations seek to direct and recontain impulses toward self-expression. If technologies are central to the invention of material cultures, these social constructions are always contested. A technological outcome, whether it be the $14.6 billion "Big Dig" in Boston or the construction of a new house, is not automatic, but negotiated. The abstract term "technology" too often has been another word for fate.[60] It is best understood neither as a hegemonic force for homogenization nor as an automatic agent of liberation, but as a complex system of tools, materials, structures, machines, and techniques under human control. Societies continually add and subtract items from their repertoire, and in using them they often come into conflict, as they construct diverse American life-worlds.

The challenge for American studies is not only to examine how technologies have been incorporated into cultures of difference, but also to prepare students to take part in the social construction of emerging technologies. Too often these are left to the private sector, as if the market alone can adjudicate the best uses of new machines. No one who has studied how computers and the Internet have been and are being used to reconstruct economic hierarchies and to remap the social sphere can see this as a laissez-faire process.[61] The use of the Internet has led to social crises and legal struggles over free speech, privacy, pornography, and e-commerce.[62]

It therefore seems appropriate, in closing, to point out another area in which the social construction of technologies will be a cultural battleground during the lifetimes of today's undergraduates. Scientists such as Aubrey de Grey at Cambridge are seeking ways to prolong human life, not merely a decade or two, but at least fifty years and perhaps as much as a thousand or more. They argue that "medicine is a branch of engineering" and expect that when cells wear out or become diseased, cloned cells or cultured stem cells can be used to replace them.[63] Based on increasing knowledge of the human genome, some have seriously begun to talk about achieving immortality. But this is not merely a wonderful achievement that can be left to private initiative, as it raises profound political questions. Will access to life-extending genetic technologies be democratically shared or more available to some racial groups and economic classes? At what age would female fertility in this population end? Would childbirth be more or less frequent? Would world population rapidly reach unsustainable levels? Would people retire at 150, or 200, or only if their bodies were breaking down? What kind of society would such a long-lived population want? Today's undergraduates may be involved in the social construction of life-extending technologies that have profound social and political consequences. If they have a deterministic view of the technologies of the past, they will be ill prepared.

Just as American colonists demanded the vote in the 1770s, they will need a voice when technological decisions are made about such matters as the technological prolongation of life, data-mining (and the right to privacy), the possible rights of intelligent machines, cyborgs, or the development of new kinds of pharmaceuticals, such as "cognition enhancing" drugs.[64] As Steven Goldman has argued, "science and technology policies have a social impact comparable to that of taxation policy in the colonial period."[65] All too often, however, new machines and processes are introduced by way of the patent system, with little or no discussion in the legislature. In the new millennium, formal politics focuses too exclusively on social programs and national security. The citizen may feel secure when terrorists are arrested, because they avowedly threaten public safety. Yet, as Mulford Q. Sibley emphasized, "one never hears of the FBI rounding up those who introduce new machines which are depopulating the countryside, depriving men of their vocations of a lifetime, destroying much of the earth, polluting the atmosphere," or otherwise transforming social and political life. "Until Americans develop the law, standards, practice, and organization for deliberate introduction or rejection of complex technology, their supposed self-government will remain largely a pretence."[66] Technological determinism may have been rejected in the academy, but in public life it remains a vigorous and misleading idea, one that tells citizens they have no agency, that they must accept a particular constructed future as though it were a single, unalterable fate. In contrast, this essay suggests that however much Americans may have feared the homogenization that new technologies made possible, they have used them for the production of difference.

Notes

I would like to thank MIT Press for allowing me to reprint portions of my argument from *Technology Matters: Questions to Live With* (MIT Press, 2006).

1. Henry Petroski, *Small Things Considered: Why There Is No Perfect Design* (New York: Vintage, 2003), 13.
2. Angela Lakwete, *Inventing the Cotton Gin: Machine and Myth in Antebellum America* (Baltimore: Johns Hopkins University Press, 2003); Judith Carney, *Black Rice: The African Origins of Rice Cultivation*

in the Americas (Cambridge, Mass.: Harvard University Press, 2002); Michael Adas, *Machines as the Measure of Man* (Ithaca, N.Y.: Cornell University Press, 1989).

3. Emily Thompson, *The Soundscape of Modernity: Architectural Acoustics and the Culture of Listening in America, 1900–1933* (Cambridge, Mass.: MIT Press, 2002).
4. Representative works include Ruth Schwartz Cowan's *More Work for Mother: The Ironies of Household Technology from the Open Hearth to the Microwave* (New York: Basic Books, 1985); David E. Nye, ed., *Technologies of Landscape: Reaping to Recycling* (Amherst: University of Massachusetts Press, 2000); Martin V. Melosi, *The Sanitary City* (Baltimore: Johns Hopkins University Press, 1999); Stephen B. Goddard, *Getting There: The Epic Struggle Between Road and Rail in the American Century* (Chicago: University of Chicago Press, 1994); Carolyn Marvin, *When Old Technologies Were New* (New York: Oxford University Press, 1988); and Susan Douglas, *Inventing American Broadcasting, 1899–1922* (Baltimore: Johns Hopkins University Press, 1987).
5. See John Staudenmaier, *Technology's Storytellers: Renewing the Human Fabric* (Cambridge, Mass.: MIT Press, 1985).
6. Cited in Langdon Winner, *Autonomous Technology: Technics-out-of-Control as a Theme in Political Thought* (Cambridge, Mass.: MIT Press, 1977), 61.
7. Nicholas Negroponte, *Being Digital* (New York: Vintage, 1995), 230.
8. For the last fifteen years more than two-thirds of all articles in the journal *Technology and Culture* have employed some form of a contextualist approach. For an overview, see John M. Staudenmaier, "Rationality versus Contingency in the History of Technology," in *Does Technology Drive History? The Dilemma of Technological Determinism*, ed. Merritt Roe Smith and Leo Marx (Cambridge, Mass.: MIT Press, 1994), 260–73.
9. Esther Dyson, George Gilder, George Keyworth, and Alvin Toffler, "Cyberspace and the American Dream: A Magna Carta for the Knowledge Age," *Release* 1.2 (August 22, 1994).
10. Cited in Winner, *Autonomous Technology*, 196.
11. Johan Huizinga, *Life and Thought in America: A Dutch Historian's Vision, from Afar and Near* (trans. from the Dutch, 1918; New York: Harper Torchbooks, 1972), 234, 237.
12. On the Frankfurt School, see Martin Jay, *The Dialectical Imagination* (Boston: Little, Brown, 1973).
13. Quoted in Jay, *The Dialectical Imagination*, 214.
14. David Reisman, *The Lonely Crowd* (1950; New Haven: Yale University Press, 2001).
15. George Grant, *Technology and Empire* (Toronto: House of Anasi, 1969), 26.
16. Jules Henry, *Culture Against Man* (New York: Random House, 1963), 15.
17. The Savio speech can be found at http://www.democracynow.org/article.pl?sid=03/11/21/1524217 (accessed June 6, 2006).
18. Theodore Roszak, *The Making of a Counter Culture: Reflections on the Technocratic Society and Its Youthful Opposition* (Garden City, N.Y.: Doubleday, 1969), 12.
19. This paragraph relies on Kenneth T. Jackson, *Crabgrass Frontier: The Suburbanization of the United States* (New York: Oxford University Press, 1985), 234–38.
20. David Riesman, "The Suburban Sadness," in *The Suburban Community*, ed. William Dobriner (New York, 1958), 375–402.
21. See the materials assembled by Peter Bacon Hales, on the transformation of the original suburb, at http://tigger.uic.edu/~pbhales/Gottscho.html (accessed June 6, 2006).
22. *New York Times*, July 13, 2003, Late Edition, Section 11, 1.
23. John B. Rae, *The American Automobile* (Chicago: University of Chicago Press, 1965), 61.
24. Ibid., 95–99. Richard S. Tedlow, *New and Improved* (New York: Basic Books, 1990), 158–81.
25. See http://www.fordvehicles.com/trucks/f150/features/ (accessed June 6, 2006).
26. This paragraph is based on an excellent paper presented by Kenneth Lepartito at the 1996 annual meeting of the Society for the History of Technology in London.
27. Regina Lee Blaszczyk, *Imagining Consumers: Design and Innovation from Wedgewood to Corning* (Baltimore: Johns Hopkins University Press, 2000), 13.
28. Ibid., 93.
29. Ibid., 229.
30. Philip Scranton, *Endless Novelty: Specialty Production and American Industrialization, 1865–1925* (Princeton, N.J.: Princeton University Press, 1997), 17.
31. Ibid., 99.

32. See, for example, William J. Lederer, *A Nation of Sheep* (Greenwich, Conn.: Fawcett, 1961); for a Marxist version of this argument, see Stuart Ewen, *Captains of Consciousness: Advertising and the Social Roots of the Consumer Culture* (New York: McGraw Hill, 1975).
33. On Taylor and Ford, see David E. Nye, *Consuming Power: A Social History of American Energies* (Cambridge, Mass.: MIT Press, 1998), 131–54.
34. Jean-François Lyotard, *The Postmodern Condition: A Report on Knowledge* (Minneapolis: University of Minnesota Press, 1984), 4.
35. Manuel Castells, *The Internet Galaxy: Reflections on the Internet, Business, and Society* (New York: Oxford University Press, 2001), 249.
36. Ibid., 250.
37. Ibid., 252.
38. Ibid., 261.
39. Daniel Miller and Don Slater, *The Internet: An Ethnographic Approach* (Oxford, U.K.: Berg, 2001).
40. George Lipsitz, *Time Passages: Collective Memory and American Popular Culture* (Minneapolis: University of Minnesota Press, 1990), 138.
41. George Ritzer, *The McDonaldization of Society* (Thousand Oaks, Calif.: Pine Forge Press, 1993).
42. Benjamin Barber, *Jihad vs. McWorld* (New York: Ballantine, 1996).
43. Francis Fukuyama, "Social Capital, Civil Society, and Development," *SAIS Review* 22.1 (Winter 2002): 23–37. For a related argument, see Thomas Friedman, *The Lexus and the Olive Tree* (New York: Anchor Books, 2000).
44. Roland Robertson, *Globalization* (Thousand Oaks, Calif.: Sage Publications, 1992). See also Arjun Appadurai, *Modernity at Large: Cultural Dimensions of Globalization* (Minneapolis: University of Minnesota Press, 1996).
45. Rob Kroes, *If You've Seen One, You've Seen the Mall: Europeans and American Mass Culture* (Urbana: University of Illinois Press, 1996), 164.
46. See http://www.aramark.com/Home.aspx?PostingID=21&ChannelID=2 (accessed June 6, 2006).
47. Alan Durning, *How Much Is Enough? The Consumer Society and the Future of the Earth* (New York: W. W. Norton, 1992), 69, 74, 68, 45.
48. Eric Hobsbawm and Terence Ranger, *The Invention of Tradition* (Cambridge: Cambridge University Press, 1992), 6.
49. Ibid., 4.
50. On technologies as national symbols, see David E. Nye, *American Technological Sublime* (Cambridge, Mass.: MIT Press, 1994).
51. Naomi Klein, *No Logo* (London: Flamingo, 2001).
52. See Michael Nevin Willard, "Séance, Tricknowlogy, Skateboarding, and the Space of Youth" in *Generations of Youth*, ed. Joe Austin and Michael Nevin Willard (New York: New York University Press, 1998).
53. John Perry Barlow, "The Future of Prediction," in *Technological Visions: The Hopes and Fears That Shape New Technologies*, ed. Marita Sturken, Douglas Thomas, and Sandra J. Ball-Rokeach (Philadelphia: Temple University Press, 2004), 181.
54. Bud Goldstone and Arloa Paquin Goldstone, *The Los Angeles Watts Towers* (Los Angeles: Getty Conservation Institute, 1997).
55. Sara Cantina, "Towers of Power," *Los Angeles Times*, October 23, 2005, Section I, 18.
56. Tricia Rose, "A Style Nobody Can Deal With: Politics, Style, and the Postindustrial City in Hip-Hop," in *Popular Culture: A Reader*, ed. Raiford Guins and Omayra Zaragoza Cruz (Thousand Oaks, Calif.: Sage Publications, 2005), 401.
57. Mike Davis, *City of Quartz: Excavating the Future in Los Angeles* (London: Verso, 1990).
58. Mike Davis, "Fortress Los Angeles: The Militarization of Urban Space," in *Variations on a Theme Park: The New American City and the End of Public Space*, ed. Michael Sorkin (New York: Hill and Wang, 1992), 158–59.
59. John Fiske, *Power Plays, Power Works* (London: Verso, 1993), 52.
60. For further discussion, see David E. Nye, *Narratives and Spaces: Technology and the Construction of American Culture* (New York: Columbia University Press, 1998).
61. Janet Abatte, *Inventing the Internet* (Cambridge, Mass.: MIT Press, 1999).
62. Sherry Turkle, *Life on the Screen: Identity in the Age of the Internet* (New York: Simon and Schuster, 1995); John Cassidy, *Dot.con* (New York: Viking, 2002); Tim Jordan, *Cyberpower: The Culture and*

Politics of Cyberspace and the Internet (New York: Routledge, 1999); and James Hughes, *Citizen Cyborg: Why Democratic Societies Must Respond to the Redesigned Human of the Future* (Boulder: Westview, 2004).

63. See http://www.sens.org/index.html (accessed June 6, 2006).
64. Rachel Fishman, "Patenting Human Beings: Do Sub-Human Creatures Deserve Constitutional Protection?" *American Journal of Law and Medicine* 15.4 (Fall 1989): 461–82; Francis Fukuyama, *Our Posthuman Future: Consequences of the Biotechnology Revolution* (New York: Farrar, Straus, Giroux, 2002); Langdon Winner, "Are Humans Obsolete?" http://www.langdonwinner.org/index.html (accessed June 6, 2006); Hughes, *Citizen Cyborg*, 30 and passim.
65. Steven L. Goldman, "No Innovation Without Representation: Technological Action in a Democratic Society," in *New Worlds, New Technologies, New Issues*, ed. Stephen H. Cutcliffe et al. (Bethlehem, Penn.: Lehigh University Press, 1992), 149.
66. Mulford Q. Sibley, "Utopian Thought and Technology," *American Journal of Political Science* 37.2 (Spring 1973): 278.

The Turn Within: The Irony of Technology in a Globalized World

Susan J. Douglas

It is the first decade of the twenty-first century. The homes of millions are pulsating with some combination of advanced communication technologies: satellite dishes or cable TV, DVD and VCR players, desktop computers, laptops, modems, CD players, streaming video, cell phones, palm pilots, instant messaging, voice mail. Supposedly we are now part of a "global village": all these technologies compose a new, electronic nervous system that radiates out around the world, connecting people and cultures in unprecedented and more intimate ways. A jolt to this nervous system in one part of world can now be felt instantly halfway around the planet. Marshall McLuhan, who coined the term "global village" in 1964, argued that these new sensory linkages, these "extensions of man," were destined to bind the world together. But McLuhan's assertion was not just about the flow of information. Embedded in it were assumptions about the emergence of a new subject position, one enabled by technology, in which people possessed (and welcomed) a more curious and empathetic global awareness of other cultures and people. "It is no longer possible," insisted McLuhan, "to adopt the aloof and dissociated role of the literate Westerner" because the new communications technologies have "heightened human awareness of responsibility to an intense degree."[1]

But if you were a resident of the United States in the early twenty-first century, did television, with its technically enabled global reach and instantaneity, hail you as a member of this so-called global village? Is the subject position this webwork of satellites, videophones, cameras, and cables seeks to constitute a globally empathetic one? After 9/11, when one would have expected the nightly news programs to provide a greater focus on international news, attention to the rest of the world was fleeting, with the exception of the war in Iraq. After a precipitous decline in celebrity and lifestyle news in the immediate aftermath of the 9/11 catastrophe, a year later the percentages of these stories in the nightly news were back to where they had been pre-9/11. In 2004, despite the war, the percentage of stories about foreign affairs on the commercial nightly news broadcasts was lower than it had been in 1997.[2] In the print media, there has been an explosion in the number of celebrity magazines—here the technologies of telephoto lenses and cell phone cameras are used to capture television stars walking their dogs or taking out the trash.

In entertainment programming, the proliferation, especially after 9/11, of nonscripted television has brought viewers into private realms—apartments, houses, resorts, or made-for-TV camps set up on remote islands—where dramas about relationships, personal behavior, and people's "confessions" urge viewers to look inward, not outward. On ABC, a bachelor sampled the wares of twenty-five very pretty women over a period of weeks before he decided which one he liked best. On MTV, five buff twenty-somethings in a preposterously swanky apartment in Las Vegas or San Diego obsessed about which one of them was inconsiderate, "a bitch," or a lout. On The Learning Channel, couples redecorated each others' houses: here people lived for window treatments and flooring options. On the Food Network, we zeroed in on the deep fulfillment that comes from mincing, dicing, and pureeing. From *The Swan* to *Queer Eye for the Straight Guy* to *The Apprentice*, the cameras zoomed in on narcissistic, consumerist obsessions in which the contestants, and we, were to focus on our bodies, ourselves. Here was the antithesis of the global village.

Because McLuhan was a technological determinist, he envisioned only a one-way trajectory for the media, independent of economic or corporate constraints. He failed to anticipate that technologies that enable us to look out beyond our borders can also encourage us to gaze at our navels, and it has turned out that the latter use is more profitable and cost effective than the former. All communications technologies are scopic technologies: as instruments of viewing, listening, and observing, they can slide our perceptions outward or inward. But they do not, cannot, do so on their own.

This essay argues that, at least in the United States, these new communications technologies have not created a global village but have, ironically, led to a fusion of ethnocentrism and narcissism, best cast as a "turn within." While I reject McLuhan's blanket "the medium is the message" aphorism, as well as his monolithic bifurcation of communications technologies as being either (and only) "hot" or "cool," I argue in this essay that communication technologies do have particular, intrinsic properties; they are simply not unidirectional. Thus, we need to consider how the "soft determinism" of technologies interacts with corporate imperatives, producing often ironic and unintended consequences. It is their scopic capabilities, their ability to zoom in or zoom out, interacting with corporate exigencies and consumer desires that can, at times, produce effects quite contrary to what pundits and the public initially thought they were likely to produce. And this, so far, has been true for McLuhan's prediction that communications technologies would create a global village.

American isolationism is nothing new, but it is striking that during this particular period, when technological capabilities and geopolitical exigencies should have interacted to expand America's global vision, just the opposite occurred. The turn within rests on four conditions: what I am calling the irony of technology; the refining of narcissism and the economies of ethnocentrism, both of which rely on narrative and journalistic story telling conventions; and the triumph of youth demographics. And while the turn within is a dominant trend promoted and reinforced by corporate media, it has not gone completely uncontested, as this essay will later note.

I want to bring together several trends not usually discussed side by side: the crisis in American journalism, particularly television journalism and the decline of the reporting of international news, the explosion in reality TV shows in the immediate post-9/11 period, and the metastasizing of celebrity culture. These trends put the lie to the supposed inevitability of faster, more portable, less expensive communications technologies producing a global village and the globally empathetic subjects who inhabit it. The relative expense of covering international stories and the ratings-driven push for "news you can use" have exacerbated ethnocentrism and parochialism in the news; the relative cheapness of producing unscripted television, in which everyday people act out or compete with each other in apartments or boardrooms, has exacerbated the self-scrutinizing narcissism of reality TV; and the comparable cheapness and profitability of celebrity media have produced a glut of PR and gossip that insist entertainment personalities are more important to focus on than anything else.

There is a powerful and underappreciated synergy between these seemingly disparate genres, between the ethnocentrism in the news and the narcissism of so much entertainment media, that propels a further spiraling within in American culture. By the spring of 2006 the reality TV boom was beginning to attenuate as the networks in particular returned to scripted programming, yet it is worth noting the important ideological work reality TV did during the immediate post-9/11 era. If television news, in particular, bears especial responsibility for squandering its ability to enhance a global awareness despite its ever-augmented capabilities to do so, reality TV, colonizing television as it did between 2001 and 2005, insisted that the most productive way to use communications technologies was to focus them on individual Americans in confined and controlled spaces hermetically sealed from foreign peoples and cultures.

I'd like to suggest that the consequences of the turn within are especially serious for young people. Let's take just one piece of evidence. In a widely reported and somewhat embarrassing survey done by the National Geographic Society in 2002, more young Americans (aged 18–24) knew that a recent season of *Survivor* was located in the South Pacific than could find either Israel or New Jersey on a map. Only 13 percent could find Iraq (although they probably didn't do worse here than most adults). But in what one would think would have been a giveaway question, fewer than half could find France, Japan, the United Kingdom, or India on the globe. Fewer than 25 percent could name four countries that officially acknowledge having nuclear weapons. Geographic illiterates, American young people came in next to last of all the countries surveyed, doing better than only youth in Mexico.[3] Of course, the United States' own geographical isolation, the underemphasis on geography in the country's schools, and the fact that the United States remains, for the time being, the world's only superpower, with others wanting and needing to know more about us than we seem to want to know about them, all contribute to this illiteracy. In *Tuned Out: Why Americans Under 40 Don't Follow the News*, David Mindich also documents young people's ignorance about current affairs, despite the passion many of them feel for promoting social justice.[4] Media industries, focusing on certain broad trends among young Americans, such as their declining newspaper readership and their concentration on personal concerns, deploy and privilege the microscopic versus the telescopic properties of communications technologies to maximize profits.

None of this scopic calibration is occurring in a vacuum. Despite the increasing mobility, portability, and reach of communications technologies that can help cultivate a global village, television news organizations chose not to use these tools in this way. Celebrity and scandal journalism dominated the 1990s—the O. J. Simpson case, the Lewinsky scandal, the JonBenet Ramsey murder—and relative peace and prosperity did not compel people to look outward. Nor is it surprising that in the wake of 9/11, some Americans wanted to hide under a collective quilt. Concerns about the consequences of outsourcing, and a forced but superficial engagement with the foreign—particularly the Middle East—may also exacerbate the isolationism that fuels cultural narcissism. But this very same context could have—and indeed, many feel *should* have—led to a turn outward. The technology is not an obstacle here. On the contrary, the new wireless, digital media provide an instantaneity and reach that McLuhan could only have dreamed of. The question is, why did just the opposite occur?

*****

Scholars have put it this way: Do machines make history?[5] This overly simplified question has been asked for decades now, in American studies, in media studies, and in the history of science and technology. And a tension has persisted between the academic response to this question and what circulates in the popular press. Consistently, most academics have argued against what has come to be called "hard determinism," in which inventions and technological systems have primary agency. But technological determinism is a rather large stream in the journalistic reservoir of how to explain societal change.[6] Too often in the press, communication technologies—television, the Internet, video cameras, cell phones—are cast as the prime movers and shakers. Television has made politics more about image than substance: television and video game violence cause school shootings, the Internet has reactivated grassroots politics, MySpace gives rise to stalkers, and so forth. And, of course, such maxims are not totally wrong.

It would be flattering, but foolish, to think that such determinism does not also at times enter academic work. And it would also be mistaken to think that scholars sometimes do not give technology its due, that in our resolve to avoid technological determinism we can underestimate how socio-technical systems help create and sustain social change. Indeed, what has come to be called "soft determinism," in which technologies are seen not as the prime movers, but as having some agency in the mix of how individuals, institutions, and political-economic systems respond to and shape technological change, has gained some credence. As Leo Marx and Merritt Roe Smith pointed out, even if technology is demoted from the primary agent of historical change to an agent whose effects are determined by socioeconomic, political, and cultural forces, the fact remains that technology may still powerfully direct the course of events. Thus, they redefine technological determinism: "it now refers to the human tendency to create the kind of society that invests technologies with enough power to drive history."[7] Yet most scholars remain quite wary of technological determinism in any form.

Witness, for example, the high ambivalence surrounding Marshall McLuhan and his various gambits in his 1964 best seller, *Understanding Media*. Here, pure and simple, communications technologies changed history, often suddenly and cataclysmically, and these media could, according to McLuhan, be easily divided up into those that were "hot" and those that were "cool." Media content was irrelevant to McLuhan; for him, the medium was the message: the technology's properties, how it engaged people cognitively and emotionally, were much more important than whatever representations or information it conveyed. McLuhan coined the term "the global village" to capture how he saw new media technologies transforming the world.

After his initial success as a media guru in the 1960s and early 1970s, McLuhan fell into serious disfavor among media studies scholars, displaced by neo-Marxist theorists who saw capitalism, not machines, as the prime shaper of media systems. The social constructivism school in science-technology studies emphasized the often complex process of negotiation and conflict among competing inventors, institutions, early adapters, and consumers that together shaped the ultimate form and uses of technology that were often a major mutation of, even at odds with, the machine's initial design and application.[8] In cultural studies, there was also an emphasis on human agency and political economy, on how people used and adapted communications technologies and their content, although admittedly in circumstances not of their own making. Thus, McLuhan's mechanistic determinism seemed simplistic and, well, historically and sociologically inaccurate. And, with the rise of ideology studies, his dismissal of media content also seemed politically naive.

The most recent rejection of McLuhan has come from the British scholar Brian Winston, who has argued against communications technologies producing "revolutions," and emphasized their long scientific gestation periods. Winston argues that "supervening social necessity" propels the diffusion of some technologies while the "suppression of radical potential," typically by self-interested corporations, thwarts the development and diffusion of others. Yet, Winston has been criticized for lumping all sorts of restraints on technological change into his notion of suppression, and in his efforts to counter McLuhanesque determinism may have underplayed some important determining features of the technologies he reviews.[9]

In fact, in recent years, and particularly with the rise of radio studies, and the impact of e-mail and the Web, some scholars, myself included, have been reconsidering McLuhan's insistence that particular media have different consequences than other media because of the inherent properties of that medium. And McLuhan became the "patron saint" of *Wired* precisely because of his vision of a technologically enabled global village and his notions about how new media cannibalize and repurpose existing media, rendering some aspects of them obsolete while amplifying others. At first the Internet's Web-based structure and many-to-many information exchange capabilities seemed to ensure a flattening of hierarchies; new, potentially more democratic relationships could result from the technology itself, it was hoped. So it is not surprising that McLuhan began to enjoy a revived following in the 1990s.

Nonetheless, while most scholars will not go all the way with his "medium is the message" aphorism, some of us have argued that a medium that denies sight to its audience, like radio, might cultivate a different sort of cognitive and psychic engagement than would television, or e-mail, which privileges print but denies tone of voice and facial cues to its users. Hence the rise of "soft-determinism." And despite the rejection of many of McLuhan's assertions, the term "global village" has become such a part of our collective common sense that it seems everyone takes it for granted: of course all these new media have created a smaller world and enhanced mutual knowledge and understanding.

I am suggesting that "the global village" is a myth, at least in the United States. There is much evidence to refute, or at least to seriously undermine, the teleological conceit that increasingly sophisticated media technologies have led automatically to increased awareness of and sympathy for other cultures. On the contrary, one could argue that the great irony of all these media extensions—satellite transmission, cable, video technology, even the Internet—is that they have, instead, promoted even more isolationist and ethnocentric views. I am particularly interested here in exploring how these communications technologies, or, more accurately, the uses they have been put to, have enabled the turn within in the United States. Drawing from the notion of "soft determinism," we can see how portability, miniaturization, low cost, the proliferation of media outlets, and speed of transmission, which indeed could have promoted the "extensions of man," have instead led to the "implosion of culture."

This leads us to one of the conditions on which the turn within rests: the irony of technology. What does this mean? While each communication technology does have its own individual properties, especially regarding which of the human senses it privileges and which ones it ignores, that can shape the transmission and reception of content in particular ways, the economic and political system in which the device is embedded almost always trumps technological possibilities and imperatives. And ideological structures trump technological systems.

Thus, communications technologies can often have the exact *opposite* consequences of what we think and hope they might be. The great irony of our time is that just when a globe-encircling grid of communications systems indeed makes it possible for Americans to see and learn more than ever about the rest of the world, Americans have been more isolated and less informed about global politics. Two historical moments stand out as absolute exemplars of this irony. The first is in the 1980s when the refinement of satellite, cable, and video technology came together to make McLuhan's "extensions of man" a technical reality but a corporate problem. The second was after 9/11 when these "extensions of man" were most needed but had their capabilities inverted. As a result, our media are not telescopes, searching outward, as McLuhan insisted. They have become microscopes, trained inside.

Of course there is an important caveat here—one does not want to substitute technological determinism with economic determinism, nor with a linear declension narrative. There has always been resistance to the one-to-many, center-to-periphery model of communications preferred by U.S. media industries. Insurgent uses of communications technologies have existed at least since the early days of wireless telegraphy, and historically such rebelliousness has become most pronounced (and destabilizing of the status quo) when corporate control has seemed most complete. Ham operators in the first two decades of the twentieth century sought to use radio for lateral communication among peers and groups in defiance of top-down efforts aimed at military and corporate control of the airwaves. In the late 1960s, counterculture disc jockeys disgusted by the hypercommercialism and banality of AM radio developed a new format and ethos, "underground" or "progressive" radio on FM. Today, bloggers seek to circumvent the gatekeepers of the mainstream media to convey alternative accounts of a variety of events, and everyday people broadcast their own videos online. Of course, these oppositional uses are then often quickly co-opted for corporate gain: witness the transformation of great swaths of the Web into strip malls and the co-optation of Napster by services such as iTunes. So corporate definitions of how to use media technology are never final, uncontested, or complete; they are, however, dominant.

Nor do I mean to idealize some allegedly mythic past in which the airwaves were filled with rich, impartial, informative international news. Scholars as diverse as Herbert Gans and Edward Herman and Noam Chomsky documented the ethnocentric, anticommunist filters that framed the news about the former Soviet Union, Eastern Europe, and Latin America.[10] The point is not to compare today with a nostalgic past that did not exist. The point is to compare what the technology permits with how it is used. Let's remember that in 1963, at the height of the civil rights movement, television news expanded from fifteen to thirty minutes so more stories could be covered in more depth. A comparable decision to expand the nightly network news in the wake of 9/11 did not occur.

In the 1970s, a series of technological advances, most notably the expanded use of geosynchronous satellites to transmit television signals and the replacement of film with video, meant that news and events from around the world could be broadcast as they were happening, live, into people's homes in real time. It was this "liveness," this "you are there" immediacy, that fueled visions of the global village and the more attuned, empathetic subject position it was seen to cultivate.

But we should also consider what the new communications technologies of the 1970s and beyond, technologies that were faster, that brought "liveness" to us, laid before Americans in our living rooms and dens.[11] The irony of technology points to the profound contradiction between the technical capabilities of these space eradicating technologies and the news values and routines that have guided their use. For example, in the United States, newsworthiness is defined first and foremost by conflict or disaster. Thus, with a few notable exceptions, since the coming together especially of satellite transmission, videotape, and cable in the late 1970s—which meant that international news could be covered and transmitted live into people's living rooms—what American viewers have had laid before their feet are famine, floods, hijackings, bloody military coups, terrorist attacks, civil wars, and genocide. Americans saw defeat. American humiliation. We watched helpless as people hung off of helicopters, struggling to get out of Saigon. The news showed us a wild, irrational-seeming anti-Americanism. Plane hijackings we couldn't control. Furious masses of people in the streets of Tehran, shaking their fists at us. Everywhere, a loss of control. Everywhere, a wound.

Later, during Desert Storm, designed, in part, to exorcise the "Vietnam syndrome," the new communications technologies brought viewers a video game war, distant and sanitized from its human toll. Simulated views from the cockpit of "smart bomb" attacks, complete with cross hairs, nighttime infrared photography of bombings, and computer graphics flash cards of the military's preferred armaments all fetishized the weaponry, placed viewers in the position of the pilots or military strategists and rendered invisible, or erased, the living targets.[12] Newsman Garrick Utley, in an article in *Foreign Affairs*, also notes that video technology and satellite transmission, with their capabilities for instant communications, "left little time for developing expertise in a specific country. Reporters became known as 'firemen,' flying from one international conflagration to the next."[13] In other words, the speed and new mobility of news reporting technology, and the graphics in the newsroom, worked in opposition to depth, and thus in opposition to global awareness and empathy.

After 9/11, despite some fleeting and pioneering explorations by Christiane Amanpour and a few others into the reasons for the attacks, the media focused almost exclusively on the heroism of American police and firefighters and repeated the assertion that the reason for the attacks was that "they hate our freedoms" rather than "probing in depth into the geopolitical situation that might have fueled the terrorism."[14] More recently, the images from the invasion in Iraq showed the initial fireworks of "shock and awe," shaky-cam images of embedded reporters whose low-definition videophone-transmitted faces seemed to come from Mars, or staged events like the toppling of the Saddam statue. Images that would have produced empathy with or sympathy for the Iraqi people were censored, as were images of the toll taken on U.S. soldiers and Iraqis alike. As the coverage evolved and the government's grip on news management faltered, Americans learned about beheadings, bombings, and an incipient civil war. So what these technologies brought over the years were, first, representations of American impotence and, then, representations of distant, supposedly triumphant conquests followed by distant, unfathomable chaos. News routines interacting with increasingly orchestrated and rigid government news management and the battle for ratings powerfully governed the scopic range of what communications technologies could show people, and thus, in fact, blunted what these technologies could permit people to see and feel.

McLuhan was writing in the 1960s when the industrial and regulatory framework of television was quite stable and appeared fixed. There were three national networks regulated by the FCC, and in exchange for having control of the nation's airways they were obliged to supply public affairs programming, much the same structure as had governed radio since 1934. For the networks, especially CBS and NBC, news divisions were a source of prestige, especially given their roles in covering the Kennedy assassination and the civil rights movement. By the mid- to late 1960s, these news divisions, using their gradually expanding arsenal of visual technologies, brought the Vietnam War, Soviet tanks rolling into Prague, and men landing on the moon into people's living rooms. It was the technology, not the institutional structures, that was changing: the conversion, by the mid-1960s, to color broadcasts, the rise of electronic newsgathering and the use of the mini-cam to bring live, on-the-spot news to people, cameras aboard spacecraft beaming images back to earth, all indeed pointed to the inevitability of a global village. McLuhan, seemingly gripped by that uncritical ideology of progress that dominated some scholarship and popular thinking in the early 1960s, could not foresee a time when corporate imperatives and financial constraints would thwart what all this new technology could actually deliver.

The high point in the irony of technology occurred not long after McLuhan's death in 1980. Just as outward-reaching media technologies continued to proliferate in the 1980s and beyond, the three broadcast networks, facing competition from the cable networks and pressure to increase their profit margins, spun off their news divisions from their entertainment divisions (which had previously partially subsidized the news), insisted that news divisions become profitable and support themselves, and, thus, downsized them. So, at the very same time—the early to mid-1980s—when the convergence of new media technologies meant that the war in the Falklands, for example, could be brought to viewers around the globe live via CNN—the broadcast networks began cutting costs in their news divisions. All the networks scaled back on international reporting and eliminated a host of foreign news bureaus. International news was replaced by less expensive entertainment news, mayhem news, lifestyle and other human-interest stories, celebrity journalism, and news about health and fitness. By 2000, new news segments devoted to personal health, such as CBS's "Health Watch" and NBC's "Life Line" were embedded news beats in the network news lineups. While those on the right and the left continue to fight over whether the news is dominated by a liberal or conservative bias, few have emphasized that the greatest bias in the news today that emerged from these decisions is the narcissism bias.

Various studies of the news media have documented the turn within in journalism. Between 1971 and 1982, foreign news in newspapers dropped from 10.2 percent to 6 percent of what journalists call the newshole.[15] In 1989, only 2.6 percent of the nonadvertising space in ten leading U.S. newspapers was devoted to international news. The percentages are dramatic as well in the newsweeklies. In *Time*, international news dropped from 24 percent to 14 percent between 1985 and 1995, and in *Newsweek* from 22 percent to 12 percent during the same time period. In 1987 *Time* featured eleven cover stories on international news; by 1997, that number had dropped to one.[16] The decline seems even more precipitous in TV news. The time there devoted to international news has, according to one study, dropped from 45 percent in the 1970s to 13.5 percent by 1995, or, in other words, a whopping 70 percent fall off.[17]

In their critique of the news media, *The News About the News*, Leonard Downie and Robert Kaiser, both top editors at the *Washington Post*, sat down separately with Peter Jennings, Dan Rather, and Tom Brokaw. They showed each anchor a broadcast from his news show in the early 1980s, just as each man had taken over as anchor. "This is amazing, truly amazing," they quote Rather as saying, responding to how few graphics and how much international news the 1981 broadcast he saw contained. All three anchors affirmed that they could never get that much international news in today. The networks had cut back on international news by cutting bureaus and foreign correspondents that are deemed too expensive.[18]

Certainly there were some historical reasons for this. In the early 1970s, the United States was still involved in Vietnam, still enmeshed in the cold war, and those two factors alone could account for the increased coverage back then. The cold war in particular provided an ongoing, us-versus-them, good-versus-evil story in which Vietnam and other international stories could be framed. Likewise, in 1979, the Iranian hostage crisis "became one of the most widely covered stories in television history"; it inaugurated *Nightline*, and Walter Cronkite, by then voted the most trusted man in America, closed every broadcast with an announcement of how many days the hostages had been in captivity.[19] It's not surprising then, one might argue, that international news coverage has dropped from such highs.

But certainly there's more to it than this. And let's examine another set of figures collected after the events in the 1970s: "total foreign coverage on network nightly news programs has declined precipitously," from nearly 3,800 minutes in 1989 to just over 1,800 minutes in 1996 at ABC (the leader) and from 3,350 minutes to 1,175 minutes at NBC.[20] In 1988, ABC featured 1,158 foreign bureau reports; by 1996, that was down to 577 reports. As media critics and disappointed journalists themselves have pointed out, the ongoing press for higher profits and lower costs continued to target international news coverage, the most expensive news to gather despite the new technology. Given that, as Garrick Utley reports, to send a correspondent and a production team plus 600 pounds of equipment to a new locale cost, in 1997, $3,000 a day plus excess baggage fees, international coverage has come to be seen as an expendable expense. In addition, many network executives believed Americans were completely uninterested in foreign news, so why bother? According to Mort Zuckerman, editor in chief of *U.S. News and World Report*, and Maynard Parker of *Newsweek*, "featuring a foreign subject on the cover of the magazine results in a 25 percent drop in newsstand sales."[21] Americans, the conventional wisdom went, wanted "news you can use."

It wasn't just commercial pressures that undercut the possibility for a global village; the entrenched narrative and filmic conventions of the network news also, with some exceptions, worked against constituting an empathetic global village subject. If ideological structures trump technological systems, one of the main ways that they do so is through dominant storytelling practices. News routines in most international stories lead to people in foreign countries being represented in highly conventional, often stereotypical ways that make them seem not at all like "us," but, as that overused word emphasizes, "other." As Herbert Gans's still classic study of the news documented, and as many developing countries have complained, foreign countries become newsworthy in the United States when they are afflicted by natural disasters, wars, coups d'etat, or terrorist attacks.[22]

Thus, the people American viewers see most frequently are victims or combatants. Those we see in the streets of Gaza or Jerusalem or Baghdad are mostly featured in long shots or medium shots, as undifferentiated members of groups or masses. They are furious protestors, masked guerillas, soldiers, or grief-stricken victims. They are "tribal," masses of them gather in the streets shaking their fists, screaming and chanting; they chop each others' limbs off; they are mute, poverty-stricken victims; they wear too many clothes, or not enough; they are antimodern. They are objects of journalistic scrutiny, lacking a subjectivity except one filled with rage or the most desperate sorrow. Occasionally, of course, viewers get a sound-bite from a person in the street, but news programs are more likely to give airtime to official spokespeople. And the sound-bites are all too brief. (The length of the average sound-bite for presidential candidates in the 1996 election was down to 7.2 seconds; why would a person on the streets of China, Sudan, or Iraq get any more?[23]) These were all snapshots, flipping on after the other, and not contextualized, embedded in narratives that made sense or garnered empathy. Why turn outward, toward them? There is nothing to understand, much to reject.

*****

At the same time that industry decisions were blunting the "global village" capabilities of communications technologies, other industrial imperatives began to privilege their narcissistic and domestic surveillance capabilities. Two dominant media trends in the first decade of the twenty-first century, the juggernaut of reality TV between 2000 and 2005, and the explosion in celebrity-based entertainment and journalism magnified the importance of personality, the interpersonal, surveillance of behavior and the body, and an even more insistent consumerism. Video technology made shows such as *COPS*, *America's Most Wanted*, and *Rescue 911* inexpensive to produce; *America's Funniest Home Videos* made the viewer the producers. MTV's *Real World*, which premiered in 1991, pioneered in using video technology to record the everyday interactions and relationship melodramas of a small group of people. But the turning point came in the summer of 2000, when *Big Brother* and *Survivor* premiered on American television, became rating success stories, and inaugurated a new programming era characterized by minimal writing and the use of everyday people instead of actors. As Susan Murray and Laurie Ouellette point out, the genre almost instantly metastasized to produce subgenres: the gamedoc (*Survivor* and *Fear Factor*), the dating shows (*Joe Millionaire* and *The Bachelor*), makeover/lifestyle shows (*Extreme Makeover, Queer Eye for the Straight Guy*), the docusoap (*Real World, Sorority Life*), talent shows (*American Idol*), and reality sitcoms (*The Osbournes, The Newlyweds*). "Not since the quiz show craze of the 1950s," they write, "have nonfictional entertainment programs so dominated the network prime-time schedule . . . By January 2003, one-seventh of all programming on ABC was reality based."[24]

Reality TV came to replace, in particular, the lineup of TV newsmagazines that had dominated various network schedules in the 1990s.[25] Programs such as *Dateline*, *20/20* and *48 Hours* offered a mix of investigative reports, typically about domestic issues, and human interest stories that often focused on individuals. While these programs did little to expand people's global horizons, they did provide a mix of hard and soft news. On unscripted TV shows, by contrast, the outside world, in the form of magazines, newspapers, books, and TV news (all of which would be part of everyday people's actual environments) is expunged from the bubble of the reality TV world. In the hands of reality TV producers and network executives, cheap, portable, interactive communications technology in the service of this genre privileges self-absorption, self-scrutiny, the intricacies of interpersonal interactions and an obsession with the private and personal at the expense of any broader public issues.

Narcissism sells, not only because it celebrates, even elevates, the quotidian aspects of our everyday lives into weighty theater, but also because narcissism—the desperate desire for the approval of others—also sells products, legions of them. Christopher Lasch's 1977 best seller, *The Culture of Narcissism*, was incredibly prescient in analyzing the role that the media, and especially advertising, played in constituting the "narcissistic personality of our time," a person utterly reliant on the approval of others, desperate to make a good first impression, filled with a deep self-loathing, terrified of aging and death, and having no core, independent self. But even Lasch didn't anticipate this televisual refining and cultivating of narcissism.

The filmic and narrative practices of reality TV are especially designed to promote intimacy and scrutiny of others and the self, quite different from the distance and objectification encouraged by the visual conventions on the nightly news. This stylistic gap also reinforces the bond between the ethnocentrism cultivated by the news and the narcissism cultivated by reality TV. So the news and its storytelling frameworks do not stand alone, they operate in relation to others that dominate in the media. As viewers, we learn who we are supposed to identify with, and who we are supposed to distance ourselves from and see as alien through both genres.

Unlike how we view the long shots of groups and crowds on the news, consider how we meet the participants in most reality shows. Of course there are the seemingly "real" group dynamic shots in the boardroom or jungle or apartment. But then the different participants are shot alone, sitting in a chair, talking to the camera, telling us their reactions, their judgments, concerns. They are subjects of their own lives, *with* inner lives. The structure of the show invites us to empathize with them, judge them, or both.

Having said that, however, let's also consider what is meant to stand for a rich inner life on reality TV. What do the participants in these shows ponder? Who dissed whom, who didn't wash the dishes, who seems sincere, who took responsibility and who shirked it, who has the hots for whom. "She's the party girl–sister I never had," opines one young woman on *The Real World* while another offers the bracing observation that "we've all been put here for a reason." The inner lives viewers get to see consist primarily of banalities, shallow personal bleatings about other people's behaviors. Inner life here consists of reactions and attitudes, in which you don't have to really know much of anything except how you feel. Inner life here is not occupied by concerns about the environment, world hunger, politics, philosophy, or the meaning of life. So, in the overall televisual experience, as we move between news and reality TV, we are invited to distance ourselves from people abroad and, instead, to insert ourselves deeply into the interpersonal relations of a group of preselected, mostly young, mostly white people whose major concerns are staged as highly narcissistic and vapid.

The point is not to single out reality TV as the latest and possibly worst example of television schlock. Rather, by exploding on the scene when it did, between 2000 and 2005, and by insisting through its televisual and narrative conventions that portable, barely noticeable cameras are best deployed, and most effectively reveal the richness of human experience, when trained on preselected people in confined, often domestic spaces, reality TV served as the validating pivot for the turn within in the news and elsewhere. The preferred discourse of the whole genre is one that celebrates a determined isolationism, a luxuriant self-absorption. It was the discourse of reality TV in particular that legitimated, after the initial aftermath of 9/11, a preference for the microscopic rather than the telescopic properties of communications technologies.

The turn within has not gone uncontested, nor is it a uniform turn. A 2002 poll conducted by the Pew Research Center for the People and the Press reported that 63 percent of respondents felt it was very important for the news to cover foreign events and that interest in international news, especially the Middle East, had increased since 2000. For college graduates, international news had become a top news subject.[26] The business press, whose readers require international news and information, provides this service to its audience. Blogs by soldiers in Iraq, freelance journalists, and academics challenge the superficiality or official line proffered by the network news. Especially in the run-up to the U.S.-led invasion of Iraq, and during the occupation and insurgency, e-mail Listservs have offered reports and commentary from international news sources, and BBC.com reported a major increase in hits from U.S. readers. The upstart network Current TV features stories by young people using small digital video cameras to cover everyday life in Iraq, dangerous methods of transport used by illegal immigrants into the United States, and the ongoing war in Afghanistan.

Thus, the turn within has contributed, at least for a subset of the U.S. population, to a hegemonic crisis for the mainstream news media. It is held in widespread disregard, with viewers on the right and the left perceiving bias, and many feeling they are not getting accurate or complete information.[27] So despite powerful industry imperatives, the life span and dominance of the turn within remains unclear, especially given the ongoing threat of terrorism and the rise of other economies, most notably those of India and China, and other nuclear powers, that may cause the spotlight to shift.

In this era in which the word *globalization* is used so constantly, and profligately, and is linked to an automatic assumption that communications technologies undergird the shrinking of the world, we need to address the scopic capabilities of these devices, and analyze which political and economic structures encourage their microscopic versus their telescopic capabilities. This calls, in part, for more policy analysis on our parts, more attention to political economy, more linking of textual analysis with that political economy. And the way state-corporate systems interact with scopic technologies changes over time, so historicizing these relationships is key.

For example, wars, colonial and postcolonial relations, economic relationships, geographic proximity and form of government, from autocratic to neoliberal, all influence different societies to different degrees regarding whether the lenses of their communications systems will zoom out and let the world in or zoom in and keep it at bay. In the United States, wars that involved large sectors of the population, from the Civil War to World War II and the Vietnam

War, deployed and often enhanced the telescopic capacities of communications technologies. Just one striking example was radio's—and CBS Radio's in particular—use and refinement of transatlantic shortwave broadcasts to bring the Blitz, or the fall of France, or D-Day directly into people's homes. Yet in the United States, unlike in Europe, geographic proximity matters for naught here, as evidenced by the virtually nonexistent attention news programs give to Canada or Mexico (except when the latter's citizens cross U.S. borders). As government regulation of the broadcast media has weakened in the United States, coverage of international affairs has also declined. And the United States, being, for the moment, the world's only superpower, has meant that others need to learn more about the United States than many feel we need to learn about them. Future research and theorizing needs to itemize, schematize, and analyze these different historical and political articulations comparatively so we can move beyond blanket technological or economic determinism.

When many of us threw McLuhan out in the 1970s, we repudiated the notion that communications technologies could make history on their own, and that rejection stands today, even given how so many communication technologies—the Internet, e-mail and cell phones—have transformed daily life. But it now appears that communication scholars were too quick in our wholesale rejection of "the medium is the message." Communications technologies do have some inherent capabilities that privilege some senses—and thus some cognitive and behavioral processes—over others. As we in American studies consider the ways technology and society affect each other in the digital age, we need to examine how these inherent capabilities are enhanced or thwarted by the institutional structures that regulate, profit from, or are even surprised and destabilized by scopic technologies.

In thinking through soft determinism and the turn within, we should keep in mind Thomas Misa's "middle-level theory" of technological determinism. Misa notes that those who adopt a more "macro" view of history and society tend to give technology a much more causal role (e.g., McLuhan), while scholars doing more "micro-level" analyses, who examine the multiple and often contingent factors shaping technological innovation and diffusion, tend to give technology itself minimal agency. "Middle-level theory" seeks to find an intermediate level of analysis in which technology is seen as both socially constructed and as society shaping.[28] It is at the middle level that we can best analyze the articulations between the scopic capabilities of communications technologies and the industrial decisions about when and why to zoom out and when and why to zoom in.

Today, the economies of ethnocentrism are trumping the telescopic capabilities of communications technologies, particularly on television. With news organizations becoming increasingly small divisions of large entertainment behemoths, the economic incentives for ethnocentric programming are overwhelming. In addition, the triumph of youth demographics means that advertisers want the eighteen to forty-nine demographic, but especially the eighteen to thirty-four demographic. Today's young people rarely watch the network news or read newspapers; in part, they don't trust the news. They see the news as not about or for them but for an older audience, and many are focused on their own lives rather than on current affairs. The mass media pander to all of this and offer fare obsessed with sex, relationships, self-surveillance, physical challenges, voyeurism, the humiliation of others, and incessant celebrity psychodramas.

Without imposing some false taxonomy on communications technologies, we should think more about which ones are telescopic, microscopic, and cinemascopic. They do not fall neatly into categories, as most are multivalent. If we think about satellite transmissions and the new videophones as being telescopic, bringing distant events and peoples into view, their zoom-in images can also isolate these events and peoples from their sociopolitical contexts. Telephoto resolution is often low. The same video technology that brought us *Big Brother* also revealed the beating of Rodney King and, through tourists' own cameras, the first news footage of the 2004 tsunami. And there may be some communications technologies, in isolation or in combination with each other, that we would consider cinemascopic, providing a wide, panoramic view and encompassing the viewer as well as the viewed. Such tableaus have existed primarily in the world of cinema, in which fictional stories about other parts of the world are meant to reveal the actual, "real" truth we don't see on the news. In other words, if we take what is most important from McLuhan—that communications technologies do have inherent properties that we need to take seriously without giving them total deterministic power—then there is still much work to do in thinking through their multiple and at times paradoxical powers.

So we must also keep in mind that the uses and effects of communications technologies are not pat; they are often, as Claude Fischer reminds us in his terrific book about the telephone, *America Calling*, contradictory. The telephone allowed for the invasion of people's privacy from the outside, which many hated; it also allowed for more immediate contact with friends and family, especially important in emergencies, which people embraced.[29] The example I know best is radio, so often cited as "bringing the nation together" during the

Great Depression and WWII. Through national entertainment programs that attracted 40 million listeners, Roosevelt's fireside chats, sporting events, and coverage of the war, of course radio enabled (much more than newspapers ever could or did) the construction of an imagined nation.[30] But radio also broadcast local programs that promoted more regional identifications sometimes in opposition to such homogenized national affiliations. Just because the networks and the advertisers who supported them (and, later, the government) wanted to construct a national market with a national self-concept and loyalties does not mean that their success was complete.

Media coverage of new communications technologies either suggests some unilinear trajectory or lapses into a utopian versus dystopian framework (the Internet will produce a thriving new public sphere; the Internet will allow child pornographers to stalk our kids). As we in American studies struggle in our own lives with the multiple, contradictory consequences of the digital revolution and with the rapid but unequal global diffusion of communications technologies, we must always remember the irony of technology, and the ongoing gaps and tensions between technological capabilities on the one hand, and the not insignificant power of ideological frameworks and corporate-state interests on the other. It is at this nexus, in this struggle and mess at the middle levels, that we will find the most interesting and important stories to tell about technology and modern life in the twenty-first century.

Notes

1. Marshall McLuhan, *Understanding Media: The Extensions of Man* (New York: Signet, 1964), 19–20.
2. "The State of the News Media 2005," Project for Excellence in Journalism, online at Journalism.org (accessed July 11, 2006).
3. Survey available at www.nationalgeographic.com/geosurvey (accessed July 11, 2006).
4. David T. Z. Mindich, *Tuned Out: Why Americans Under 40 Don't Follow the News* (New York: Oxford University Press, 2005).
5. The classic article here is Robert L. Heilbroner's "Do Machines Make History?" *Technology and Culture* 8 (July 1967): 335–45.
6. Leo Marx and Merritt Roe Smith, "Introduction," in *Does Technology Drive History?* ed. Leo Marx and Marrit Roe Smith (Cambridge, Mass.: MIT Press, 1994), xi.
7. Marx and Smith, "Introduction," xiv.
8. The flagship book here is *The Social Construction of Technological Systems: New Directions in the Sociology and History of Technology*, ed. Wiebe Bijker, Thomas P. Hughes, and Trevor Pinch (Cambridge, Mass.: MIT Press, 1989).
9. Brian Winston, *Media, Technology, and Society: A History from the Telegraph to the Internet* (New York: Routledge, 1998).
10. Herbert Gans, *Deciding What's News* (New York: Vintage Books, 1979); Edward S. Herman and Noam Chomsky, *Manufacturing Consent* (New York: Pantheon, 1988).

11. Robert Stam, "Television News and Its Spectator," in *Film and Theory*, ed. Robert Stam and Toby Miller (New York: Blackwell, 2000).
12. Susan J. Douglas, "Camouflaging Reality with Faux News, Clever Decoys," *In These Times*, February 13–19, 1991.
13. Garrick Utley, "The Shrinking of Foreign News," *Foreign Affairs*, March-April 1997, 4.
14. S. Elizabeth Bird, "Taking It Personally: Supermarket Tabloids after September 11," in *Journalism After September 11*, ed. Barbie Zelizer and Stuart Allen (New York: Routledge, 2002), 145. See also Michael Traugott and Ted Brader, "Explaining 9/11" in *Framing Terrorism: The News Media, the Government, and the Public*, ed. Pippa Norris, Montague Kern, and Marion Just (New York: Routledge, 2003).
15. James F. Hoge, "Foreign News: Who Gives a Damn?" *CJR*, November-December 1997. The newshole is the amount of time or space devoted to reporting the news, as opposed to the entire paper or broadcast, which also includes advertising.
16. W. Lance Bennett, *News: The Politics of Illusion* (New York: Longman, 2003), 14–15.
17. Hoge, "Foreign News."
18. Leonard Downie Jr. and Robert G. Kaiser, *The News About the News* (New York: Alfred A. Knopf, 2002), 111.
19. Melanie McAlister, *Epic Encounters: Culture, Media, and U.S. Interests in the Middle East, 1945–2000* (Berkeley: University of California Press, 2004), 198.
20. Utley, "The Shrinking of Foreign News," 2.
21. Hoge, "Foreign News."
22. Gans, *Deciding What's News*, 35–37.
23. Bennett, *News*, 34–35.
24. Susan Murray and Laurie Ouellette, eds., *Reality TV: Remaking Television Culture* (New York: New York University Press, 2004), 3–4.
25. I am grateful to Amanda Lotz for this point.
26. "Public's News Habits Little Changed by September 11," Pew Research Center for the People and the Press, online at people-press.org/reports/display.php3?ReportID=156 (accessed July 11, 2006).
27. See the Pew Center's Study, *Media: More Voices, Less Credibility*, at http://people-press.org/commentary/display.php3?AnalysisID=105 (accessed July 11, 2006).
28. Thomas J. Misa, "Retrieving Sociotechnical Change from Technological Determinism," in *Does Technology Drive History?* ed. Marx and Smith, 115–41.
29. Claude S. Fischer, *America Calling: A Social History of the Telephone to 1940* (Berkeley: University of California Press, 1994).
30. See Susan J. Douglas, *Listening In: Radio and the American Imagination* (Minneapolis: University of Minnesota Press, 2004).

Say It Loud, I'm Black and I'm Proud: African Americans, American Artifactual Culture, and Black Vernacular Technological Creativity

Rayvon Fouché

> The actual *beginnings* of our expression are post Western (just as they certainly are pre-western). It is only necessary that we arm ourselves with complete self knowledge[;] the whole technology (which is after all just *expression* of who ever) will change to reflect the essence of a freed people. Freed of an oppressor, but also as [Askia] Touré has reminded we must be "free from the oppressor's spirit," as well. It is this spirit as emotional construct that can manifest as expression as art or technology or any form.
>
> Amiri Baraka[1]

"Say it loud, I'm black and I'm proud." The rhythmically pulsating refrain of the James Brown song and the title of his 1969 album publicly vocalized the African American desire to reclaim, recover, and articulate self-claimed black identity and expression. Not surprisingly, the song became an anthem in black America during the late civil rights movement. A few years before the release of this album, Stokely Carmichael clearly articulated the meaning of black power that James Brown referenced in his song. In the same-titled book, *Black Power*, Carmichael defined black power as "a call for black people in this country to unite, to recognize their heritage, to build a sense of community. It is a call for black people to begin to define their own goals, to lead their own organizations and to support those organizations. It is a call to reject the racist institutions and values of this society."[2] At a most basic level, Carmichael was calling for African Americans to gain control of their existences within the United States, as well as abroad, and to understand that there is something special, unique, and valuable about cherishing, nourishing, and supporting black people, black cultures, and black communities. In a similar way, Amiri Baraka, in the essay "Technology & Ethos," was calling for black people to rethink their relationships with technology and take action to make technology more representative of black culture.[3] More important, Baraka was arguing that through black technological utterances rooted within black cultures, black communities, and black existences—or what I would call expressions of black vernacular technological creativity—technology would be more responsive to the realities of black life in the United States.

Carmichael and Baraka represent two of many critical black voices that have pointed out difficulties black people have encountered searching for a place of space within American society and culture. Yet, the commentary by Baraka is an unusual break from the traditional lines of criticism. In "Technology & Ethos" Baraka exposed the fact that of the many people, organizations, and institutions that have participated in derailing black struggles for power and equality, technology is infrequently part of the discussion. Currently technology—even with the ever-growing volume of technological critiques—is publicly understood to change society positively by making life more healthy, productive, and efficient, thus better. Americans are continually bombarded with seemingly endless self-regenerating progressive technological narratives. In this capitalist-supported tradition, the multiple effects that technology has on African American lives go underexamined. This uplifting rhetoric has helped obfuscate the distinctly adversarial relationships African Americans have had with technology.

In the article "Technology Versus African Americans" Anthony Walton contends that "the history of African-Americans since the discovery of the New World is the story of their encounter with technology, an encounter that has proved perhaps irremediably devastating to their hopes, dreams, and possibilities."[4] Technology such as the ships that transported African slaves to the "New World," the overseers' whips, cotton cultivation, "Jim Crow" rail cars, segregated buses, inner-city public housing, and voting machines have contributed, directly or indirectly, to the subjugation of African American people. Historically, technology has been a potent form of power in material form that has politically, socially, and intellectually silenced African American people, and in the worst cases rendered them defenseless and invisible. Cornel West has called this affect the black diaspora problematic of invisibility and namelessness. This problematic constructs "black people as a problem-people rather than people with problems; black people as abstractions and objects rather than individuals and persons; black and white worlds divided by a thick wall (or a 'Veil') . . . black rage, anger, and fury concealed in order to assuage white fear and anxiety; and black people rootless and homeless on a perennial journey to discover who they are in a society content to see blacks remain the permanent underdog."[5]

Technology as material oppression is not the only way to consider African American technological experiences. As interesting as this mode of analysis can be for thinking about the technological control of African Americans, it strips black people of technological agency. It inherently closes down discussions about the ways African American people consume and use technology, and conceals the reasons that black people produce meanings for technological artifacts, practices, and knowledge that regularly subvert the architectured, or constructed, meanings of technology.[6] A major limitation of this perspective is that it does not embrace the ways that African American people acquire technological agency by being resourceful, innovative, and most important, creative.

Studies of African American creativity often center on the vernacular. The black vernacular tradition is primarily associated with the production or performance of music, dance, literature, visual art, and sport.[7] By recasting African American artistic and aesthetic creativity as American "survival technology," Joel Dinerstein's *Swinging the Machine* presents a case for engaging the technological.[8] Techno-dialogic, Dinerstein's term to explain how "the presence (or 'voice') of machinery became integral to the cultural production of African American storytellers, dancers, blues singers, and jazz musicians," highlights the creative interplay between modern industrialization and black expressive culture.[9] Dinerstein effectively relies on Mikhail Bakhtin's dialogic theory of language to discuss how, during live artistic performances, black "musicians brought the power of machines under artistic control and thus modeled the possibility of individual style within a technological society."[10] As much as this essay embraces the intent of his work, especially his efforts to theorize "a cultural tradition of resistance to technology in African American expressive culture,"[11] it aims to extend black creativity by proposing the concept of black vernacular technological creativity to describe the ways African American people interact with material forms and affects of technology.

Black vernacular technological creativity is characterized by innovative engagements with technology based upon black aesthetics, or, in Albert Murray's terminology, the "technology of stylization."[12] This differs from Dinerstein's approach in that black vernacular technological creativity is a process of engaging material artifacts as opposed to performing black-informed expressive or aesthetic representations of technology. Yet it is similar, in that black vernacular technological creativity results from resistance to existing technology and strategic appropriations of the material and symbolic power and energy of technology. These maneuvers enable African American people to reclaim different levels of technological agency. Some resistant responses and technological appropriations are stronger, more politically motivated, and culturally embedded than others. As a result, black vernacular technological creative acts—spanning the continuum from weaker to stronger—can be seen in three ways: redeployment, reconception, and re-creation.[13] Redeployment is the process by which the material and symbolic power of technology is reinterpreted but maintains its traditional use and physical form, as with blues musicians extending the perceived capability of a guitar without altering it. Reconception is the active redefinition of a technology that transgresses that technology's designed function and dominant meaning, as in using a police scanner to observe police activities. Re-creation is the redesign and production of a new material artifact after an existing form or function has been rejected, as in the case of DJs and turntablists developing new equipment. In developing this framework, the goal is not to make evaluative statements or privilege one type of black vernacular technological creativity but to express multiple ways that African Americans as culturally and historically constituted subjects have engaged the material reality of technology in America.

To explore black vernacular technological creativity, familiar ways of examining the nature of technological experiences have to be rethought. As effective as existing approaches to the study of technology are for understanding technological developments by members of dominant cultures in Western society, they are lacking in their abilities to handle the creation, development, and use of technology by those racially marginalized. Since these theories aim to assess technological activity by dominant groups, they are limited in their ability to address the wide variety of technological experiences that fall outside of the realm of dominant cultural experiences. Unfortunately, many existing social theories of technology also do not address the significance of absence—in specific, the significance of the theoretical absence of African Americans from technological decision making and what these absences can tell us about the nature of technology in America. In many cases, perception of what "counts" as technological activity is deeply intertwined with deleterious representations of the racialized other. In other words, technological activities that cannot be effectively categorized within the dominant canon of science and technology fall to the wayside.

One cannot expect African Americans, who have traditionally been relegated to peripheral sites within American society and culture, to interact with technological products analogously to the members of the dominant American culture. It is the misconception of fair and equal Americanness—reeking of the value-neutrality of technology—that is highly responsible for the systematic disregard for technological activities that are peripheral to the dominant society's. By understanding that the locations of black people within American society are the historical by-products of a businesslike effort to fix racism within

American culture, a new set of questions to explore technology and African American lives emerges. Questions contemplating how technology has been "raced" throughout American history, as well as how to understand and see African American technological agency, are essential for a broader conception of the complex nature of race in the creation, production, and use of technology. New questions will produce a more textured understanding of the roles that black people have played as producers, shapers, users, and consumers of technology within American society and culture.

American Culture, African Americans, and Theories of Technology

Historically, technology has been one of the defining elements of American society. Individual American technological ingenuity, from Samuel Morse to Bill Gates, has been a hallmark of American culture and an important factor in building a financially prosperous nation.[14] However, questioning the ways technology and American culture have been co-produced has been a twentieth-century enterprise. Works by Siegfried Giedion, S. Colum Gilfillian, William Fielding Ogburn, and Lewis Mumford began a more thorough analysis of the social components of technological development.[15] By midcentury, Abbott Payson Usher, Lynn White, and Leo Marx began to recontextualize our understanding of the nature of technology in American society and culture.[16] Marx's *The Machine in the Garden*, which began as an article first published in *New England Quarterly*, marks an important turning point in study of technology and American culture.[17] Though reworked and extended by John F. Kasson's *Civilizing the Machine* and David E. Nye's *American Technological Sublime*, *The Machine in the Garden* would become, as Jeffrey Meikle has argued, the "starting point for all attempts to understand the complex connections among developing technologies, their representations in text and image, and the multiple realities of American cultural experience."[18] In addressing the connections between technology and nature, technology and progress, technology and inventive/engineering achievement, and technology and nationalism, Marx eloquently presented a set of themes that subsequently would shape and direct the future study of technology and American culture.

By the late twentieth century, the connections between technology and American culture Marx presented would become central to the emerging field of science and technology studies. Many researchers began to ask, with a critical edge, if technology is truly autonomous within the societies they inhabit and whether or not technological changes drive social changes.[19] This research endeavored to reexamine how individuals and groups (re)shape and (re)construct meanings for new and existing technology.[20] Eventually more effective tools, such as those exhibited in the systems approach, the social construction of technology (SCOT), and actor-network theory emerged to unpack the complex interconnections between technology and American society. The systems approach is clearly displayed in *Networks of Power*, Thomas Hughes's transnational study of the developments of electrical distribution systems in Berlin, London, and New York. This approach "analyzes technology as heterogeneous systems that in the course of their development acquire a technological momentum that seems to drive them in a specific direction with a certain autonomy."[21] Wiebe Bijker and Trevor Pinch articulated SCOT in the article "The Social Construction of Facts and Artefacts."[22] SCOT focuses on the development of technological artifacts as relevant social groups negotiate to "close," or stabilize, the meanings of these artifacts. In *Science in Action* Bruno Latour outlined a dominant version actor-network theory.[23] This model argues that for a technology—an artifact, a practice, or constitutive knowledge—to be successful, a seamless web, or network, of "durable links tying together humans and nonhuman entities," or actants, must be created.[24]

The systems approach, SCOT, and the actor-network theory met when Bijker, Hughes, and Pinch edited the highly influential *The Social Construction of Technological Systems*.[25] This volume illustrated the common ground between the three methods of analysis. John Law, in the article "Technology and Heterogeneous Engineers," pointed out the three main similarities. "First, they concur that technology is not fixed by nature alone. Second, they agree that technology does not stand in an invariant relation with science. Third . . . they assume that technological stabilization can be understood only if the artifact in question is seen as being interrelated with a wide range of nontechnological and specifically social factors."[26]

But, this is about as far as their commonalties can be taken. Differences arise when researchers use these methods to determine the agents of technological change and development. SCOT locates the power to shape technology within human actors' social interests, whereas the systems approach has no such loyalty to overarching human technological agency. Technology can shape society as well as other technology. Technology can potentially have momentum, be autonomous, and be deterministic. Actor-network theory goes one step further to say that there are no inherent differences between human and nonhuman actors. They are all "actants" and should be treated symmetrically.

As productive as these approaches have been for the study of dominant technological voices, products, and experiences, they have been equally unproductive for those traditionally marginalized within American society and culture. Feminist technological critic Judy Wajcman has argued that these methods of investigating the social and cultural implications of technology overlook "the ways in which technological objects may be shaped by the operation of gender issues," and how technological developments are "shaped by a set of social arrangements that reflect men's power in the wider society."[27] Wajcman's work exploring gendered components of technology has clearly broadened our understanding of the ways gender imbalances have influenced our technological world.[28] The inattention to gender issues that Wajcman indicated is reflected in the comparable indifference to racial issues in analyzing the creation, development, production, and distribution of technology. As result, these approaches theoretically shut down discussions about black technological experience.

For example, Thomas Hughes's systems approach is not particularly applicable to the experiences of African American inventors of the late nineteenth and early twentieth centuries. There could not have been black system builders similar to Alexander Graham Bell, Thomas Edison, or Henry Ford. Since late-nineteenth-century America reverberated with renewed enthusiasm for overt racism—beginning at the end of Reconstruction and culminating in the United States Supreme Court's *Plessy v. Ferguson* decision in 1896—most people of African heritage found the traditional avenues followed by inventor-entrepreneurs closed. African American technologists, engineers, and inventors could never reach the position to wield the power and resources necessary to construct a technological system. Granville Woods, the most successful black inventor of this period, attempted to invent and develop systems of railway locomotion and inductive communication. Due to the institutionalized racism within technical communities, the hope of being a system builder never materialized for him.[29] The lives and experiences of the most promising African American inventors reveal that black people could construct only the components of a system and never the system itself.

Actor-network theory is also limited in its ability to handle culturally embedded racism. Anthropologist David J. Hess contends that the actor-network approach "is not very good at explaining why some actors are excluded from the game and why the playing field is not level . . . for this reason categories such as race, class, gender, colonialism, and industrial interests tend to be absent from actor-network analyses."[30] Moreover, African Americans traditionally have not possessed the power to gather large interrelated groups "of disparate elements of varying degrees of malleability," and heterogeneously engineer.[31] It is even more difficult to talk about how marginalized people participate in seamless weblike networks of interaction when often they are not allowed into the web. More often than not, members of the dominant American culture enrolled black people against their will, as in slavery and forced segregation. Actor-network theory neither accounts for the processes of how the dominant American culture oppressed and purposely marginalized a segment of American society, nor how it reinvented the margins of society for its own supremacist and colonizing purposes.[32]

The first version of SCOT was not particularly applicable to the marginalized status of African Americans. For SCOT, relevant social groups, which can be "institutions and organizations . . . as well as organized and unorganized groups of individuals," are deemed the most important factor in shaping and designing a technology for a specific meaning or purpose.[33] Since African Americans historically have been denied basic human rights, participation in the larger American processes of social, cultural, and technological development has been extremely limited.

Stewart Russell's critique of SCOT is right on the mark in regard to race. Russell indicates that SCOT's discussion of closure, or the adoption of a technological feature, is highly misleading since it has overtures of "consensual acceptance" which may not always be the case. To this end Russell writes, "a group which stands to lose out from an imposed technological change . . . will probably not have been consulted, and certainly will not 'see the problem as being solved.'"[34] Langdon Winner's critique of social constructivist theories of technology cogently addresses the issue of power. Winner argues "insofar as there exist deeper [*racial*,] cultural, intellectual, political, or economic origins of social choices about technology or deeper issues surrounding these choices, the social constructivists choose not to reveal them."[35] As a result, SCOT can be seen as reinforcing racist stereotypes and supporting the constructed inferiority and marginalization of black people by building a program of investigation that obscures black Americans' ability to shape technology products.

A reassessment of SCOT by Trevor Pinch and Ronald Kline, and the focus on users of technology rather than producers, created new opportunities to examine African Americans and technology.[36] The finer language that defines a relevant social group as a fluid assemblage of individuals who share a common meaning of an artifact opens up interpretive flexibility to acknowledge and consider a multitude of coexisting technological meanings for a variety of social groups and creates an opportunity to study how African Americans, and other marginalized peoples, create their own relevant social groups that decide which technologies work for them and how to use them. However, as productive as this approach can be, SCOT still is limited by overlooking structural factors such as institutional racism, regional discrimination, economic disparity, and a host of other factors that have led many forms of African American technological creativity to be categorized as inferior.[37]

The creativity of black people often has been framed in this pejorative way. For instance, African American artists, writers, and musicians such as Jacob Lawrence, Langston Hughes, and Miles Davis, at different moments in history, have been constructed as primitive, Africanized, and animalistic. Interestingly, this primitivized creativity is not extended to African American people when it comes to the technological realm, because primitives cannot be technologically sophisticated. In the technological realm, creativity by African Americans is regularly dismissed as cleverness, instead of being interpreted as smart, ingenious, or innovative.

In 1858, when plantation owner Oscar J. E. Stuart attempted to patent the double cotton scraper that his slave Ned invented, Attorney General Jeremiah S. Black firmly determined that "a machine invented by a slave, though it be new and useful, cannot, in the present state of the law, be patented. I may add that if such a patent was issued to the master, it would not protect him in the courts against persons who might infringe it."[38] The Patent Office made it abundantly clear that inventions by slaves were not worthy of patent protection. The implication was that slaves were not human beings and definitely not citizens. With this reasoning, it was inconceivable for the Patent Office to grant a Negro a patent.[39] Even when American society began to accept the fact that African Americans did invent, black inventors were framed in a negative context. Some even went to the extent of denying that black people possessed American ingenuity. A white attorney expressed this perception by making the following comment in a patent rights battle involving a black inventor: "It is a well-known fact that the horse hay rake was first invented by a *lazy negro* [*sic*] who had a big hay field to rake and didn't want to do it by hand."[40] The stereotype of black inferiority, fueled by the Uncle Remus trickster character of American folklore, implied that African American people invented in order to sustain poor work habits rather than to produce creative solutions to existing problems.[41]

The acknowledgment of the margins, and those activities, people, and institutions that exist and thrive there, are fully absent from all of these approaches, because they consistently focus on winners of technological controversies and why certain technology has won. It would appear as if African Americans, throughout American history, did not have the ability to make technological decisions of their own and have led lives in which technology was foisted upon them. By overlooking the implications of race, many theories of technology limit the examination of African American technological experiences. Yet, it is not just these theories that have shaped black relationships to technology. Technology, as well as racist ideologies that have been built into the material structure and form of technology, has also been used to do racist work.

Race-ing Technology

In discussing *Time* magazine's cover photo of O. J. Simpson after his arrest, John Fiske discloses how technology can be used to reinscribe and maintain negative representations of African Americans within the dominant American culture. Moreover, Fiske's work alludes to the ways technology is "raced" and embedded with racialized politics. On June 27, 1994, *Time* magazine editors selected Matt Mahurian's "photo-illustration" of O. J. Simpson for the cover. Unbeknownst to readers, the image was technologically darkened. This came to only light after *Newsweek* printed the same unaltered image on its cover. O. J. Simpson was not black enough, so by enrolling computer-related technology to produce the valued racialized image, *Time*'s staff reconstituted his visual representation to meet the dominant American culture's perceived expectations of a black felon. The magazine's editor exhibited *Time*'s insensitivity to American racial identity politics when he commented that the image "lifted a common police mug shot to the level of art, with no sacrifice of truth."[42] Yet to a large number of African Americans who disapproved of the altered image, there was a significant sacrifice of truth. The editor did not see anything wrong with technologically darkening O. J. Simpson's image, because O. J. Simpson was being presented as a criminalized black man. The implication here is that blackness, regardless of hue, carries the same meaning: black = criminal. In this way technology is being used to dislocate racism; that is, "racism is dislocated when it is apparently to be found only in the behaviors of a racial minority and never in those of the white power structure."[43] This dislocated racism can be more clearly seen in the editor's rationale for darkening the image.

> The harshness of the mugshot—the merciless bright light, the stubble on Simpson's face, the cold specificity of the picture—had been subtly smoothed and shaped into an icon of tragedy. The expression on his face was not merely blank now, it was bottomless. This cover, with the simple, non-judgmental headline "An American Tragedy," seemed the obvious right choice....First, it should be said (I wish it went without saying) that no racial implication was intended, by TIME or by the artist. One could say that it is racist to say that Blacker is more sinister, and some African Americans have taken that position in the course of this dispute.[44]

The racism tied up in this photographic manipulation is dislocated onto African American people and represented by black anger. The result of these types of interactions is that black people are perceived to overreact and misconstrue certain acts as racist, while simultaneously the fictitious racelessness of whiteness is stabilized by the denial of the power of race in America.

As critical as Fiske is of those who use the technology, he does not address the problematic technological politics this case presents. For instance, this technologically modified image is considered to be devoid of racism because of the assumed value-neutrality of technology in conjunction with the disavowal of the American fetishism of black criminality. In a sense, the technology mediated a re-representation of the "true" image, to "fix" it and make it "right." Thus, to make it "right" was to make it darker. Those technologically transmogrifying O. J. Simpson's image did not acknowledge what Franz Fanon calls the process of "epidermalization."[45] That is, in the most basic sense, much of the information that people read, interpret, and use, that results in differential power and racial relations, is inscribed within skin, skin color, and the body.

In bringing together race, technology, and representation, Richard Dyer in *White* has examined how photographic technology was created from a racialized perspective. Dyer exposes photographic and cinematographic whiteness, and the ways in which producers of the related technology have used white skin tone and color as their reference. Dyer writes that film

> stocks, cameras, and lighting were developed taking the white face as the touchstone. The resultant apparatus came to be seen as fixed and inevitable, existing independently of the fact that it was humanly constructed. It may be—certainly was—true that photo and film apparatuses have seemed to work better with light-skinned peoples, but that is because they were made that way, not because they could be no other way.[46]

As this image-replicating technology developed, the white facial skin tone was used as the "standard" to determine the success or failure of each technology. To this end, Dyer argues, "experiments with, for instance, the chemistry of photographic stock, aperture size, length of development and artificial lighting all proceeded on the assumption that what had to be got right was the look of the white face."[47] With the racially valenced underlying design principle that white facial skin tone was normal and everything else was abnormal, photographic technology was constructed to privilege and perfect representations of whiteness. The way this has played out for nonwhite people, in specific for African Americans, is that photographing black people is frequently considered to be a "problem." This is another instance in which we can see the dislocation of race and racism. Instead of viewing the technology as the problematic, we see blackness as the problem. Alterations are made onto the black body (as in reflective makeup, oiling, stronger lighting), rather than the technology itself.

Through these alterations, the misperceived value-neutrality of the technology and how the technology is raced can continue to be hidden. As effectively as Dyer points out the racialized nature of photographic technology, he neither addresses the material consequences these acts of potential technological violence can have on black people, nor does he discuss how racialized technology continues to maintain and sustain white hegemony.

In the article "Do Artifacts Have Politics?" Langdon Winner presents an approach to address race, racism, and technology as well as how technology can be racialized. Winner describes how New York City public works builder Robert Moses designed several Long Island overpasses, leading to certain recreational facilities, to prohibit access by undesirable others. Winner writes that "automobile-owning whites of 'upper' and 'comfortable middle' classes, as he [Moses] called them, would be free to use the parkways for recreation and commuting. Poor people and blacks, who normally used public transit, were kept off the roads because the twelve-foot-tall buses could not get through the overpasses."[48] Thus, one man was able to embed his racial ideology within these technological artifacts, thereby racializing them during their construction and eventual use. Winner's example is not without its critics, but his work challenges us to think about how to critique technological activities from a racial vantage point.[49] By acknowledging that technological artifacts, practices, and knowledge can be racialized, Winner opens up a larger discussion about racial politics and technology in American culture.

What can move us forward in understanding issues of race, racism, African American experiences, and technology is to not focus on and give too much explanatory power to "authorial intentions" in order to produce the definitive meaning of a technology.[50] We must "engage the ambivalence of artefacts" and the multiple meanings this ambivalence creates.[51] By acknowledging the tensions between discordant discourses and accepting nondominant communities as legitimate locations from which to explore the nature of technology within American culture, we can embrace the complexity and contradiction in technology and societies. Specifically for African Americans, this approach validates asking how African Americans see, view, feel, understand, and interact with technology from their own perspective. When we redirect our position of analysis for African Americans, we can begin to ask questions that address the ways African Americans often view technology differently than others. The move away from the object, to the person or the community, creates new opportunities to study the ways those marginalized engage technology within their everyday lives.

In this regard, black intellectuals Martin Luther King Jr. and Amiri Baraka articulated similar, yet different, visions of why, when considering the place of technology in the world, we should focus on the individuals and community rather than on material objects. As with Winner, their brief writings on technology can open new avenues to explore why and how African Americans through black vernacular technological creatively can redeploy, reconceive, and re-create technology.

Black Vernacular Technological Creativity

Martin Luther King Jr., in his posthumously published book *Where Do We Go from Here: Chaos or Community?*, began to critically examine the deepening divide between morality and technology. In the chapter titled "The World House," King expressed concern about the ways he saw modern science and technology and freedom revolutions shaping the emerging global community. He saw a great deal of change, but he also was unsure if we had the capacity to embrace each other as equal human beings, break the tradition of human exploitation, and use technology to bring communities together rather than destroy them. King wrote, "We must work passionately . . . to bridge the gulf between scientific progress and our moral progress. One of the great problems of mankind is that we suffer from a poverty of spirit which stands in glaring contrast to our scientific and technological abundance." He signaled his concern that "the richer we have become materially, the poorer we have become morally and spiritually" and cautioned that "when scientific power outruns moral power, we end up with guided missiles and misguided men."[52] King pressed for a revolution in values not only cultural but also technological. He argued that technological development did not have to be oppositional to a global moral vision. He was also troubled by what he saw as the potential for technology, if used inhumanly, to exploit individuals, communities, and societies. King was calling for "a shift from a 'thing'-oriented society to a 'person'-oriented society," and declared that "when machines and computers, profit motives and property rights are considered more important than people, the giant triplets of racism, materialism, and militarism are incapable of being conquered."[53]

Of course King's critique was situated within the context of sixties-era protest, which demanded the reassessment of the expanding military industrial complex.[54] But his critique also was situated within the context of the emerging technological medium of television. King, and the civil rights movement, effectively appropriated the power of television. Sasha Torres contends that television, as a technology of representation, powerfully displayed the civil rights protests to the larger, primarily white, American society and altered the way that white America saw and viewed African Americans.[55] The technologically mediated televisual representation of terrible events such as the attack on the Selma-to-Montgomery marchers as they crossed the Edmund Pettis Bridge changed how many people saw the movement. White audiences began to sympathize with the civilized marchers, who were being brutalized as they peacefully demonstrated for their rights. Torres shows that television can be a fertile location to ask questions about technology in relation to American race relations, politics of representation, and African American life. However, this appropriation does not specifically represent black vernacular technological creativity. Civil rights activists were not actively engaging the technology of the television to alter the way they were presented to American society. Thus, as much as television recontextualized the civil rights movement, the change in black televisual representation was quite serendipitous.

White audiences' newly found sympathy for the nonviolent civil rights protesters did not carry over to the Black Panther Party for Self Defense. The potent images of gun-toting Huey P. Newton, Bobby Seale, and a host of other Black Panther Party members redeployed the gun and precipitated an important reversal of its technological power. Guns were instruments traditionally used to control—in the loosest sense of the word—black bodies. But the Black Panther Party members inverted this power. They redeployed guns as effective and visible artifacts to create a sense of fear among many white Americans, the same fear that many African Americans had felt for generations. This appropriation by the Black Panther Party of the material and symbolic power of the gun, against those who had used it so powerfully to subjugate African Americans, enabled them to claim power African Americans infrequently access.[56]

Following in this black nationalist tradition, Amiri Baraka would level a critique of technology from a black perspective. In "Technology & Ethos," Baraka wrote that "machines (as Norbert Weiner said) are an extension of their inventor-creators. That is not simple once you think. Machines, the entire technology of the West, is just that, the technology of the West . . . .Political power is also the power to create—not only what you will—but to be freed to go where ever you can go—(mentally and physically as well). Black creation—creation powered by the Black ethos brings very special results."[57] Baraka felt that the West had long ago gone down the wrong path in attempting to technologize humanity rather than humanizing technology. In his opinion, the Western technological tradition of creating "technology that kills both plants & animals, poisons the air & degenerates or enslaves man" was misguided.[58]

Baraka was equally interested in probing what could happen and the questions that could be asked if black people had technological power and became agents of technological change:

> Think of yourself, Black creator, freed of european [*sic*] restraint which first means the restraint of self determined mind development. Think what would be the results of the unfettered blood inventor-creator with the resources of a nation behind him. To imagine—to think—to construct—to energize!!! How do you communicate with the great masses of Black people? How do you use the earth to feed masses of people? How do you cure illness? How do you prevent illness? What are the Black purposes of space travel?[59]

In a sense he was asking how black people could express their own creativity and design technology that would represent their social, cultural, and technological aesthetics. He would get to the heart of this question through an analysis of a typewriter.

> A typewriter?—why shd [sic] it only make use of the tips of the fingers as contact points of flowing multi directional creativity. If I invented a word placing machine, an "expression-scriber," *if you will*, then I would have a kind of instrument into which I could step & sit or sprawl or hang & use not only my fingers to make words express feelings but elbows, feet, head, behind, and all the sounds I wanted, screams, grunts, taps, itches, I'd have magnetically recorded, at the same time, & translated into word—or perhaps even the final xpressed thought/feeling wd not be merely word or sheet, but *itself*, the xpression, three dimensional—able to be touched, or tasted or felt, or entered, or heard or carried like a speaking singing constantly communicating charm. *A typewriter is corny!!* The so called fine artist realizes, those of us who have *freed* ourselves, that our creations need not emulate the white man's, but it is time the engineers, architects, chemists, electronics craftsmen, ie film too, radio, sound, &c., that learning western technology must not be the end of our understanding of the particular discipline we're involved in. Most of that west shaped information is like mud and sand when you're panning for gold![60]

Baraka clearly stated that the typewriter—a technology designed by someone who did not see the world from a black perspective—could not fit his aesthetic sensibilities. He used the typewriter to ponder what the results of black technological creativity would look like if black people were freed from Western technological domination. But Baraka, like King, would cautiously ask, "What is our spirit, what will it project? What machines will it produce? What will they achieve?" He demanded that black technological expression be humanistic, which in his words "the white boy has yet to achieve."[61]

A more recent example of what Baraka alluded to can be seen with the creative technological reconception at Black Liberation Radio. In 1986 DeWayne Readus (soon to be known as Mbanna Kantako) began Afrikan Liberation

Radio (which became known as Black Liberation Radio in 1988 and is now Human Rights Radio) in his apartment located in the John Hay Homes public housing complex in Springfield, Illinois. Readus was already well known locally for his activism regarding public housing issues.[62] The radio station began as a means for the John Hay Homes Tenants Rights Association to organize the residents and began weekly broadcasts in January 1988.[63] Initially, the station mainly aired mixes, rap, reggae, political and social commentary, and occasionally listener phone calls. The station had a total power of one watt, and, due to the segregated nature of Springfield, most of Springfield's black population could receive the broadcast. The event that changed the position of Black Liberation Radio with Springfield's black community was the broadcast of what would become known as the Gregory Court massacre.

The events began to unfold on March 19, 1989, when the emotionally unstable Douglas "Dougie" Thomas held his girlfriend, Karen Lambert, and her sister, Nicole, hostage. During the forty-two-hour standoff, Thomas's family members wanted to talk to him, convince him to let the women go, and surrender. The Springfield police apparently were minimally responsive to the family's requests and eventually entered the apartment by force after shooting two canisters of tear gas through a window. In the end, Thomas shot the two women and himself; only Nicole was not fatally injured.[64] In the following week there was a great deal of disagreement between the black residents and the police regarding the order of the events that resulted in the deaths of Dougie Thomas and Karen Lambert. It was unclear to many if Thomas had shot himself and the Lambert sisters before the tear gas or because of the tear gas. After this incident, Black Liberation Radio led the questioning of Springfield's police and began to broadcast police activity and air the resident encounters with police brutality. Soon Black Liberation Radio was harassed for its activism, as well as for broadcasting without an FCC license.[65]

Readus/Kantako said his station was a form of electronic civil disobedience. Thus, he clearly knew what he was doing. His technologically driven response was not an accidental by-product of his hobby. He understood this set of technological objects as a potent means of regaining power and a voice within an oppressive local system. More important, Black Liberation Radio rearticulated the politics of surveillance in this African American community. The station supported an "inverted 'neighborhood watch,'" and observed "the police [and city officials] as the [violent and] threatening intruder."[66] Black Liberation Radio creatively reconceived surveillance technology to surveil the surveillants. As a result of this technologically and culturally rooted inversion of power, African Americans living in the Hay Homes renegotiated their relationships with the oppressive dominant power structure. This powerful reconception of a set of technological products of surveillance was based on the needs and desires of a black community.[67]

In similar ways to the Black Panther Party and Black Liberation Radio, resistance has been a motivating factor for musicians in the reconception of technological artifacts, practices, and knowledge.[68] One example of this can be seen with DJs and the act of scratching. Scratching is the purposeful manual manipulation of an LP recording in the reverse direction of the spinning turntable to produce a "scratching" noise. Depending on the speed, duration, and the music already inscribed on the LP record, scratching can produce a plethora of sounds. When DJs began scratching, they subverted the fundamental meaning constructed for record players as well as for that of the LP records. What is significant about this basic maneuver is that it drastically diverges from the principal meaning embedded in the technological network associated with records and record players: to listen to prerecorded sound/music. DJs were thus able to creatively reconceive the technological products associated with recorded music and the knowledge associated with their functions based on their own black/ethnic musical sensibilities.[69]

The sonic and cultural priorities that led these musicians to reconceive recording equipment began to exert a broader influence as the popularity of hip-hop music exploded in the 1980s and 1990s. Initially, existing technology was incapable of reproducing the desired sounds. Musicians such as Herbie Hancock, who embraced the tonal flexibility of synthesizers, would often have to "hack" them to produce sounds like those exhibited in his Grammy award-winning single "Rockit".[70] Others, like Eric Sadler—one of the producers of Public Enemy's incendiary hip-hop album "Fear of a Black Planet" (1990)—who desired to reproduce a gritty, dirty, and for him, more authentic sound, had to rely on a different approach. During an interview, Sadler explained why he preferred to work in a less than pristine studio. "One of the reasons I'm here in this studio is because the board is bullshit. It's old, it's disgusting, a lot of stuff doesn't work, there are fuses out . . . to get the old sound. The other room, I use for something else. All sweet and crispy clear, it's like the Star Wars room. This is the Titanic room."[71] Even though he had access to a much newer studio, he specifically wanted to use this seemingly inferior equipment because it allowed him to create a rich, rough, bass-heavy sound that emulated the "old sound" of records from the 1960s and 1970s that he valued. What he called the "sweet and crispy clear" sound produced by the newer equipment simply did not fit his aesthetic, sonic, or cultural priorities.

As hip-hop became an important part of American culture, and represented an extremely lucrative market, the music industry came to embrace the technological tweaks of early hip-hop musicians and directly supported the development of equipment designed specifically to tap into this market. DJ legend and hip-hop pioneer Grandmaster Flash was instrumental in re-creating a set of new technological objects and practices that addressed black cultural needs. Moreover, Grandmaster Flash's engagement with technology, like Readus/Kantako's, was not an accident. He had a history of technological innovation. He commented that it was his "love of technology, and specifically electronic equipment, that got me into DJing over 20 years ago. I remember stepping to the packed schoolyard jam with my equipment and records in hand ready to debut my new innovation, The Quick Mix Theory. Like a mad techno-scientist, I had spent months holed up in my room testing dozens of needles, sampling sounds and perfecting my newest experiment."[72] His technological rhetoric acknowledges that he understood he was re-creating technology based on his own personal aesthetics as well as using scientific methods to develop his technique.[73] A newer extension of his technologically rooted creativity can be seen in Rane Corporation's Empath mixer. Grandmaster Flash played a key role in this device's technical design, and in a 2003 interview he spoke of his often contentious, but ultimately successful, working relationship with Rane:

> The items on a mixer that you touch the most were too far away and other items that you touch weren't there. So when I made that phone call to Rane, I told them that . . . I did have a problem with some things. So when I had conversations with [Rane's director of sales] Dean Standing, all my frustrations of 25 years were coming up. They finally said, "Flash, what do you want with the mixer?" And I just flooded them with what I wanted. I met with [Rane engineer] Rick [Jeffs] and that was probably the closest thing to a fistfight that it could possibly get. With his genius, he'd say, "Flash, but it's not normally done this way." And I'd say, "But you must!" He'll say, "The mixer doesn't have enough room for that." And I'd say, "Well, you gotta squeeze it." He said, "What's going to be the output format?" And I told him XLR, quarter-inch, and RCA! He'd come back with, "Why don't we do two of the three," and I'd say no. As I gave him my wish list, he'd have to keep going back to the schematic diagram and make it work.[74]

Flash overrode the reservations of the engineers to produce one of the most innovative DJ mixers on the market today. In the end, whether it is the valorization of old equipment, the subversion of the phonograph through scratching, or the collaboration between turntablists and the music industry, the vernacular technological creative innovations of hip-hop musicians have deeply imprinted black cultural aesthetics, priorities, values, beliefs, and sensibilities on the dominant culture. I think Tricia Rose says it best when she writes:

> Rap technicians employ digital technology as instruments, revising black musical styles and priorities through the manipulation of technology. In this process of techno-black cultural syncretism, technological instruments and black cultural priorities are revised and expanded. In a simultaneous exchange rap music has made its mark on advanced technology and technology has profoundly changed the sound of black music.[75]

Within the exploration of techno-black cultural syncretism and black vernacular technological creativity lies the potential to end the silence surrounding African American technological experiences.

Conclusion

African American technological experiences need to be studied to alter the current discourse of American technology, rather than to multiculturalize our narrow understanding of technology in America. With new multicultural and multiracial approaches to understanding the nature of technology and American culture, traditional narratives can no longer produce, contain, and maintain the explanatory power they once possessed. To develop a more thoughtful analysis of African American technological experiences, we need to think differently about the questions we ask and the tools we use to answer those questions. Technological knowledge must be interrogated, because it is inextricably intertwined with relations of power that are regularly applied to regulate black existences. Stuart Hall writes, "Knowledge linked to power not only assumes the authority of 'the truth' but has the power to make itself true. All knowledge, once applied in the real world has real effects, and in that sense at least, 'becomes true.'"[76] Following from Hall, it can be said that what we know about the relations between black people and technology primarily comes from dominant subject positions that unfortunately tell us more about how African Americans are controlled and regulated than about how black people engage technology from their own locations within American culture. The existing approaches used to understand technology in American society and culture overlook racialized power and conflict when they reduce everything to various forms of negotiation. This is not to implicate or label social theories of technology as forms of epistemological imperialism in the manner in which Edward Said writes about orientalism; but Said's thoughts are insightful.[77] In writing about colonialism and imperialism, Said inveighs that "both are supported and perhaps even impelled by impressive ideological formations that include notions that certain territories and people require and beseech domination, as well as forms of knowledge affiliated with domination."[78] Just as the intellectual work of which Said writes is tainted from the very beginning, social theories of technology are besmirched by similar dominant cultural efforts that are intended to maintain domination, but are concealed within the rhetoric of flexibility and freedom.

To gain a deeper understanding of black vernacular technological creativity, it is vital to examine the experiences of African Americans from where they stand in American society and culture rather than from the dominant position reflecting back on black lives. Black vernacular technological creativity is rich in historical value and replete with rebellion, resistance, assimilation, and appropriation in forms we would often not recognize and in places we are not accustomed to looking. It is from this space that we can see how black people reclaim a level of technological agency by redeploying, reconceiving, and re-creating material artifacts in their world. By focusing on black vernacular technological creativity and engaging in uncovering the multiple layers of black communities and their interactions with technology, we can avoid making the "they are all the same" essentialization of the marginalized mistake regarding African Americans.[79]

Technology is often thought of as a value-neutral "black box" for inputs and outputs. Critical studies of technology have opened the black box, but there are many hidden compartments that still need to be explored. To access these concealed compartments, or the "blackness" in the black box, we need to reassess and expand our study of technology to examine how racially marginalized people, such as African Americans, interact with technology and how technology mediates multiple African American experiences with racism. To address African Americans and technology, we must think about the ways in which black people see race and racism—important realities of everyday black existence. This is difficult because race and racism, in relation to technology, have always been hidden in a mysterious place of "unlocation."[80] By uncovering African Americans creating technological artifacts, practices, and knowledge that have become parts of the American material and technological cultures, black people will become visible metaphorically and materially. This work will enable black people to move out of the shadows, lift the veil, remove the mask, and solidify and develop decidedly positive technological representations and existences for African Americans within American society and culture.

Notes

1. Imamu Amiri Baraka, "Technology & Ethos," in *Raise, Race, Rays, Raze: Essays Since 1965* (New York: Random House, 1971), 157.
2. Stokely Carmichael and Charles V. Hamilton, *Black Power: The Politics of Liberation in America* (New York: Random House 1967), 44.
3. Baraka, "Technology & Ethos," 155–57.
4. Anthony Walton, "Technology Versus African-Americans," *Atlantic Monthly* 283.1 (January 1999): 16.
5. Henry Louis Gates Jr. and Cornel West, *The Future of the Race* (New York: Alfred A. Knopf, 1996), 84.
6. Ronald Kline, "Construing 'Technology' as 'Applied Science': Public Rhetoric of Scientists and Engineers in the United States, 1880–1945," *Isis* 86 (June 1995): 197; Leo Marx, "The Idea of 'Technology' and Postmodern Pessimism," in *Technology, Pessimism, and Postmodernism*, ed. Yaron Ezrahi, Everett Mendelsohn, and Howard P. Segal (Amherst: University of Massachusetts Press, 1994), 14; Judy Wajcman, *Feminism Confronts Technology* (College Station: Pennsylvania State University Press, 1991), 14–15.
7. Gena Dagel Caponi, ed., *Signifyin(g), Sanctifyin', and Slam Dunking: A Reader in African American Expressive Culture* (Amherst: University of Massachusetts Press, 1999), 1–41.
8. Joel Dinerstein, *Swinging the Machine: Modernity, Technology, and African American Culture between the World Wars* (Amherst: University of Massachusetts Press, 2003), 22.
9. Ibid., 126.
10. Ibid., 130.
11. Ibid., 25.
12. Albert Murray, *Stomping the Blues* (New York: McGraw-Hill, 1976), 90.
13. Black vernacular technological creativity is related to the ways Ron Eglash has written about technological appropriations by marginalized peoples. See Ron Eglash, Jennifer L. Croissant, Giovanna Di Chiro, and Rayvon Fouché, eds., *Appropriating Technology: Vernacular Science and Social Power* (Minneapolis: University of Minnesota Press, 2004).
14. Ruth Schwartz Cowan, *A Social History of American Technology* (New York: Oxford University Press, 1997).
15. Sigfried Giedion, *Mechanization Takes Command: A Contribution to Anonymous History* (New York: Oxford University Press, 1948); S. Colum Gilfillian, *The Sociology of Invention* (Cambridge, Mass.: MIT Press, 1935); Lewis Mumford, *Technics and Civilization* (New York: Harcourt, Brace, 1934); William Fielding Ogburn, *Social Change, with Respect to Culture and Original Nature* (New York: B. W. Huebsch, 1922).
16. Abbott Payson Usher, *A History of Mechanical Invention* (Cambridge, Mass.: Harvard University Press, 1954); Lynn Townsend White, *Medieval Technology and Social Change* (Oxford: Clarendon Press, 1962); Leo Marx, *The Machine in the Garden* (New York: Oxford University Press, 1964).
17. Leo Marx, "The Machine in the Garden," *New England Quarterly* 29 (March 1956): 27–42.
18. John F. Kasson, *Civilizing the Machine: Technology and Republican Values in America, 1776–1900* (New York: Grossman, 1976); David E. Nye, *American Technological Sublime* (Cambridge, Mass.: MIT Press, 1994); Jeffrey L. Meikle, "Leo Marx's *The Machine in the Garden*," *Technology and Culture* 44.1 (January 2003): 159.
19. Donald MacKenzie and Judy Wajcman, *The Social Shaping of Technology: How the Refrigerator Got Its Hum* (Philadelphia: Open University Press, 1985).
20. Wiebe Bijker, "Sociohistorical Technology Studies," in *Handbook of Science and Technology Studies*, ed. Sheila Jasanoff, Gerald E. Markel, James C. Petersen, and Trevor Pinch (Thousand Oaks, Calif.: Sage Publications, 1995), 229–56.
21. Bijker, "Sociohistorical Technology Studies," 250.
22. Thomas P. Hughes, *Networks of Power: Electrification in Western Society, 1880–1930*, (Baltimore: Johns Hopkins University Press), 1983; Wiebe E. Bijker and Trevor Pinch, "The Social Construction of Facts and Artefacts, or, How the Sociology of Science and the Sociology of Technology Might Benefit from Each Other," *Social Studies of Science* 14 (August 1984): 399–441.
23. Bruno Latour, *Science in Action: How to Follow Scientists and Engineers through Society* (Cambridge, Mass.: Harvard University Press, 1987).
24. Donald MacKenzie, *Knowing Machines: Essays on Technical Change* (Cambridge, Mass.: MIT Press, 1996), 13.
25. Wiebe E. Bijker, Thomas P. Hughes, and Trevor Pinch, eds., *The Social Construction of Technological Systems* (Cambridge, Mass.: MIT Press, 1987).

26. John Law, "Technology and Heterogeneous Engineers: The Case of Portuguese Expansion," in *The Social Construction of Technological Systems*, ed. Bijker, Hughes, and Pinch, 112–13.
27. Judy Wajcman, "Feminist Theories of Technology," in *Handbook of Science and Technology Studies*, ed. Jasanoff, Markle, Petersen, and Pinch, 204.
28. Wajcman, *Feminism Confronts Technology*; Judy Wajcman, *Technofeminism* (Cambridge: Polity, 2004).
29. Rayvon Fouché, *Black Inventors in the Age of Segregation* (Baltimore: Johns Hopkins University Press, 2003), 26–81.
30. David J. Hess, *Science Studies: An Advanced Introduction* (New York: New York University Press, 1997), 110.
31. Law, "Technology and Heterogeneous Engineers," 113.
32. Russell Ferguson, "Invisible Center," in *Out There: Marginalization and Contemporary Cultures*, ed. Russell Ferguson, Martha Gever, Trinh T. Minh-ha, and Cornel West (Cambridge, Mass.: MIT Press, 1990) 9–14.
33. Bijker and Pinch, "The Social Construction of Facts and Artefacts," 30.
34. Stewart Russell, "The Social Construction of Artefacts: A Response to Pinch and Bijker," *Social Studies of Science* 16 (May 1986): 337.
35. Langdon Winner, "Upon Opening the Black Box and Finding It Empty: Social Constructivism and the Philosophy of Technology," *Science, Technology, & Human Values* 18.3 (Summer 1993): 370–71; emphasis added.
36. See also Ronald Kline and Trevor Pinch, "Users as Agents of Technological Change: The Social Construction of the Automobile in the Rural United States," *Technology and Culture* 37.4 (October 1996): 763–95.
37. Hans K. Klien and Daniel Lee Kleinmann, "The Social Construction of Technology: Structural Considerations," *Science Technology and Human Values* 22.1 (January 2002), 28–52.
38. Norman O. Forness, "The Master, the Slave, and the Patent Laws: A Vignette of the 1850s," *Prologue*, 25th Anniversary Edition (1994): 50.
39. Steven Lubar, "The Transformation of Antebellum Patent Law," *Technology and Culture* 32 (October 1991): 932–59.
40. "A Colored Man's Invention," *New York Recorder*, February 13, 1892; emphasis added.
41. Joel Chandler Harris, *Uncle Remus, His Songs and His Sayings; the Folk-lore of the Old Plantation* (New York: Appleton, 1881).
42. James R. Gaines, "To Our Readers," *Time*, July 4, 1994, 4.
43. John Fiske, *Media Matters: Everyday Culture and Political Change* (Minneapolis: University of Minnesota Press, 1994), 272.
44. Ibid., 273, quoted from Gaines, "To Our Readers," 4.
45. For epidermalization see Frantz Fanon, *Black Skin, White Masks* (New York: Grove Press, 1967), 11; Stuart Hall, "The After-life of Frantz Fanon: Why Fanon? Why Now? Why *Black Skin, White Masks*?" in *The Fact of Blackness: Frantz Fanon and Visual Representation*, ed. Alan Read (Seattle: Bay Press, 1996), 12–37.
46. Richard Dyer, *White* (New York: Routledge, 1997), 90.
47. Ibid.
48. Langdon Winner, "Do Artifacts Have Politics?" *Daedalus* 109.1 (Winter 1980).
49. Exchanges between Bernward Joerges, Steve Woolgar, and Geoff Cooper have critiqued Winner's argument in certain contexts, but the strengths of the racialized aspects of this argument still remain; Bernward Joerges, "Do Politics Have Artefacts?" *Social Studies of Science* 29 (June 1999), 411–31; Steve Woolgar and Geoff Cooper, "Do Artefacts Have Ambivalence? Moses' Bridges, Winner's Bridges, and other Urban Legends in S&TS," *Social Studies of Science* 29 (June 1999), 433–49; Bernward Joerges, "Scams Cannot Be Busted: Reply to Woolgar and Cooper," *Social Studies of Science* 29 (June 1999), 450–57.
50. Bernward Joerges, "Do Politics Have Artefacts?" 423.
51. Woolgar and Cooper, "Do Artefacts Have Ambivalence?" 443.
52. Martin Luther King Jr., *Where Do We Go from Here: Chaos or Community?* (New York: Harper & Row, 1967), 171–72.
53. Ibid., 186.
54. Howard Brick, *Age of Contradiction: American Thought and Culture in the 1960s* (Ithaca, N.Y.: Cornell University Press, 2001); Robert A. Rhoads, *Freedom's Web: Student Activism in an Age of Cultural*

Diversity (Baltimore: Johns Hopkins University Press, 1998); Theodore Roszak, *The Making of a Counter Culture: Reflections on the Technocratic Society and Its Youthful Opposition* (Garden City, N.Y.: Doubleday, 1969); Kirkpatrick Sale, *SDS* (New York: Random House, 1973).

55. Sasha Torres, *Black, White, and in Color: Television and Black Civil Rights* (Princeton, N.J.: Princeton University Press, 2003).
56. The Black Panther Party members were not the first black people to redeploy the power of the gun in the twentieth century. Black activist Robert F. Williams powerfully advocated black armed resistance in the late 1950s and early 1960s (Timothy B. Tyson, *Radio Free Dixie: Robert F. Williams and the Roots of Black Power* [Chapel Hill, N.C.: University of North Carolina Press, 1999]).
57. Baraka, "Technology & Ethos," 155.
58. Ibid., 157.
59. Ibid., 155–56.
60. Ibid., 156–57.
61. Ibid., 157.
62. Doug Pokorski, "Hay Homes Group Again Seeks Bus Service," *The State Journal-Register* (Springfield, Ill.), October 23, 1985, sec. local 10; "Readus to Head Hay Tenants Group," *The State Journal-Register* (Springfield, Ill.), February 23, 1986, sec. local 35.
63. Mary Nolan, "FCC Turns Off Hay Homes Radio Station," *The State Journal-Register* (Springfield, Ill.), April 12, 1989, sec. local 9.
64. Jay Fitzgerald, Mike Matulis, Cathy Monroe, Michael Murphy, and Jacqueline Price, "Hostage Ordeal Ends," *The State Journal-Register* (Springfield, Ill.), March 20, 1989, sec. local 1; Jay Fitzgerald, "With 2 Dead the Questions Remain," *The State Journal-Register* (Springfield, Ill.), March 21, 1989, sec. local 1.
65. Nolan, "FCC Turns Off Hay Homes Radio Station"; Ron Sakolsky and Stephen Dunifer, eds., *Seizing the Airwaves: A Free Radio Handbook* (Oakland, Calif.: AK Press, 1998), 117–20.
66. Fiske, *Media Matters*, 273.
67. In light of the racialized nature of American culture, surveillance has become a technology of whiteness. This is primarily because black people have been constructed as the "others" in American society. In part due to skin color, the "difference" of blackness (or brownness) is regularly the object and target of surveillance. That which is not normal is surveilled, and black people within American society and culture have traditionally been the abnormal against which normal is judged.
68. Trevor Pinch and Karin Bijsterveld, "Sound Studies: New Technologies and Music," *Social Studies of Science* 34 (October 2004): 635–48; Mark Katz, *Capturing Sound: How Technology Has Changed Music* (Berkeley: University of California Press, 2004); Joseph Schloss, *Making Beats: The Art of Sample-Based Hip-Hop* (Middletown, Conn.: Wesleyan University Press, 2004); Jonathan Sterne, *The Audible Past: Cultural Origins of Sound Reproduction* (Durham, N.C.: Duke University Press, 2003); Timothy D. Taylor, *Strange Sounds: Music, Technology, and Culture* (New York: Routledge, 2001); Paul Théberge, *Any Sound You Can Imagine: Making Music/Consuming Technology* (Hanover, N.H.: Wesleyan University Press, 1997).
69. David Albert Mhandi Goldberg, "The Scratch in Hip-Hop: Appropriating the Phonographic Medium," in Eglash et al., *Appropriating Technology*, 107–44.
70. Siva Vaidhyanathan, *Copyrights and Copywrongs: The Rise of Intellectual Property and How It Threatens Creativity* (New York: New York University Press, 2001), 149–51.
71. Trisha Rose, *Black Noise: Rap Music and Black Culture in Contemporary America* (Hanover, N.H.: University Press of New England, 1994), 77.
72. See http://www.grandmasterflash.com/ (accessed June 8, 2006).
73. Jeff Chang, *Can't Stop Won't Stop: A History of the Hip-Hop Generation* (New York: Picador, 2005), 112–13.
74. Jim Tremayne, "With a Hot New Mixer on the Market and a Revitalized Career in Motion, DJ Pioneer Grandmaster Flash Finds that Necessity Is Still the Mother of Invention," *DJ Times*, March 2003.
75. Rose, *Black Noise*, 96.
76. Stuart Hall, ed., *Representation: Cultural Representations and Signifying Practices* (Thousand Oaks, Calif.: Sage Publications, 1997), 49.
77. Edward Said, *Orientalism* (New York: Pantheon Books, 1978).
78. Edward Said, *Culture and Imperialism* (New York: Knopf, 1994), 8.
79. Cornel West, "The New Cultural Politics of Difference," in *Out There*, ed. Ferguson et al., 19–36.
80. Heidi Mirza, ed., *Black British Feminism* (New York: Routledge, 1997), 5.

EXPOSITION
BOOSTER POSTCARD
WHAT EVERY ONE SHOULD KNOW
CANAL STATISTICS.
Length of Canal 50½ Miles.
Culebra Cut Depth 300 Feet.
Locks in Pairs 12.
Locks Usable Length 1000 Feet.
" Width 110 " .
Canal Force Actually At Work 35.000
Canal Co Formed By The French 1879.
" Work Begun " " " 1881.
Eight Years Work " " " Cost $300.000.000
Work Begun By U.S. May 4 1904.
U.S. Paid French Canal Co $40.000.000
U.S. " Republic of Panama $10.000.000
Cost of Canal Total Over $375.000.000
Date of Completion Jan 1st 1915
A Yearly Saving of Millions
to the Shipping World.
SEATTLE
PORTLAND
P.P.I. Exposition
San Francisco
1915
LOS ANGELES
SAN DIEGO
Meeting of the Atlantic & Pacific,
THE KISS
OF THE OCEANS 1915.
PANAMA PACIFIC INTERNATIONAL EXPOSITION
SAN FRANCISCO CALIFORNIA.
COPYRIGHTED
C. A. de Lisle Holland

Imperial Mechanics: South America's Hemispheric Integration in the Machine Age

Ricardo D. Salvatore

> The nations all await
> Our coming, with their portals opened freely;
> They are Republics, each our sister State,
> And down the world, from Mexico to Chili,
> The railway's mutual commerce should pulsate.
>
> —F. D. Carpenter, "From Zone to Zone," in H. R. Helper, *The Three Americas Railway* (1881), 284, 286

> The airplane, more than any other means of transportation, has overcome Latin America's geography. It was the airplane that finally conquered the towering mountains, bottomless swamps, and dense jungles. More than rails or roads or rivers, planes helped solve our neighbor's transportation problems.
>
> —Delia Goetz, *Neighbors to the South* (1941), 116

The rise of the machine as the dominant representation of U.S. superiority and supremacy in the Western Hemisphere corresponded with the transition in the form of conceptualizing empire: from Manifest Destiny in the mid-nineteenth century to Benevolent Informal Empire (Good Neighbor Policy) in the fourth decade of the twentieth century. The railroad, the automobile, and the airplane represent different moments in the evolution of U.S. technology and were used, at each moment, to deploy U.S. claims to technological and cultural superiority. In addition to affirming U.S. dominance, these technological "marvels" were crucial to envisioning ways of integrating Central and South America into the sphere of U.S. commerce and influence. Subsequent imaginaries of transportation systems uniting the three Americas—the Pan-American Railroad, the Pan-American Highway, and Pan American Airways—constituted projections of U.S. machine civilization into the terrain of inter-American relations. I call these visions "transportation utopias" to emphasize their fictional and compre-

Figure 1.
The Kiss of the Oceans postcard, ca. 1911 © Blue Lantern Studio/Corbis.

hensive nature and also to underscore the displacement of machine magic to the nonplace of "Pan-America," a virtual space created at the conjuncture of foreign policy principles, business expansionism, and knowledge enterprises.

"Imperial mechanics" refers to the interrelationship between imaginaries of empire and conceptions of machine-civilization. My use of this concept responds to the necessity to connect the two notions in a way that allows us a better understanding of U.S. hegemonic conceptions and policies with regard to Latin America. "Imperial mechanics" is a cognitive and rhetorical displacement of "hard machines" (railroads, automobiles, airplanes, canals) into the terrain of Pan-Americanism and informal empire, a field usually associated with the "soft machine" of inter-American conferences, political pronouncements about peace and security, bureaucratic bargaining, and hemispheric propaganda. This essay shows the extent to which U.S. visions of hemispheric integration were built upon notions of connectivity, circulation, and modernity that promoters and policymakers considered embedded in machines.

The construction of the Panama Canal (1904–1914), an engineering achievement of gargantuan proportions, marked the ascent of the United States to a position of international leadership. A Herculean mechanical force that separated the continent into two, creating a passage between the Caribbean Sea and the Pacific Ocean, gave the United States enhanced naval control of the "American Mediterranean" and heightened the intensity of international navigation and commerce by reducing transportation distances and costs.[1] Paradoxically, this engineering marvel was presented as a monument to hemispheric unity and cooperation. Displaced into the terrain of inter-American relations, the Panama Canal was expected to render an additional service: to present to the Latin American neighbors U.S. technological superiority as a mechanical, visible fact. It was a spectacular machine that authorized the United States to speak as the new hemispheric hegemon.

This essay examines the role of these crucial mechanical constructs—"spectacular machines" and "transportation utopias"—in the making of imaginaries of hemispheric integration that functioned as the foundations of Pan-Americanism. They were two moments of a broader use of "hard machines" to reconfigure inter-American relations, at a time in which U.S. foreign-policymakers embraced a notion of hemispheric hegemony based upon a combination of selective interventions in Central America and the Caribbean (formal empire) and the expansion of U.S. investment, commerce, and influence through expertise in South America (informal empire). In the first section, I present the Panama Canal as an experiment in spectacular politics that was expected to teach objective lessons in governance and development to the republics of

Central and South America. In the second section, I examine the three "transportation utopias" that, between 1890 and 1945, stimulated the fiction of a united hemisphere in moments in which the countries of Latin America lagged behind in transportation development. In the final section, I focus on the 1942 journey of two American motorists in South America and use this journey to reflect upon the tensions between the dream of continental transitability and the reality of political fragmentation and economic backwardness.

Hard and Soft Machines in U.S. Hemispheric Policy

The U.S. informal empire in Latin America can be seen as a "soft machine," intensive in representations—a machine that produced, processed, and disseminated texts and images in order to accommodate older notions of self and other to the new conditions of capitalist accumulation (mass consumer capitalism and its technologies of representation).[2] It was this complex representational machine that generated a "rediscovery" of the region—of its urban and consumer modernity, of its antiquity, of its "Indianness," of its aberrant social inequality. World fairs, films, photographic exhibits, travel narratives, geographic manuals, and other representations brought into the U.S. sphere of visibility a subcontinent that had remained until then relatively unknown.

This soft machinery of empire organized practices and technologies of observation and representation (photography, travel writing, statistical handbooks, consumer surveys, etc.) to make available to the "American public" the diversity and difference of "Spanish America." Representational machines were fundamental for the construction of new knowledge about South/Latin America. Indeed, the construction of an informal empire in South America was a collective enterprise dependent upon representations. In South America in particular, where military invasion, territorial annexation, and direct colonization were not viable alternatives, the imperative of conquest had to be replaced by persuasive arguments about expertise, technology, and markets.

Since the Spanish-American War (1898), U.S. curiosity for "Spanish America" grew in exponential fashion. Early in the 1890s the International Bureau of the American Republics put together a series of handbooks, statistical manuals, and geography readers that brought the Central and South American republics to the attention of U.S. readers. Starting in 1909, John Barrett, director of the Pan-American Union, disseminated the idea of South America as a "continent of opportunities." He was followed by an army of business prospectors, journalists, travel writers, Spanish teachers, archaeologists, geographers, and historians, all eager to put in print their impressions of

South America. By a process of accumulation, they built the textual and visual foundations of U.S. engagement with the subcontinent.

The extension of informal empire to the lands beyond the "American Mediterranean" required an immense accumulation of representations. Beyond diplomatic tours, Pan-American conferences, or company towns, the Pax Americana existed in maps, paintings, geography books, novels, and natural history exhibits.[3] The building of informal empire was a process of cultural engagement that constructed simultaneously the object of desire and the reasons for extraterritorial influence. Business-based mechanical metaphors were important in formulating early-twentieth-century visions of hemispheric integration. The notion of Pan-Americanism prevalent in the period 1906–1925 entailed a "mechanics of commercial penetration" that integrated U.S. business and government forces into a coordinated effort to conquer South American markets and spread the "American way of life" among the new republics.[4] The "practical Pan-Americanism" predicated by John Barrett presented inter-American relations as a great machine in which American firms and governmental agencies cooperated—as different departments of a big corporation do—in disarming anti-Americanism, disseminating U.S. technologies, and creating the institutional framework in which the forces of trade and investment could best function.

Businessmen and the leaders of the Pan-Americanist movement agreed: to conquer South American markets a vast commercial and cultural offensive was needed, one predicated upon better knowledge of the region and the demonstration of the American way of life to South Americans. At the 1919 Pan-American Commercial Conference, speakers presented the need to disseminate among Latin Americans the superiority of U.S. road-building methods, told participants of the effectiveness of "publicity films" to present U.S. firms and products more effectively to South Americans, and informed the public of the establishment of the syndicated news agencies (U.P. and A.P.) that would integrate the newspapers of the region.[5] "Pan-America" was in actuality an enhanced territory for the free flow of goods, technical assistance, news, and business solutions in ways that turned upside down old diplomatic conceptions of the Pax Americana.

Implicit in the message of Pan-Americanism was an imperative to transmit to Latin Americans the superiority of U.S. technology and mass culture. In this sense, it could be said that the soft machinery of empire served to disseminate "America," a nation-culture self-fashioned in terms of its command of "hard machines." The hegemony of the U.S. informal empire depended crucially on the transmission of "American standards" of journalism, radio broadcasting, filmmaking, and advertising, as well as on the creation of imitative demands for automobiles, refrigerators, toothpaste, and typewriters.

In this essay, I focus on the "hard machines" that gave support to the imaginaries of inter-American integration. I start with the analysis of a "colossal machine" built to expand commerce and generate awe among the U.S. southern neighbors. The Panama Canal stood as the symbol of the new U.S. position in the world as well as a spectacular "objective lesson" about governance and development. It was expected that, in view of this engineering marvel, the Latin American republics would pay attention to what their older sister had to say and would act accordingly. The message was clear: at the time in which Europe was embarking in a devastating total war, the United States was taking over the banner of "Western civilization." Latin American nations had to understand this relocation in the leadership of Western civilization and accept in good terms what the U.S. had to offer: investment funds, new production and transportation technologies, sanitation and educational expertise, economic advice, and the peaceful arbitration of (Latin American) disputes.

This spectacular machine, I suggest, failed to impress Latin Americans. More enduring and appealing were imaginaries of hemispheric integration built upon new means of transportation. Machines emblematic of U.S. modernity (the railroad, the automobile, and the airplane) were endowed with the capacity of integrating the Americas, overcoming the obstacles imposed by geography. Pan-American railways, highways, and airways presented the possibility of increased commercial and cultural interaction among cities, nations, and peoples of the United States and Latin America. By mechanical means, U.S. promoters tried to shorten the time and distance that separated the two societies and cultures.

Colossal and Spectacular Machines (the Panama Canal)

From the start, the Panama Canal was conceived as a technological marvel that would open the gates of South America, a region that until then had remained trapped in the European sphere of influence. Through this "path between the seas," raw materials and foodstuffs would flow out of South America and reach U.S. and European markets. The Pacific countries in particular would gravitate into the orbit of U.S. commerce and navigation. The canal was an "American" achievement of epic proportions, an engineering solution to the four-centuries-old search for an "isthmian route" across Central America.[6]

The Panama Canal was one of the greatest engineering feats of all time. The sheer size and complexity of the project bedazzled contemporary observers. The

Figure 2.
Group of tourists in front of Panama Canal, Gatun, Panama, ca. 1910 © Underwood & Underwood/Corbis.

lock chambers could contain ships larger than the Titanic. Each lock was taller than a six-story building. No bridge or dam built in the United States had used such a volume of reinforced concrete.[7] The engineering design was unprecedented. The gates opened and closed with an electrical system operated from a giant switchboard. The canal used the power of falling water to generate the electric energy needed by its motors and towing locomotives. The giant chain that stopped ships before the gates was impressive, as were the emergency dams.[8]

The canal was an experimentation ground for "American technology." It was a test of the resistance of giant concrete structures, of the efficacy of centrally controlled electric systems, of the preciseness of hydraulic calculations, of the safety and productiveness of combination-machines to excavate the earth, pour concrete, and move materials. It was a site where new materials (the latest steel alloy) and new devices (a giant electric switchboard, the great locks) could be implemented experimentally. And it was the place for fine-tuning public health technologies for the control of malaria and yellow fever.

Promoters of the canal expected the shortening of transportation time to attract trade from the Pacific states and to divert traffic previously carried through the Magellan Strait.[9] Frederic Haskin saw the Panama Canal as an extension of the Mississippi River traffic and imagined the day when the steel products from Pittsburgh and the grain from Kansas would reach all ports of the world.[10] Despite the criticism of railroad promoters and naval imperialists, the project galvanized public support, because it promised important gains in terms of economic growth, national defense, and international image for the United States.[11]

With time, the operation of the canal surpassed the most optimistic expectations in terms of total volume of traffic. But it failed to integrate the economies of the American hemisphere. The canal benefited the United States more than it did the rest of the Americas.[12] In 1934–1935, 60 percent of the cargo transported across the canal corresponded to U.S. accounts, and over 33 percent of total traffic consisted of commerce *internal to the United States.*[13] The Atlantic countries of South America did not make much use of the canal. Only the Pacific nations of South America, holding modest shares of the region's import-export trade, benefited from the new passage.[14] Thus, the main economic effect of the canal was to bring closer the two coasts of the United States.[15] Shipments across the canal reflected the increasing specialization of Latin American nations as exporters of raw materials. Through the canal passed coffee and rubber from Brazil, sisal from Yucatan, nitrates from Chile, tin from Bolivia, sugar from Cuba, and oil from Peru.[16]

The strategic value of the canal to the United States was clear to most observers. Because the canal served to move ships from the Pacific to the Atlantic Ocean more rapidly, it improved significantly the capacity of the U.S. Navy to patrol the seas. The Canal Zone was, after all, a U.S. military bastion. It had about 8,000 troops, landing fields, and wireless stations, and could be used to supply warships with coal, food, and medicines. During WWI, crucial supplies to the allied armies went through the canal, including important volumes of Chilean nitrates used for explosives.[17] Placed under the jurisdiction of the War Department, the Canal Zone served during WWII as a key U.S. military enclave.[18]

The canal was a gigantic transplantation of "industrial America" into the Central American jungle. Its management and design benefited from the U.S. experience in railroad building and other great engineering projects. The relocation of industrial-age technology produced spectacles of mechanical power that awed contemporary observers. Cranes swung in the air enormous buckets of cement; large steam shovels deposited tons of materials in trains, seventy-five of them running daily; the filling of Gatum Lake took 26 million gallons of water. The cableways, the towing locomotives—all provided spectacles of overwhelming modernity.

The canal created a new location for rooting pronouncements about U.S. technological superiority. Here was a concrete example of American technology to be seen and admired by Latin Americans, an engineering work that provided an objective lesson in politics and development, a visible test of the powers of a machine civilization in the Tropics. U.S. visitors to the region presented the canal as a major force that could activate the dormant economies of South America.[19] To John Barrett, director of the Pan-American Union, the canal was a catalyst for a "commercial awakening" of South America.[20] Business prospector and journalist James A. Collins attributed to the "canal effect" sanitation improvements, railroad building, the development of ports, and the more intensive exploitation of natural resources.[21] Besides seeing new economic opportunities, Collins anticipated a profound change in the mentality of South Americans:

> It is not only the Canal itself. That is big and impressive in its engineering and operation. As the Latin American rides through, and sees his ten-thousand-ton liner lifted up and sucked down by the locks like a toy boat, you can feel the mental change taking place inside him. Being supersensitive to impressions, it is a transformation bordering on the spiritual.[22]

Spectacular politics was the mode of self-fashioning of the informal American empire. When, from Washington, President Wilson pressed the button that activated the detonator that blew off the last piece of earth connecting the two tunnels at Culebra Cut, he was also demonstrating a basic connection between technology and imperial governance. From the distance, he could finalize a work that would give his nation the renown of an advanced industrial power, capable of colossal engineering works. The canal provided a mechanical "example" designed to impress the Latin Americans neighbors. W. Blanchard, geography professor at the University of Illinois, wrote in 1937:

> The story of a great torrential river made over into a navigable waterway, of a great mountain ridge grooved and bridged by a great artificial lake, of a pest hole transformed into a healthful dwelling place, of all this accomplished without the least suspicion of graft or corruption—such an accomplishment is one of which any nation may well be proud. . . . In addition, it has been *an impressive lesson for our Latin American neighbors*—an example of what may be accomplished for peaceful intercourse among the nations.[23]

The completion of the canal was celebrated as a national achievement at the 1915 Panama-Pacific Exposition in San Francisco. F. Haskin anticipated this fair as a moment of communion in which Americans would gather to celebrate their nation's unity together with the completion of Columbus's quest.[24] U.S. technology had cut the continent in two in order to unite the world through commerce. Fair organizers offered visitors a miniature version of the Panama Canal, to be gazed upon from an elevated position. The exposition's "aeroscope" allowed people to see the technological achievement in the jungle as if it were within reach.[25]

The canal was also a lesson in government. In the midst of WWII, the colossal machine in the jungle became the expression of a political principle: the inevitable triumph of democracy over dictatorship. International law expert Norman Padleford saw the canal as a demonstration that democracies were able to carry out technological achievements that required a scale of cooperation generally associated with totalitarian regimes and coercive labor.[26] The canal displayed the powers (capital, management, social labor, science) generated by a democracy in the hinterlands of empire.

In addition, the canal was conceived as an engineering work that could generate "good will," as a cog in the larger machinery of inter-American cooperation. In February-March 1912, Secretary Philander Knox visited the countries that would be most immediately affected by the canal to bring the good news of enhanced commerce, peace, and cooperation.[27] To Knox, the canal signaled the dawn of a "new day" that would remove Central American and Caribbean nations from their "isolation," open exceptional commercial possibilities, and bring general prosperity to the area. This would put an end to the "malady of revolution and financial collapse" that had plagued the region since its independence.[28]

To the extent that the magic of this machine worked, the intervention threat embedded in the Roosevelt Corollary would no longer be needed. With the end of the cycle of revolutions and recurrent debt crises, the Caribbean and Central American nations would finally escape from the menace of European imperialism. The mechanical enclave would in this way release the United States from a concern that had preoccupied its statesmen since the times of Jefferson. Rather than a manifestation of "doux commerce" or of Smithsonian optimism about the power of markets, Knox's speeches revealed the belief that a piece of engineering work could act as ancillary to foreign policy.

The Panama Canal was at the same time a material expression of the original Monroe Doctrine, and a validation of the wisdom of that doctrine. It was a machine with an ideological purpose.[29] Besides opening the gates to Pacific-Atlantic navigation, the canal would reinforce the providential mandate that had entitled the United States to speak for Latin America, interpreting its claims for peace and prosperity and carrying them into effect. This enunciatory capacity was predicated upon the powers of machine civilization. It was this arrangement of knowledge and technology that gave Secretary Knox the possibility to foretell the strengthening of hemispheric cooperation.

> Even now, it is a great bond between us. In the future I perceive it will be a common heritage binding together the nations of this hemisphere with a force no power can break, and while it has in Providence been given to us of the North to state and interpret it [the Monroe Doctrine], it has never been invoked to the detriment of the people of the South or operated to their hurt.[30]

A monumental mechanical solution took the wind off the ideology of Manifest Destiny, authorizing new narratives of inter-American transitability, mutual knowledge, and cooperation. In the face of a technological development that had cut the continent into two, prior imaginaries of imperial penetration lost actuality and persuasiveness. The older obsession with penetrating the subcontinent through its "river arteries" receded into the backstage of imperial strategies.

In spite of the enthusiasm conveyed by many U.S. commentators, it was clear that not many Latin Americans passed through the canal or knew of its existence.[31] Accustomed to traveling to Europe, the sons and daughters of Latin America's "high society" had little chance to see either the Panama Canal or the Canal Zone.[32] Leading mercantile houses in South America continued to use the traditional Atlantic routes to Europe. The U.S. press did all it could to promote the opening of the canal as a major national achievement, but the almost simultaneous beginning of the war in Europe displaced the attention of readers, both in the United States and in Latin America.[33] Some notable figures of the Latin American literary elites showed scorn toward U.S. culture and materialism. The Nicaraguan vanguardists criticized the Panama Canal as the embodiment of imperialist designs and of a materialistic culture.[34] However, by and large, the canal passed unadverted to the southern neighbors. The Colossal Machine failed to impress most Latin Americans, simply because it was too far from sight. By contrast, mechanical constructs that symbolized transitability and the union of a fragmented subcontinent could appeal to South Americans more effectively.

Transportation Utopias

In the age of Pan-Americanism, three "transportation utopias" dominated U.S. thought about hemispheric integration. All three responded to the imperative of exporting to Latin America innovations in transportation that had brought about economic progress and national integration in the United States. These marvelous machines (the railroad, the automobile, and the airplane), it was expected, could connect Latin America to U.S. markets and, in this way, carry to the undeveloped republics the comforts and modernity that structured

American life. To an extent, these "transportation utopias" developed out of disenchantment with an earlier project (pertaining to the era of Manifest Destiny) that considered penetrating the ex-Spanish colonies through waterways. The machine-age produced imaginaries of hemispheric (overland) integration built around specific "hard" machines. These machines, it was assumed, would bring greater connectivity, transitability, and visibility to the region. "Transportation utopias" were part of the imperial vision that projected to modern machines the solutions of problems that dollars, diplomacy, and gunboats had been unable to solve.

River Arteries of Empire

In the mid-nineteenth century, a series of naval explorations sought to extend the frontiers for U.S. commerce. Those destined to South America intended to prove the feasibility of opening up the subcontinent through its "river arteries." Waterways were considered natural avenues for the penetration of "progress" into new or uncharted territories. To test the navigability of the great South American rivers, the U.S. Navy carried out two surveys in the 1850s. Lieutenants Herndon and Gibbon explored the Amazon River basin in 1851–52, while Lieutenant Page navigated the River Plate and Paraná River in 1853–55.[35] These surveys brought back favorable expectations about the region's progress, as well as great misgivings about the obstacles to overcome.

In the Amazon basin, the development of trade depended crucially on improved transportation. Herndon and Gibbon imagined "a vast internal waterway system laced through the Tropics, awaiting only freedom of navigation, steamboat fleets, new settlement, and commercial enterprise."[36] In their transportation utopia, U.S. technology and enterprise could bring "progress" to the Amazon, linking the region directly with U.S. ports on the Gulf of Mexico and in the Mississippi Valley. A natural complement of the U.S. economy, South America would also welcome the resettlement of southern plantation owners. The River Plate expedition also brought back promissory expectations about regional development. Page reported the expansion of coastal shipping and the establishment of meat-salting plants along the Paraná River and the liberal spirit that animated the new governors.[37]

Great obstacles, however, prevented a rapid development of international trade and transportation in the Amazon and the River Plate regions. The Brazilian state objected to the idea of declaring the Amazon and its tributaries international waterways and, contrary to expectations, granted exclusive navigation privileges to a Brazilian company for thirty years. In the Plate-Paraná river system, the poor navigability of small streams in northern Argentina and Bolivia and the open hostility of the Paraguayan president made steam navigation quite difficult.[38] Only in a small section of the Plata-Paraná rivers, steam navigation combined with European immigration promised to bring about "development."[39] With time, overburdened by institutional and natural hurdles, the "river arteries" theory of commercial integration gave way to mechanical metaphors.

The Pan-American Railway

The first great mechanical utopian possibility to emerge out of the Pan-Americanist movement was the idea of an intercontinental railway.[40] The project of connecting all the capitals of the American continent through railways was first exposed at the 1889–90 meeting of the American republics in Washington. Secretary Blaine strongly endorsed the idea, calling Latin American nations to cooperate in the design and construction of such a railway system. In his welcoming speech to the delegates of the first Pan-American Conference, Blaine said:

> We believe that we should be drawn together more closely by the highways of the seas and that at no distant day the railway system of the North and South will meet upon the Isthmus and connect by land routes the political and commercial capitals of all Americas.[41]

Later, the U.S. Congress established an International Railway Commission, headed by Henry Gassaway Davis, an influential railroad builder, and approved funds for the surveys the project required. The Army Corps of Engineers carried out the surveys, the reports were published in five volumes during the years 1895 through 1899. In 1901, with the support of Latin American delegates, Davis managed to create a permanent Pan-American Railway commission that included philanthropist Andrew Carnegie and three representatives from Mexico, Peru, and Guatemala.[42]

In 1903, Charles Pepper, journalist and Cuba expert, was commissioned to travel to South America to evaluate the feasibility of the Pan-American Railway. His report was quite optimistic. Pepper found Mexico extending its railways to the border with Guatemala and Argentina building a railway into Bolivian territory. The Chilean congress had approved the construction of a trans-Andean railway that would unite Valparaiso to Buenos Aires. Peru had earmarked new fiscal resources for railroad development. And several South American republics were in the process of settling their boundary disputes, something Pepper considered crucial for the completion of the transcontinental railroad.[43]

Pepper's optimistic expectations proved unrealistic. Financial constraints and political suspicion made Latin American governments stall the construction of key links in the network.[44] In U.S. foreign-policy circles the project gradually lost support.[45] By the time of the Buenos Aires meeting (1910), Davis could show only meager results. The Mexican section was still forty miles short from the Guatemalan border. Central American nations had made almost no effort in railroad building. South of Panama the prospects were better; Peru, Ecuador, and Argentina had made some progress, but still 4,200 miles remained to be built.[46]

Lacking crucial connections,[47] the existing railways could not yet integrate the Americas. During WWI more important priorities pushed the project of a Pan-American Railway onto the backstage.[48] Even enthusiastic supporters of the project had to admit that the dream of traveling on train from New York to Buenos Aires had to be abandoned. The most that the New York passenger could expect was to reach Panama.[49] By the time of the 1928 Havana conference, many delegations started to abandon the idea of a transcontinental railway, favoring instead the construction of highways.[50] Like other machines, the Pan-American Railway had become obsolete.

For almost forty years, this transportation utopia had sustained the expectation of a North-South hemispheric integration.[51] As early as 1907, John Barrett, one of the most prominent promoters of commercial Pan-Americanism, imagined that sanitation and railroads could radically transform the reality of Latin America. In the near future, he thought, a New Yorker could travel by train to any Central American capital. If adequate transfers of American technology were instrumented, even the malaria-infested and isolated areas of Central America could become sites of economic progress.[52]

> This is sure to come some day . . . the whole Mosquito Coast and the rest of the Caribbean shore of Central America will be busy with prosperous commercial entrêpots, which, in turn, will be connected by railroads with all parts of the hitherto impenetrable jungle, as well as with the mountain capitals and towns. In fact, I look to see, during the next twenty years, a transformation in Central America which will astonish the world and make it difficult to realize that, in 1907, it was commonly regarded as a terra incognita.[53]

The dream of railroads awaking a dormant Latin America resurfaced after the completion of the Panama Canal.[54] In actuality, rail transportation failed to connect all countries of the region. Barrett's vision of "prosperous commercial entrêpots" in the jungle remained illusory. When railroads connected the lands of Central America, it was for the benefit of the United Fruit Company. But the prospect of railroads uniting hot and inhospitable lands had significant weight in the imagination of American statesmen and businessmen.

Various attempts to carry out this utopian project—railroads in the jungle—ended in formidable failures. The story of the Madeira-Mamoré Railway is a case in point. The early attempt made by Phillip and Collins in 1878 had to be abandoned due to the ravaging effects of malaria on the labor force.[55] A Baldwin locomotive rusting in the jungle stood as the salient monument to this failure.[56] From 1909 to 1912, P. Farquhar took over the project and, with the help of tropical medicine, imported laborers, and alliances with local tribes, was able to complete the project. But Farquhar's imagined business-empire in the jungle never came to full fruition. Malaria continued to devastate the region, and the world crisis of rubber in 1913–14 made Farquhar's finances tremble. Worn out by negative publicity in the Brazilian press, the American entrepreneur gave up his concession in 1930.[57] With time, this spectacle of displaced modernity gradually sunk into in the Brazilian imagination as a mechanical ghost (*trem fantasma*) that evoked destruction and decadence.[58]

The Pan-American Highway

The idea of a system of highways to connect the three Americas emerged from the Pan-American Conference at Santiago, Chile, in 1923.[59] The following year engineers from nineteen Latin American nations gathered in Washington to study the problem and suggest viable alternatives for the project. Congresses on highways in Buenos Aires (1925), Havana (1928), and Rio de Janeiro (1929) perfected the plan, making Latin American governments responsible for the surveys and the investments that would make the interconnection possible.[60]

Since its inception, this road utopia stimulated a series of dreamscapes about commercial integration, the extension of U.S. tourism, the exportation of U.S. culture, and the possibility of apprehending the regions' realities and problems. An intercontinental road created the impression that large markets for U.S. manufacturers would open in Latin America and that the region's raw materials would flow into the United States. Through modern highways, the comforts of American life would flow into Latin America. William Rudolph wrote in 1943:

> The Pan American Highway and its feeders should provide a new and sound basis for interchange between one American nation and another, between the farm and the city, between the mine far up in the mountains and the world outside, between the desert which has salt and the tropical farmlands which have all varieties of food. It is not alone a symbol; it is a distinct advance, which should provide the stepping-stone for extending the comforts of modern life to all parts of the American continent.[61]

When the war closed Europe to commerce, it became necessary to seek new sources of raw materials and markets as well as alternative destinations for U.S. tourism. Cuba and the Hawaiian Islands were already incorporated in the American imagination as places of sexual pleasure and entertainment.[62] Developing alternative tourist meccas in Central and South America required important improvements in overland transportation.[63] Good roads could make accessible beautiful beaches, awesome jungles, Inca ruins, and Spanish colonial architecture. The possibility of an intercontinental highway ushered in great expectations. Motorist and photographer H. Lanks wrote in 1939:

> It is very possible for Panama to become a great Riviera, attracting pleasure seekers from all over the world. Surely it will become a great Mecca for American tourists as soon as the Pan American Highway is opened through Mexico and all Central America. Then, the great middle class American tourist can experience foreign travel and comfortable adventure from his own car over the 3,305 miles from the American border to the Canal.[64]

Penetrating Latin America through roads was functional to an imperial vision that presented technology as an instrument for hemispheric integration. Selling cars and building roads were means to exhibit to Latin Americans the achievements of a technologically superior society.[65] Hence the project of an intercontinental highway encapsulated both foreign-policy dreams and business objectives. It projected into a mechanical package (automobiles + highways) the expectation of the American middle class and the political dreams of its leadership. At last, "Latin" and "Anglo-Saxon" America could be united by a technology that was representative of American mass consumer culture.[66]

Like the previous transportation utopia, the Pan-American Highway never saw its completion. While the roads from the U.S. Midwest and California to the border towns of Laredo and Nogales were completed by 1936 and 1940, paved roads into Mexican territory stopped 750 miles short of Mexico City.[67] The same was true of Central American roads, where the *fruteras* still controlled the means of transportation. In the 1940s, Latin American governments invested more in domestic roads, but did not pay much attention to completing the Pan-American Highway.[68] Road builders' optimism about an engineering project that would unite a fragmented continent contrasted with the actual possibilities of Latin American governments to finance these roads.

This was not a case of cultural resistance against this emblem of U.S. modernity, the automobile. On the contrary, the Brazilian and Argentine middle classes had developed by the 1920s a strong affection toward the automobile. Car races, automobile clubs, and car exhibitions signaled the early development of a "car culture" in the Southern Cone.[69] The lack of road connectivity in South America was the result of uneven regional development. In 1942 Paul Pleiss and Herbert Lanks drove their car around South America to prove the feasibility of completing the Pan-American Highway.[70] Their journey demonstrated that the idea of a Pan-American Highway was far from being a reality: many interruptions prevented the travelers from even circumventing the subcontinent. The American motorists concluded that only the "adventurous motorist" should try to replicate their feat. The "heart of South America" was still unreachable by car.

Pan American Airways

The last transportation utopia, centered on the airplane, resulted from the combination of private entrepreneurship and national defense policies. The rivalry between the United States and Germany in the field of commercial aviation prepared the terrain for the emergence of a transcontinental flying company. U.S. policymakers, concerned about German influence in Latin America, considered the operation of SCADTA, a German aviation company based in Colombia, as a security threat to the Panama Canal. Consequently, they tried to push it out of business. In 1927, when SCADTA sought landing rights in Panama (the same year that Lindbergh crossed the Atlantic with his *Spirit of St. Louis*), the U.S. Air Force decided to give full support to an American aviation company that could block German influence. The name of this company was Pan American Airways.[71]

In 1928 Juan Trippe entered into the business. With the support of the U.S. Air Force (from which he got the license) and the financial backing of J. P. Morgan and other leading bankers, Trippe built a successful airline company that linked the capitals of South America.[72] His successful business strategies produced a connectivity that both railways and roads had failed to attain. By ordering aircrafts in advance of the competition, securing landing rights from Latin American governments, and preventing the emergence of competitive companies, Pan American Airways grew rapidly to become the region's premier airline.[73] By 1930, only two years after its start, Pan Am controlled forty-eight ground stations and had flight lines over 19,200 miles.[74]

Starting with a mail service between Key West and Havana, Trippe soon expanded services across all of South America. "Panagra," a subsidiary of Pan Am, controlled the Andean routes. The company's regular international flights connected all the eastern coast of South America (Panama, Quito, Guayaquil, Lima, Arequipa, Antofagasta, and Santiago). In addition, the company offered flights from Santiago to Buenos Aires and from the latter location to Salta, Argentina, and La Paz, Bolivia. All these connections by air were completed before 1934.[75] On the eastern side of the continent, Pan Am's Brazilian subsidiary (Panair do Brasil) provided connections along the eastern coast from Rio de Janeiro all the way to Panama.

Pan Am was a clear case of a monopoly with a technological edge. Its large "flying boats," at first a motive for amazement, soon became a common sight in the landscape of South American airports. The election of amphibian planes was based upon the notion that in order to overcome the mountainous terrain of Andean nations, airplanes had to follow water routes, landing on lagoons, rivers, or seaports. This was particularly true of the lines scheduled on the eastern coast. Pan Am contributed also to the process of technological transfer, developing radio and weather stations along its scheduled routes.[76] A vehicle for modernity in transport and communications, Pan American Airways represented, perhaps more than other U.S. companies, the hegemony of U.S. capital over the skies of Central and South America. Through its subsidiaries, Pan Am controlled most of the region's air traffic, almost without competition.[77]

A showcase for the possibility of conducting foreign policy through private enterprise, Pan Am Airways soon became a symbol of inter-American goodwill. Carrying medicines in time of war, transporting delegates to the Pan-American conferences, keeping flight schedules in the midst of revolutions and coups, the company acquired a reputation for neutrality and international cooperation. In addition, the company collaborated with U.S. scientific expeditions to demonstrate the importance of aerial photographs in archaeology, geography, and other disciplines. In 1930, for instance, Pan American Airways assisted the Carnegie Institution in a project that promised an aerial mapping of Maya sites. In five days a Skirosky amphibian plane commanded by Charles Lindbergh overflew Tikal, Uaxactum, Chichen Itzá, and Petén searching for other potential archaeological sites.[78]

Pan American Airways facilitated and intensified business connections. U.S. company executives and engineers and upper-class Latin Americans traveled on the company's scheduled routes to conduct business in the region. Panagra planes carried oil-drilling equipment into the midst of the Ecuadorian jungle. Airplanes, the ultimate machine, could overcome the obstacles imposed by South America's geography—the Andean mountains, the great rivers, and the jungles—and penetrate into the "heart" of the subcontinent. Furthermore, the airplane presented Latin American business communities and ruling elites with a new spectacle of modernity and with a new level of comfort.[79] But it failed to integrate the three Americas in the way that Secretary Blaine, John Barrett, or Henry G. Davis had envisioned. Pan American Airways realized the utopia of uniting the continent for the "air traveler," leaving unresolved the region's transportation backwardness. Most of the region's cargo and passengers continued to be carried on trains, roads, and ships, if not on animals' backs.

Fragmented South America: American Motorists on the Ground

On the ground, things look quite different. The utopia of uniting a continent by mechanical devices falls apart quite easily in the face of uneven regional development, insufficient government finances, and distinct national priorities accorded to transportation improvements. On the ground, the fragmentation of South American geography, economies and societies disrupts the connectivity and cooperation envisioned by the engineers of transportation utopias. The story of two U.S. motorists traveling across South America demonstrates the illusory nature of continental connectivity and the hybridity of the region's transportation systems.

On New Year's Day of 1942, Paul Pleiss and Herbert Lanks embarked on a long car journey around South America.[80] Their goal was to chart automobile routes in the region and to assess the transitability of the existing sections of the Pan-American Highway. Their journey, sponsored by the American Automobile Association, was to give American motorists an up-to-date impression of the state of the roads in South America, with the view of expanding private tourism. Their car, "Silver," adorned with flags of all American nations, was itself a message of inter-American "goodwill." Pleiss and Lanks traveled from Caracas (Venezuela) to Guayaquil (Ecuador) following the Simón Bolivar highway. From the northern coast of Peru they reached La Paz (Bolivia), passing through Cusco. Next, they drove alongside the Pacific Coast to the town of Temuco in southern Chile, then crossed to Argentina, reached the southern tip of the continent (Rio Gallegos), and returned north to Buenos Aires. From there, they drove to Montevideo and Porto Alegre, arriving at their final destination, Rio de Janeiro, on May 15, 1942.

After driving 13,600 miles, the American motorists found the Pan-American Highway to be still an undeveloped project. Driving around the subcontinent was possible, but only after overcoming many obstacles and dangers. Wide paved roads were the privilege of a few countries. Oil wealth allowed Venezuela the luxury of road-building machines, paved highways, and a relatively large number of cars.[81] Good stretches of paved roads could also be found in the richest countries of the Southern Cone (Argentina, Chile, and Uruguay) in areas

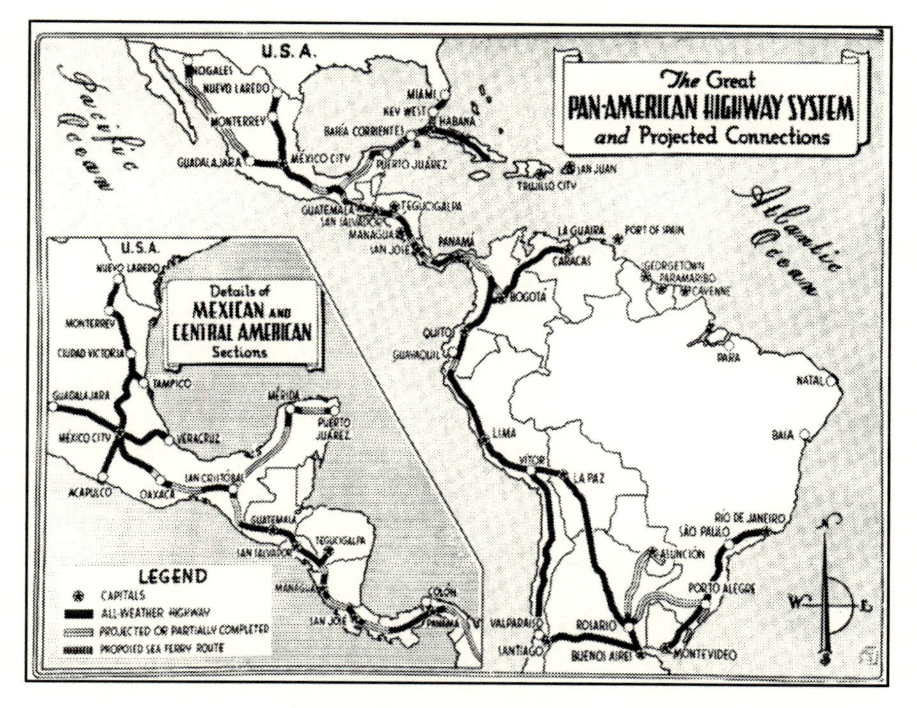

Figure 3.
Map of the Pan-American Highway System, ca. 1942 © Bettmann/Corbis.

that produced agricultural commodities for European markets. The same was true of the Peruvian coast, where roads connected sugar and cotton haciendas.[82] Within and around foreign enclaves our travelers found good paved roads, built by oil or mining companies. This was the case in the oil-rich area of Lake Maracaibo and in the Peruvian northern coast, near Talara.[83]

Outside of the oil districts or the rich export areas, most roads were unpaved, earth consolidated with gravel. Rains made the roads slippery, gusty winds covered the car with sand, swollen rivers interrupted the drive for day, and rugged terrain eroded the car's springs.[84] In Colombia, mountain roads were narrow, with one-direction traffic. On the post roads of Peru, motorists had to cross ancient rock bridges and swollen rivers, driving always under the menace of sliding rocks. In high altitudes, Silver's motor heated up quite frequently. On muddy roads travelers had to tie chains around the tires. In Bolivia, cross-ditches intended for drainage forced Pleiss and Lanks to stop frequently.[85] In Uruguay, the motorists turned to the coast and drove along the beach because the road was flooded.

On several occasions along their journey, the American travelers had to board their car on a ship, simply because the roads were impassable. This happened at Guayaquil and later at Rio Grande do Sul.[86] At other points of the journey, the road simply disappeared. The two men found this situation on the border between Ecuador and Peru, twenty-five miles south of Riobamba, and again on the border between Peru and Chile, south from Tacna.[87] In Argentina, after crossing the Chilean border, there was no road to reach Bariloche. So, the American motorists drove along open country, passing through a private sheep ranch.[88] In Patagonia, South of Lake Buenos Aires, the road suddenly ended, giving way to tracks easily erased by the wind.[89]

On the ground, the connectivity promised by the Pan-American Highway, the grand design of inter-American policy, was only fictional. With so many obstacles, interruptions, and difficulties, the journey from La Guaira to the Magellan Strait did not seem appropriate for the average American car owner and tourist. The Pan-American Highway existed only in the minds of statesmen and in the blueprints of road builders, as something to be realized in the future. In the meantime, South America remained disconnected, fragmented by bad or nonexistent roads, plagued by unresolved border disputes, and ruled by governments unconcerned with highway development. The itinerary Lanks and Pleiss followed—surrounding South America along roads close to the coast, rather than penetrating the interior—reflected South America's backwardness in overland transportation. The existing roads—a mere collection of tracks, post roads, and *caminos*—did not constitute an interconnected system of transportation.[90]

The vision of South America crisscrossed by American-style roads and motorists did not look viable. Existing roads did not allow foreigners to penetrate into the "heart" of South America. Only the "adventurous motorist" could reach into the interior of the Orinoco, and only through a "combination of car roads, river boats, and native buses."[91] From Peru, our travelers tried to venture into the affluent rivers of the Amazon, but the lack of roads made their attempt impossible. The road leading to Cusco, the archaeological capital of South America, was exceedingly difficult: it was narrow and made of stone; slides of rocks blocked traffic at several points.[92] The automobile, the quintessential representative of U.S. consumer culture, was of little help in bringing Latin American nature and culture closer to home.

Nonetheless, the motorists' journey afforded a unique opportunity for gauging regional "progress." Like prior travelers, Pleiss and Lanks saw in South America an intriguing mixture of modernity and tradition, a region where American mechanical modernity stood in tension with "Spanish" and "Indian" cultural formations. Implants of "Americanness" appeared here and there. The travelers encountered a few isles of "automobile culture" along their journey. In

Argentina, modern gasoline stations belonging to the Argentine Automobile Club (ACA) were distributed along the road. Near the Bolivian border there were signs of car races organized by the ACA. Similar clubs operating in Peru and Chile provided maps and information to motorists. A thriving truck-driving business had developed in Patagonia.[93] In Peru and Venezuela, there were plenty of stores selling gasoline at quite reasonable prices. And repair shops and stores selling auto parts could be found everywhere.[94]

In areas of capitalist export agriculture, machine-civilization had made great inroads. In the coastal plantations of Peru, cotton, sugarcane and rice were cultivated with the help of "stationary steam-engines" imported from England.[95] The same was true at the mining and petroleum enclaves. Talara, an oil company town in northern Peru, looked like any other "American industrial town," with its combination of machines, whiteness, and orderly spatial design.[96] At Oroya, Pleiss and Lanks found the largest copper smelting plant in the continent. Here U.S. machinery and management had eroded all traces of South American difference. "At Oroya *we found a little America*, isolated in the mountains, where the American employees of Cero de Pasco and their families live, between twelve thousand and thirteen thousand feet above sea level."[97]

But the marks of the unmodern ("Spanish traditions" and "Indianness") were also quite visible. "Old Spain" appeared once and again, in the cockfights at Carabobo, in the girls covered with *mantillas* at Valencia, and in the red-tiled roofs of other Venezuelan towns. Similar remnants of "Spanish glory" could be seen at Lima and, to a lesser extent, at La Paz. But more impressive and novel was the other South America. As the travelers entered Ecuador, "Indo-America" became more evident. Indigenous peoples walked alongside the road, packs of llamas crossed the roads, and here and there, ruins pointed to the existence of an ancient culture. In Peruvian territory, the American motorists stopped at various Andean market towns, took pictures of indigenous peasants, and had a taste of Andean music and religious processions. The signs of the colonial experience were mixed with the signs of "antiquity." At the temple of Pachacamac, the American motorists stopped to witness an amazing spectacle: cars speeding past the prehistoric ruins; against them, the cars looked like small dots.[98] Ancient ruins made the motorists aware that they had entered the territory of the Twantinsuyo, and this increased their respect for the Peruvians.[99]

The long journey around South America provided encounters with "primitive" technology as well as surprising moments of mechanized production and native engineering. In the Venezuelan highlands, Lanks and Pleiss found the ancient method of terrace cultivation, stone walls that protected the crops from the livestock, and peasants who threshed their grain with the help of donkeys. Further south, they saw "natives" twisting sisal ropes by means of homemade wooden machines.[100] In northern Ecuador, peasants were "working in the fields, plowing with oxen or winnowing the grain by hand."[101] Surrounding Lake Titicaca were indigenous peasants plowing the land using a stick with a blade, known since "Inca days."[102] These "primitive" methods of husbandry and agriculture coexisted with marvels of "ancient" architecture and engineering, such as the road in the high pass of Infiernillo (Peru) and the swinging bridges restored and strengthened with steel cables.[103] The travelers' praise turned superlative at the contemplation of Inca architecture. They marveled at the "mathematical precision" of Inca walls at Cusco and Sacsayhuaman.[104] Contrasting with the greatness of Peruvian past civilizations were contemporaneous indigenous peasants, who lived under quite "primitive" and "rude" conditions. The land of the Inca ruins was also the territory of contemporary poverty and extreme social inequality.

Lanks's narrative makes evident the expectations as well as the limitations of this particular transportation utopia centered on highways. Well-paved roads were capable of bringing immediate progress to the surrounding areas: opening up land, developing previously untapped resources, and creating wealth. This was particularly visible in areas where U.S. companies had developed models of "industrial towns." Their journey also made clear that American machine-civilization—and its corresponding way of life—had already penetrated the tropics, the Andes, and the pampas. But these enclaves constituted exceptions in the landscape of South America. Preindustrial methods and primitive roads still dominated in most areas of the subcontinent, making the travel of the American motorist a journey into a remote past. "Motoring" South America generated the impression of an unmanageable hybridity, in which elements of Spanish and indigenous cultures coexisted in tension with selective elements of Euro-American modernity. The smokestack of Cerro de Pasco—polluting the air and destroying the vegetation—stood as a monument to the intrusion of American technology in the world of an ancient indigenous culture. Packs of llamas crossing a Peruvian mountain road showed clearly that the great transportation utopia (the Pan-American Highway) had brought almost no benefit to the Andean peasantry. In the end, the difficulty of connectivity in the Americas was a problem of uneven development and extreme social inequality.

Conclusion

Scholars of U.S.–Latin American relations have often treated the world of engineers, machines, and machine-culture as disconnected from U.S. policies toward the region, assuming, perhaps too hastily, that these policies are based upon ideas and principles generated within the government and foreign-policy communities. By using the term "imperial mechanics" I want to suggest that foreign-policy principles and imaginaries of hemispheric integration were built upon conceptions of machines emerging from the engineer and business worlds. "Transportation utopias" and "spectacular engineering" constitute two moments of a broader use of machines to reconfigure inter-American relations. These were projections of U.S. machine-civilization onto the territory of hemispheric foreign policy.

"Hard machines" facilitated the construction of visions of a hemisphere united under U.S. supremacy. The Panama Canal constituted a grandiose display of American labor, technological ingenuity, sanitary control, and labor management. It stood for the achievements of a great democratic republic working in inhospitable terrain (the "jungle"). The canal represented also the possibility of "South American progress"—the economic opportunities opened by flows of U.S. trade, investment, and technological assistance—and, consequently, a vehicle for the diffusion of "goodwill" among the sister republics south of Panama. The webs of commerce, it was expected, could pacify national spirits, put an end to local revolutions, and diminish the animosity against the Colossus of the North. This was perhaps the biggest magic of the machine: to act as an ancillary to foreign policy, confirming in the end the wisdom of the Monroe Doctrine and the correctness of "American" perspectives on governance and development.

"Transportation utopias" helped to imagine the unity of a continent separated by national rivalries, boundary disputes, and quite diverse national communities. Blueprints for intercontinental railways and highways brought about great expectations about the transitability and connectivity of the hemisphere. Modern transportation systems could overcome the barriers imposed by nature (mountains, deserts, jungles) and bring the products and cultures of South America closer to the United States. Americans projected in these machines, through the intensification of tourism and commerce, the possibility of shortening the cultural distance that separated "Latin" from "Anglo" America. Trains, roads, and airways could bring more frequent contacts among the two Americas, generating the mutual understanding that was deemed crucial for the establishment of inter-American cooperation. The airplane added to these imaginaries the notion of enhanced, hemispheric visibility. From the air, it was easier to gauge the region's natural resources, map its archaeological sites, and get a comprehensive notion of the region's potentialities and shortcomings.

Ironically, it was a private entrepreneur (Juan Trippe) supported by U.S. financiers and the defense establishment that carried into practice the greatest dream of all: to penetrate the jungles, mines, valleys, and cities of the subcontinent with only a few hours of airplane flying. The realization of the third "transportation utopia"—in part the result of innovative "flying boats" and of the greater reliability of radio messages—left the question of lower transportation costs and regional development unsolved. But it reinforced U.S. self-awareness about its technological primacy. The narrative of two U.S. motorists who circum-motored South America in 1942 shows the difficulties in realizing the ideal of an American continent united by a system of highways. Their automobile journey demonstrated how the region remained a patchwork of different transportation technologies—an area still fragmented by particularities, rivalries, and financial constraints. It is quite interesting that this narrative was published *after* Juan Trippe managed to connect all South American capitals by air. Whereas Pleiss and Lanks's journey suspends for a moment the dreamscape of an unobstructed conquest of the subcontinent by American motorists and tourists, Juan Trippe's successful enterprise enhanced the visibility of informal empire.

The simultaneity of Andeans carrying their cargoes on the backs of llamas and middle-class Latin Americans flying in airplanes to their pleasure destinations reminds us of the limitations of machines in overcoming uneven development and inequality. But the efficaciousness of "transportation utopias" should not be measured by the degree to which they were brought into effect. For their chief role in the making of U.S. foreign policy was symbolic: to sustain the fiction of a homogeneous landscape (Latin America) ready to "be connected" to a superior technological civilization (the United States). Modern transportation technologies enabled the fiction that the flows of commerce, investment, and labor would enhance welfare and sustain inter-American cooperation. Like the Panama Canal, the intercontinental railways and highways stood as a visible representation of the possibilities of inter-American cooperation through development. Embedded in them were foreign policy principles. In the 1930s FDR multilateral Pan-Americanism appeared as a possibility when airline connections made South American skies part of the U.S. hinterland. Trippe's empire of the skies made the original (unilateral) Monroe Doctrine obsolete, enabling its replacement for a more benevolent and participatory conception of empire: the Good Neighbor Policy.

Clearly, the transition to multilateralism and hemispheric common defense were the result of new geopolitical imperatives created by the impact of the Great Depression and the emergence of Nazism. But in the meantime, "transportation utopias" helped to maintain the fiction of a united hemisphere, connected by rail, roads, and airplanes. Other technologies played an important role in the development of U.S. policies toward Latin America. Modern means of communications, such as the radio and news services, were crucial in the construction of "hemispheric cooperation" during the years of the Good Neighbor Policy. Photography and motion pictures contributed also to place Latin America under the sphere of U.S. visibility and scrutiny.

The term "imperial mechanics" connects the disconnected worlds of foreign policy, business enterprise, and technology, suggesting that notions of "hard" and "soft" machines informed U.S. foreign policies in Latin America. Pan-Americanism was the "soft machine" predicated upon the assumption of superiority achieved in the terrain of "hard machines." Barrett's "practical Pan-Americanism" was one way of understanding and building the machinery of informal empire. This initiative called for forms of hemispheric integration that did not depend on commercial or customs agreements. Similarly, transportation utopias projected an integration that rested only upon the assumption of U.S. technological superiority and the willingness of Latin American republics to join U.S. transportation initiatives. The Pan-Americanism that developed in the first four decades of the twentieth century combined the projection of these two types of machines onto the imagined land of a united hemisphere.

If dominion over the seas was at the core of the construction of the British Empire, the U.S. hegemony over South America was imagined as dominion in the fields of technology and expertise that would open transcontinental flows of goods, capital, and information over a land mass integrated by roads, railways, and commercial flights. Thus, South America's integration with a more advanced machine-civilization was rooted on two guiding principles: first, the notion that U.S. technological transfer to South America could awaken the economies and societies of the region; second, the idea that a land mass, fragmented by centuries of colonialism and national political rivalries could be united by modern transportation technologies. What Simón Bolivar had failed to achieve in 1826 and James Blaine projected as a vision in 1889—the idea of hemispheric integration—was replaced by new mechanical imaginaries that promised connectivity, circulation, and modernity to a region burdened by uneven development and fragmentation.

Notes

The author thanks the reviewers for their useful comments. He has benefited from the excellent research assistantship of Juan Pablo Scarfi. Funds for the research were facilitated by the Argentine SECyT, program Pict 12,205.

1. On the way, the Panama Canal was represented at the Panama-Pacific Exhibition (1915); see Bill Brown, "Science Fiction, the World's Fair, and the Prosthetics of Empire, 1910–1915," in *Cultures of United States Imperialism*, ed. Amy Kaplan and Donald Pease (Durham, N.C.: Duke University Press, 1993), 129–63.
2. See Ricardo D. Salvatore, "The Enterprise of Knowledge: Representational Machines of Informal Empire," in *Close Encounters of Empire: Writing the Cultural History of U.S.-Latin American Relations*, ed. Gilbert Joseph, Catherine LeGrand, and Ricardo Salvatore (Durham, N.C.: Duke University Press, 1998), 69–104.
3. Ibid., 71.
4. Ricardo D. Salvatore, "Early American Visions of a Hemispheric Market in South America," in *Transnational America: The Fading of Borders in the Western Hemisphere*, ed. Berndt Ostendorf (Heidelberg: C. Winter, 2002), 45–64. On the cultural dimensions of U.S. economic foreign policies, see Emily Rosemberg, *Spreading the American Dream: American Economic and Cultural Expansion, 1890–1945* (New York: Hill and Wang, 1982). For the displacement of the American dream to the periphery, see John Bruce-Novoa, "Offshoring the American Dream," *CR: The New Centennial Review* 3.1 (Spring 2003): 109–45.
5. Ibid., 59–63.
6. Frederic Haskin understood this U.S. accomplishment as the fulfillment of the need to find a passage to Asia, created by the closing of the Turks' conquest of Constantinople. Frederic J. Haskin, *The Panama Canal* (Garden City, N.Y.: Doubleday, Page, 1913), 3.
7. David McCullough, *The Path Between the Seas: The Creation of the Panama Canal, 1870–1914* (New York: Simon and Schuster, 1977), 590–92.
8. Ibid., 590–610.
9. Emory R. Johnson, "The Panama Canal in Its Commercial Aspects," *Bulletin of the American Geographical Society* 35.5 (1903): 481–91.
10. Haskin, *The Panama Canal*, 348.
11. George E. Church et al., "The Panama Canal in 1908: Discussion," *The Geographical Journal* 33.2 (February 1909): 177–80.
12. Ten and a half months after the canal's opening, 1,088 vessels had passed through it. Forty percent of its traffic was internal to the United States. G. G. Huebner, "Economic Aspects of the Panama Canal," *American Economic Review* 5.4 (December 1915), 816–29.
13. Carol Y. Mason and Adagrace Rowlands, "Panama Canal Traffic," *Economic Geography* 14.4 (October 1938): 325–37.
14. Eastbound, bulky commodities such as petroleum and cotton from Peru and nitrates from Chile competed with the sugar shipped from the Philippines and Hawai'i, the lumber of the Pacific states, and the oil shipped from California. Mason and Rowlands, "Panama Canal Traffic."
15. W. O. Blanchard, "The Panama Canal: Some Geographical Influences," *The Scientific Monthly* 45.6 (December 1937): 494–502.
16. Helen M. Strong, "Changes in Entrêpot Markets for Tropical and Other Exotic Products," *Annals of the Association of American Geographers* 15.4 (December 1925): 180–86.
17. Blanchard, "The Panama Canal."
18. Norman J. Padleford, "The Panama Canal in Time of Peace," *American Journal of International Law* 34.4 (October 1940): 601–37.
19. Haskin, *The Panama Canal*, 352–55.
20. John Barrett, *Panama Canal: What It Is, What It Means* (Washington, D.C.: Pan-American Union, 1913).
21. James H. Collins, *Straight Business in South America* (New York: D. Appleton, 1920).
22. Ibid., 272.
23. Blanchard, "The Panama Canal," 502; emphasis added.
24. Haskin, *The Panama Canal*, 378.
25. Brown, "Science Fiction, the World's Fair, and the Prosthetics of Empire."
26. Padleford, "The Panama Canal in Time of Peace," 636.

27. The tour included Panama, Costa Rica, Nicaragua, El Salvador, Guatemala, Honduras, Venezuela, Puerto Rico, Santo Domingo, Haiti, Jamaica, and Cuba.
28. "Secretary Knox's Visit to Central America," *The American Journal of International Law* 6.2 (April 1912): 493–98.
29. "When the canal is opened and the ships of all countries of the world come sailing through the Carib [*sic*] seas, the peculiarity of our position with its special requirement will be accentuated and the wisdom of that doctrine be confirmed again and specially." Ibid., 494.
30. Ibid., 495.
31. "Every Latin American who passes through the Canal Zone goes back home enthusiastic about doing things in the big-scale way," commented Collins. James H. Collins, *Straight Business in South America?*, 282.
32. Good portraits of upper-class Latin Americans traveling to Europe and the United States can be found in Ingrid E. Fey and Karen Racine, *Strange Pilgrimages: Exile, Travel, and National Identity in Latin America, 1800–1990s* (Wilmington, Del.: SR Books, 2000).
33. McCullough, *The Path Between the Seas*, 609.
34. Poet Pablo A. Cuadra ("Poemas nicaragueneses 1930–1933") condemned the American adoration of "the machine, the great hotels, the dance halls, the movies, tourism—in a word, the superficial luxuries of a material civilization." David E. Whisnant, *Rascally Signs in Sacred Places* (Chapel Hill: University of North Carolina Press, 1995), 155.
35. Responding to the interests of southern traders who wanted to search for alternative markets, the "Amazon expedition" tried to find another Mississippi south of Capricorn. J. Valerie Fifer, *United States Perceptions of Latin America, 1850–1930* (Manchester, U.K.: Manchester University Press, 1991), 6–29.
36. Ibid., 11.
37. These findings reflected the optimism of the American Geographical and Statistical Society, which claimed in 1852 that "at least one quarter of the whole of South America is now, for the first time, within the reach of our enterprise." Ibid., 13.
38. Ibid., 11–17.
39. These navigation utopias developed in conjunction with plans to build overland East-West railroads in ways that would replicate the expansion of railroads in the U.S. West. In time, as British investment crowded the field of railroad development, U.S. railroad builders began to speak of the "failure" of railroads in developing the natural resources of South America.
40. The origin of the idea remains disputed. In 1881 Hinton Helper, a critic of slavery (author of the "Impending Crisis of the South," 1857) and promoter of southern development via the conquest of markers in South America, put together a collection of essays titled *The Three Americas Railway*.
41. Quoted in John Anthony Caruso, "The Pan American Railway," *Hispanic American Historical Review* 31.4 (November 1951): 612.
42. Caruso, "The Pan American Railway," 613–20.
43. Ibid.
44. In 1896 U.S. army engineers had estimated that the Pan-American Railway would have an extension of 10,400 miles from New York to Buenos Aires. Of these, 6,700 miles were already in operation, leaving 3,700 miles to be constructed. By 1904 only 460 miles had been actually built. "The Pan-American Railway," *Bulletin of the American Geographical Society* 36.8 (1904): 469.
45. Despite the enhancement of the commission to include more Latin American representatives and Carnegie's open support of the project, the idea began to languish when the U.S. government decided to pass the responsibility for its development to private initiative. Caruso, "The Pan American Railway," 623–28.
46. Ibid., 629–30.
47. The railway had no connections between Central America and Colombia, between Colombia and Ecuador, between Ecuador and Peru, and between Uruguay and Brazil, to mention only the most salient gaps.
48. After Davis's death in 1916, the project was forgotten, only to be revived by Argentine engineer Juan Briano in the late 1920s.
49. Robert S. Platt, "Central American Railways and the Pan-American Route," *Annals of the Association of American Geographers* 16.1 (March 1926): 12–21.
50. Caruso, "The Pan American Railway," 637.

51. From the beginning, U.S. consuls in Latin America promoted the idea of an intercontinental railroad as one that would bring about prosperity and confraternity to the countries of the Americas. See "Intercontinental Railway of the South," *Monthly Bulletin of the Bureau of the American Republics* 2.5 (November 1894).
52. John Barrett, "Resourceful Central America," *The American Review of Reviews* (July 1907): 3.
53. Ibid., 4.
54. In 1914, Barrett projected a more complex view of the region's development, incorporating railways to more general flows of investment and migration from the United States. John Barrett, "The Pan-American Era," *Saturday Evening Post*, October 10, 1914, 2.
55. Attacks by indigenous peoples (the Caripunas) to the company's camp persuaded the engineers that the project was unviable.
56. This sad story of American entrepreneurship in the Brazilian jungle reached the U.S. public through Neville B. Craig's book *Recollections of an Ill-Fated Expedition to the Headwaters of the Madeira River in Brazil* (Philadelphia: J. B. Lippincott, 1907).
57. For a complete, though optimistic, view of Farqhuar's entrepreneurial adventures in Brazil, see Charles A. Gauld, *The Last Titan: Percival Farquhar, American Entrepreneur in Latin America* (Stanford, Calif.: Institute of Hispanic American and Luso-Brazilian Studies, 1964).
58. Francisco Foot Hardman, *Trem Fantasma: A modernidade na selva* (Sao Paulo: Companhia das Letras, 1988).
59. Paul Pleiss, "A Motor Journey Round South America," *The Geographical Journal* 101.2 (February 1943): 77.
60. Herbert C. Lanks, "The Pan American Highway, II," *The Scientific Monthly* 49.5 (November 1939): 417–30.
61. William E. Rudolph, "Strategic Roads of the World: Notes on Recent Developments," *Geographical Review* 33.1 (January 1943): 129.
62. See Rosalie Schwartz, *Pleasure Island: Tourism and Temptation in Cuba* (Lincoln: University of Nebraska Press, 1997); and Jane C. Desmond, *Staging Tourism: Bodies on Display from Waikiki to Sea World* (Chicago: University of Chicago Press, 1999).
63. Mexico was an exception in this regard. Because of its proximity, it offered U.S. tourists the possibility of visiting "primitive cultures" to release temporarily the pressures of machine-civilization. Helen Delpar, *The Enormous Vogue of Things Mexican* (Tuscaloosa: University of Alabama Press, 1992), 55–164.
64. Lanks, "The Pan American Highway," 428.
65. Transnational advertising companies contributed to this process of technological diffusion. In the late 1920s the strategic alliance between J. Walter Thompson and General Motors expanded to Brazil and Argentina the images of comfort and possibilities associated with the automobile. Ricardo D. Salvatore, "Yankee Advertising in Buenos Aires: Reflections on Americanization," *Interventions: International Journal of Postcolonial Studies* 7.2 (July 2005): 216–35.
66. Walt W. Rostow presented the automobile age as the end-stage of development, something he called "the era of mass consumption." *The Stages of Economic Growth* (1959; New York: Cambridge University Press, 1965). Other scholars attributed to the design of automobiles the power of mobilizing the American imagination toward the aesthetics of modernity. Ray Batchelor, *Henry Ford: Mass Production, Modernism, and Design* (Manchester, U.K.: Manchester University Press, 1994).
67. Pleiss, "A Motor Journey Round South America," 77.
68. In the northern tip of the network, environmentalists raised concerns that the road could destroy the forests and affect indigenous lifestyles. Hence, the so-called Darien gap was never completed. P. J. K. Burton, "The Impact of the Pan American Highway," *The Geographical Journal* 139.1 (February 1973): 49–50.
69. Richard Downes, "Autos over Rails: How U.S. Business Supplanted the British in Brazil, 1910–28," *Journal of Latin American Studies* 24.3 (October 1992): 551–83.
70. Pleiss, "A Motor Journey Round South America."
71. Wesley P. Newton, "International Aviation Rivalry in Latin America, 1919–1927," *Journal of Inter-American Studies* 7.3 (July1965): 345–56.
72. The story of Pan American Airlines is well documented in Matthew Josephson, *Empire of the Air: Juan Trippe and the Struggle for World Airways* (New York: Harcourt, Brace, 1944).
73. The success of connecting the hemisphere through the air was also the result of the good financial support and a crucial alliance with W. Grace Co.

74. H. Case Willcox, "Air Transportation in Latin America," *Geographical Review* 20.4 (October 1930): 587–604.
75. William A. Krusen, *Flying the Andes: The Story of Pan American–Grace Airway* (Tampa, Fla.: University of Tampa Press, 1997), chaps. 2–4.
76. Krusen, *Flying the Andes*, 56, 49, and 113–17.
77. Carl H. Pollog, "Commercial Aviation in the American Mediterranean," *Geographical Review* 27.2 (April 1937): 255–68.
78. Alfred V. Kidder, "Five Days over the Maya Country," *The Scientific Monthly* 30.3 (March 1930): 193–205.
79. On the spectacular use of the airplane, see Eric Paul Roorda, "The Cult of the Airplane among U.S. Military Men and Dominicans during the U.S. Occupation and the Trujillo Regime," in *Close Encounters of Empire*, ed. G. Joseph, K. LeGrand, and R. Salvatore (Durham, N.C.: Duke University Press, 1998), 269–310. W. Krusen narrates how South American politicians used Panagra planes to make their campaign tours. Krusen, *Flying the Andes*.
80. Pleiss was the driver of the expedition; his companion, Herbert C. Lanks, took care of the photographs. Lanks later published an illustrated report of their journey in *By Pan American Highway through South America* (New York: Appleton-Century, 1942).
81. "Of course, it is oil that is now putting Venezuela on the highway map of South America." Lanks, *By Pan American Highway*, 8–9.
82. Of the 1,000 miles of the Pan-American Highway that passes through Peruvian territory, only 200 miles were paved; the rest were surfaced with clay. Ibid., 86.
83. Lanks, *By Pan American Highway*, 15–16, 84.
84. In the Peruvian *sierra*, the *carretera central* remained unpaved, impassable during the rainy season.
85. Ibid., 77, 119, and 165–67.
86. Pleiss, "A Motor Journey Round South America," 70, 76.
87. Lanks, *By Pan American Highway*, 127.
88. Pleiss, "A Motor Journey Round South America," 75.
89. Lanks, *By Pan American Highway*, 163.
90. Important links in the highway system were missing. A crucial connection uniting the Panama Canal Zone with the Colombian roads had not been completed. The same was true about the vast region north of Rio de Janeiro.
91. Lanks, *By Pan American Highway*, 8.
92. Machu Picchu could not be reached by car yet. Tourists had to take a train alongside the Urubamba valley and then continue ascending on horseback. Ibid., 99–102, 110.
93. Native truck drivers, proud members of an honored and well-paid profession, shared with U.S. motorists the hospitable and egalitarian sociability of rural taverns.
94. Driving along the Chilean northern coast, in the rough road crossing the nitrate area, their car ("Silver") broke down. The hero of American technology suffered a severe case of "metal fatigue." The springs, the rear axle, the steering system, and other core mechanical parts had to be repaired. Pleiss, "A Motor Journey Round South America," 73.
95. Ibid., 89–90.
96. Ibid., 83–84.
97. Ibid., 96–97; emphasis added.
98. Ibid., 93.
99. At Ollantaytambo they stopped to peek into the interior of a peasant home to describe the owner's humble possessions. Ibid., 111.
100. Ibid., 21, 23.
101. Ibid., 52.
102. Ibid., 106, 122.
103. Ibid., 97, 103–4.
104. Ibid., 109.

Precision Targets: GPS and the Militarization of U.S. Consumer Identity

Caren Kaplan

> Here I believe one's point of reference should not be to the great model of language (*langue*) and signs, but to that of war and battle. The history which bears and determines us has the form of a war rather than that of a language: relations of power, not relations of meaning.
>
> —Michel Foucault, "Truth and Power"[1]

For most people in the United States, war is almost always elsewhere. Since the Civil War, declared wars have been engaged on terrains at a distance from the continental space of the nation. Until the attacks on the World Trade towers and the Pentagon in September 2001, many people in the United States perceived war to be conflicts between the standing armies of nation-states conducted at least a border—if not oceans and continents—away. Even the attacks of September 11 were localized in such a way as to feel as remote as they were immediate—watching cable news from elsewhere in the country, most U.S. residents were brought close to scenes of destruction and death by the media rather than by direct experience. Thus, in the United States, we could be said to be "consumers" of war, since our gaze is almost always fixed on representations of war that come from places perceived to be remote from the heartland.

Digital communications and transnational corporate practices are transforming the modes, locations, and perceptions of nationalized identities as well as the operations of contemporary warfare. Certainly, war is consumed worldwide by global, as well as national, audiences. Indeed, if the conflicts of the present age cannot be described as between nation-states but as between the extra- or transnational symbols of political, religious, and cultural philosophies or ideologies, drawing on national identity becomes a more challenging task. Yet, conditions specific to the United States need to be explored in relation to the network of discourses, subjects, and practices that make up our nation and its government. The United States still signifies a coherent identity, if only as the enemy or perpetrator of attacks against people outside its national borders or as the defender of borders that are perceived by many of its residents as too porous and insecure. Situating the cultural, political, and economic workings of the United States within transnational conditions aids our understanding of the ways in which national identity operates as a powerful enhancement to contemporary globalization. The issue is not the difference between national and international subjects of study but the mystification of the national such that its identifications with global capital disappear from view, leaving behind patriotic articulations of security, prosperity, and freedom.

When U.S. President Dwight Eisenhower and his speechwriter Malcolm Moos coined the term "military-industrial complex" in 1961, they described a moment poised between the aftermath of the world wars and the advent of the conflicts to come in which the U.S.-based armaments industries could combine their influence with those in the military and the government, who would come to gain from such an alliance. Eisenhower argued that this kind of war corporatism could tip the hallowed liberal balance between defense and social programs, leading to a war economy without end. Over the last forty years, the hybrid form of governmentality that Eisenhower delineated in his speech materialized as Congress, industry, and the military created a culture of cooperation that overcame any internal tensions to produce a normalization of what could be construed as conflict of interest, or even cronyism. As the work of James Der Derian, Tim Lenoir, Jennifer Terry, and others demonstrates so powerfully, for people in the United States war is not at all elsewhere but is, in fact, deeply imbricated in everyday life as a "military-industrial-media-entertainment network."[2]

Who becomes a militarized subject through this network in the United States today? Two primary ways in which militarization operates in U.S. contemporary culture are the pervasive use of Geographic Information Systems (GIS), the primary model of data collection, sorting, and storage in use for over thirty years, and the practice of so-called target marketing, a geographically based form of classifying neighborhoods through subsets of demographic information. The same year that Eisenhower critiqued the military-industrial complex, scientist Jonathan Robbin founded General Analytics Corporation (GAC—the forerunner of Claritas, Inc.) to explore the industry potential of a new science—geodemography, the use of the computer to identify and map subsets of the U.S. population by zip code and neighborhood. Throughout the 1960s and into the 1970s, GAC/Claritas linked geography to demography by drawing on the statistics and classification system used in the biological sciences to categorize plants and animals by species and by fine-tuning the zoning capacities of U.S. Postal Service zip codes. By the 1970s, Claritas was applying multivariate regression analysis to census and marketing survey data and "target marketing" could be said to be in full swing.[3]

Geodemography and target marketing could come about, however, only as an outgrowth of GIS. The power of GIS lies in its ability to link information and inquiries of various kinds to location. The flexibility of its analytical capacities can be attributed to the ways in which the system combines layers of information, including visual material, to answer complex questions in increasingly precise registers. As John Pickles has argued, GIS "contributes to a (re)placing of the 'visual' and the 'spatial' at the center of social life through its role as an element in the restructuring of global, regional, and local geographies, the assertion of new disciplinary codes and practices, and the constitution of new images of earth and society."[4] It is difficult to imagine the Web-based Internet with all of its graphic interfaces without the cultural shift engendered by GIS. Indeed, it could be said that the centrality of geographical images in information sciences helped to create the visual logics of contemporary U.S. subjectivity.

The development of GIS exemplifies the era of the military-industrial complex. Its emergence required computer research, geo-mapping, photography, and satellite programs—a process that involved academic, government, military, and commercial participation. The science behind GIS is not limited by nationality. Most histories point to the development of "Canadian GIS" (CGIS) in 1967, the system invented by that country's Department of Energy, Mines, and Resources to inventory land use and geographical information, as the first fully realized "system."[5] The power and resources of the transnational technoscience that the United States and the U.S.S.R. "raced" to secure were fully available to the U.S. military and research universities during the cold war. As the United States rushed to militarize space and extend the range of weapons that could be used for deterrence or for waging attacks on competing superpowers, computer science and satellite programs burgeoned. The geographic identification, sorting, and surveillance offered by GIS produced new commercial, military, academic, and governmental needs. Combined with the remote sensing capacities of new satellite systems that could generate continuous images of the earth's surface, GIS provided an affirmation of the "whole earth" ethos that was coming to characterize the cultural zeitgeist in the United States during the 1960s and 1970s while offering fresh possibilities for security and surveillance as the cold war alliances shifted and reconfigured under the pressures of new conditions and crises.[6]

I am particularly interested in the temporal and discursive overlap of these two technologies, GIS and geodemography, with a third: the global positioning system (GPS). All three emerge in the postcolonial era of globalization with all of its attendant tensions and negotiations between national and transnational culture. GPS originated as a military technology—a system of satellites launched by the U.S. Department of Defense in the early 1970s[7]—that offered precise ground locations for both defensive and offensive purposes.[8] The offensive purposes most famously enabled by GPS were the navigation of the weapons systems during the first Persian Gulf war in 1990–91. Since that time, and in connection with a complicated process of partial declassification and cooperative ventures between civilian, governmental, military, and commercial interests, GPS has become a ubiquitous consumer technology available in cars, watches, and PDAs. GPS has become integrated into the agriculture and transportation industries, law enforcement, and innumerable other commercial, municipal, and federal applications (it crops up regularly in discussions of border security).

In this article, I am inquiring into the conditions that produce U.S. militarized consumer and citizen subjects in relation to technologies that link geography, demography, remote sensing, and contemporary identity politics (including geopolitics). These subjects can be understood to be the "targets" of two seemingly distinct contexts and practices: the target of a weapon and the target of a marketing campaign.[9] In both cases, something or somebody has to be identified, coordinates have to be determined with available technologies, and the target has to be clearly marked or recognized in time and space. GIS provides the model for databases as well as the representational logic for both warfare and marketing, while GPS offers enhanced precision in locating such targets through accurate positioning. Geographically based location technologies that draw on discourses of precision make possible the subjects of both consumption and war.

"Where Am I? Ask a Satellite"[10]

> The technology is already here. Drivers using the Global Positioning System (GPS) have an option to jump in their cars, plug in area maps, and know exactly where they are. Pilots, charting their own courses, can savor free flight, saving time and fuel as they go, and sailors can navigate harbors in high-tech mode. Up-to-date hikers are already augmenting the compasses in their gear with transistor-size GPS receivers, and truck drivers are both tracking and being tracked. It will not be long before universal GPS coordinates serve as postal zip codes. Business cards will list not only telephone and fax numbers and e-mail addresses but will give precise latitude and longitude coordinates for our home and business addresses.
>
> —L. Casey Larijani, *GPS for Everyone*[11]

Commercial and civilian GPS publications tend to characterize GPS as an advance in human society on the order of the discovery of fire or the antibacterial properties of soap. An introductory GPS textbook from 1996 tells us that GPS is the "ultimate achievement of humankind's urge to know where [one is], at extraordinarily high levels of precision."[12] At the heart of this overheated assertion is the belief that human beings are urgently concerned with where they are and where they are going. Most important, technological assistance in the direction and interpretation of these processes is required. From "the clay tablets of the Mesopotamians some 5,000 years ago" to the GPS-enhanced watch or cell phone of the corporate executive in contemporary U.S. society, the truism of the desire to know where you are is presented as absolute and unquestioned.[13] Regardless of political perspective, U.S. discourse on GPS throughout the 1990s and into the next century assumes that "maps hold some primal attraction to the human animal."[14] The "lure" of maps, presented as timeless and cross-cultural, is presented as a foundational attitude of civil society—access to mapping (especially technologically enhanced mapping) is a hallmark of democracy. From the end of the first Persian Gulf war and throughout the 1990s, as GPS increasingly became part of the popular imaginary of location and navigation in the United States, enthusiastic endorsements of "GPS for Everyone" offered precise positioning for the masses.[15] As a 1994 article in the *Wall Street Journal* put it, GPS is "An Answer to the Age-Old Cry: Where on Earth Am I?"[16] A piece in *Rolling Stone* in 1992 trumpeted, "Lost in America—Not!"[17]

The proliferation of ads, press releases, and media spots (such as coverage in tabloid TV and print media on celebrity use of GPS) throughout the 1990s and into the next century focused on location—where you are—but linked closely to that designation was almost always something existential: *where* you are reveals *who* you are. For example, a 1995 article on in-car navigation systems in *Popular Science* boasted: "Real Men Don't Ask Directions."[18] A software review in 1993 advised: "Find Yourself with GPS MapKit SV."[19] For North Americans, the marketing of this novel technology emphasized personal empowerment and self-knowledge linked to speed and precision (save time, increase efficiency, avoid getting lost). Buried in the promotional hype of the emerging technology was the kind of conventional paradox of hegemony with which middle-class consumers of digital electronics are now quite familiar in the new millennium: the digitalization of information about yourself that you provide voluntarily to enhance your "lifestyle" also brings you into networks of surveillance. Who you are, geographically, is a target—of marketers, governments, identity thieves, hackers, and so on.[20]

When civilians use commercial digitalized navigational assistance based on GPS, then, they are participating in the expansion of mapping into more extreme relational contexts, which has the effect of intensifying unequal social relations. The digital mingling of position and identity into target subjects underscores the martial and territorial aspect of mapping throughout the modern period. Maps are always subjective representations; their parameters and spatialized views reflect the needs and interests of those who intend to use them. While the history of maps stretches back into the earliest recorded representations, the rise of print culture, the spread of capitalism, and the desire to chart the mobile circulations of modern culture created a specific practice of mapping.[21] New nation-states required maps of redrawn borders. Maps became indispensable to track armies in war. By the turn of the nineteenth century, the convergence of aerial perspectives made possible by aviation and the relatively new technology and art of photography intensified the visual logic of mapping to the degree that it became possible, and even an advantage, to conduct war from the air.[22] Thus, the legacy of geography, war, and aerial perspective are writ large in GPS. When people turn to satellites to tell them where they are, they mobilize these histories. At the same time, these technologies of location situate consumers within the mythologies of individual empowerment and precision that advertisers employ to market the idea that one must always be locatable.

The Ultimate Achievement: The Myth of Precision

> In order to release a bomb so that it will hit the target, the exact point in space must be determined.
>
> —Albert L. Pardini, *The Legendary Norden Bombsight*[23]

Like many technologies in use in daily life, GPS was created as a result of military research and development. While its commercial and industrial applications bear little resemblance to practices of war, the ways in which GPS operates in daily life are extremely similar to its original purpose, targeting through precise positioning.[24] Yet precision is a relative concept. Oceans have been navigated with a sextant and the naked eye, but the destined port of harbor does not have to be observed within terms that we would think of as "precise" today. The imaginary properties of precision adjust to the means available to achieve them. Thus, the powerful association of GPS with precision marks a nexus point in discourses of modern technoscience, especially those linked to aviation and remote sensing. The quest to pinpoint precisely the object of sight, as when aiming a weapon, emerges as a collaborative goal for the military-industrial complex only if the right tools are available to meet the dominant cultural, economic, and political imperatives.

GPS exemplifies the belief in precision as a required element in armaments, especially in bombardment, and the militarization of space. According to military historians, the entire rationale for GPS development was linked to the demands of precision in missile guidance. Standard histories point to the checkered experience of aerial bombing raids during the world wars and after to demonstrate the importance of accurate targeting, especially for bombardments at night or in poor weather conditions. However, Donald MacKenzie argues that the desire for bombing precision is neither natural nor inevitable but the product of "a complex process of conflict and collaboration between a range of social actors including ambitious, energetic technologists, laboratories and corporations, and political and military leaders and the organizations they head."[25] MacKenzie's research demonstrates that a technologically determinist discourse of accuracy or precision marks the attitudes of both the political Right and Left in debates about military technologies during the period between the two world wars. How precision came to dominate discourses of military strategy in the period before World War II through the Vietnam War and beyond to the first Persian Gulf war is a complicated tale of the competing claims on resources between the branches of the U.S. armed forces as well as the growing power of what should really be termed the *governmental*-military-industrial complex.[26] Above all, the mystique of precision became the underlying rationale for the founding of an air force separate from the navy (which had its own flight craft and pilots) and for the organizing of U.S. national defense and offensive warfare on the principles of airpower.

The rise of airpower as a military strategy is linked to the belief, passionately argued in the aftermath of WWI's previously unimaginable number of civilian as well as military casualties, that precision bombing would be a more humane practice than previous strategies of ground wars. Intrinsic to the argument for aerial bombardment are the key European Enlightenment precepts of distance, precision, and the truth-value of sight. Each of these concepts itself requires an underlying belief in the mastery of technology and the superiority of information systems that privilege vision. Nothing brought these disparate discourses and ideologies together more effectively than the development of enhanced bombsights in WWII. The ability to target selected sites on the ground from a machine traveling at rapid speeds through uncertain weather at heights great enough to remain safe from enemy detection and attack was not easy to achieve. The U.S. military itself was divided on the subject of airpower and the necessity of a separate air force branch.[27] Moreover, given the technological constraints, it was not clear that the moral high ground that precision bombing seemed to offer was achievable. Navigational and computational errors, inaccurate intelligence, weather interference, and human and technological failures often sent bombs awry, killing innocent civilians while destroying nonmilitary sites and structures. Nevertheless, the impression prevailed that U.S. precision bombing was far superior to its obverse strategy: tactical or saturation bombing, a technique that focused on destroying the morale of the civilian population in enemy territory through wide-scale devastation and terror.[28]

Aerial bombardment during WWI had consisted of dropping armaments by hand with "no bombing sights, no aiming points, and no true bombs."[29] As the world geared up for the next war, entire industries were pressed into the quest for high-tech solutions to the perceived need for precision—both to better the record of aerial bombardment and to protect the lives of the airmen. With the new bombsights developed for WWII, popular belief held that a bombardier's precision was increased such that he could "drop a bomb into a pickle barrel."[30] While this claim to precision was often contradicted by evidence, the bombardier became a heroic, even iconic, figure in popular perception.[31] As Conrad Crane argues, accurate daylight bombing, with its precision mystique, called upon "traditional," favored American characteristics such as marksmanship, fair play, and other "frontier" stereotypes, adding to its strategic appeal to planners and the public alike.[32] The precise aim of the bombardier (in truth, more the result of skilled mathematical calculation and new technologies than the classic "line of sight" attributed to great marksmen) became legendary. The development of the Norden and Sperry bombsights, along with the engineering of the B-17 long-range airplane (known as the "Flying Fortress"), brought daylight precision bombing into the policy and strategy of modern warfare as an integral component of airpower.

The bombsights developed for use in WWII were designed to address the problem of hitting a stationary target from a moving vehicle, the airplane. As Albert L. Pardini, one of the most devoted chroniclers of the Norden bombsight, explained the process, "a falling bomb, in order to hit the target, must be released at the correct distance back from the target so that it will not fall short or over."[33] Factors such as gravity, true airspeed, air resistance, and wind had to be accounted for as well. Since the Norden bombsight used gyroscopes to hold the optical system in place during the movements of the plane through the air, it was viewed as a huge advance over the aerial surveillance and bombardment used in WWI. The Norden bombsight and its counterparts transformed problems of time and space into principles of geometry—offering the latest science and technology to the bombardier, who was responsible not only for delivering the payload of armaments as accurately as possible but for getting the entire crew back safely by reducing costly repetitions and extra runs. It was highly valued by the air crews who used it and became the subject of its own mythologized reputation (complete with fan clubs, exhibitions, Web sites, etc.).[34]

The use of the Norden bombsight during WWII has been trumpeted as the technology that made airpower possible. Yet, the United States developed a doctrine of airpower, based on precision and the capacity of long-range bombers to reach their targets, that was not, in fact, able to be sustained under actual conditions of war. As Crane relates, elements such as weather, inadequate training, defensive fire, and camouflage resulted in only 14 percent of the bombs dropped by the Eighth Air Force during the first half of 1943 to hit within 1,000 feet of their targets (rather than the 90 percent advertised by the top brass at the beginning of the war).[35] The predicted precision of bombing raids, as Murray points out, was based on "unrealistic" situations: "on bombing ranges in the south-western United States, in conditions of perfect visibility, with bombers dropping on an individual basis and with no hostile anti-aircraft or enemy fighters."[36] Thousands of U.S. and allied flight crews lost their lives in pursuit of daylight targets when they came under sustained defensive fire in the theater of war.

Despite these significant problems in execution, the airpower mystique of precision presented a cleaner, neater image than the wholesale, destructive blasting of terror bombing or cannon bombardment. The airmen who died either were incinerated or decimated in the air on impact or in the ruins of their aircraft if they hit the ground—arguably less of a cumulative visual calamity than the thousands of corpses moldering in the trenches of World War I. If, as Sven Lindqvist argues, Europe and the United States had learned from their colonial experiences that terror was the best way to devastate morale and crush opposition, the emergence of the doctrine of airpower and precision bombing proved to be the only acceptable rationale for the kind of large-scale attacks necessitated by wars between industrial powers in the modern period.[37] Thus, the doctrine of precision, while pervasive, could not overlay completely the fact that morale or terror bombing was an accepted practice of the U.S. military and its allies. The firebombings of Dresden and Tokyo, therefore, were but dress rehearsals for the decision to drop atomic bombs on Nagasaki and Hiroshima. While the use of the atomic bomb appeared to shift the logic from precision to mass, total terror and destruction, it is with the development of GIS and satellite-guided missile systems that characterize later wars that airpower and the mystique of precision return with a vengeance.

The Precision of "Space Power" and the First Persian Gulf War

> One of the major advantages that planners like Schwarzkopf possessed was excellent intelligence assets. The environment of the desert campaign lent itself to the acquisition of accurate information about the enemy. In space, reconnaissance satellites like the KH-11 and the lacrosse radar-imaging satellite provided untouchable (by Iraqi forces) coverage of the battlespace. Other satellites carried the critical communication channels and equally important were those satellites that allowed the global positioning system (GPS) to work. GPS allowed unprecedented levels of accuracy concerning battlefield navigation that was so vital in maneuver warfare.
>
> —Alastair Finlan, *The Gulf War 1991*[38]

Precision returns as a popular discourse in military-industrial society when positioning technologies made possible by the satellite systems that were launched in the 1970s and '80s offered new standards of accuracy.[39] Most histories of the Persian Gulf War make the point that this was the first war to make extensive use of satellite technology. Not even fully operational when the war began in 1990, GPS quickly took pride of place in the pantheon of satellite-assisted technologies that the U.S. military and its allies used in the conflict. Combining multispectral imagery from US LANDSAT remote sensing and GPS, commanders had access to detailed maps that could be updated quickly and accurately. Approximately 4,500 GPS receivers were used in the war, winning over troops on the ground and pilots in the air.[40] Shifting the scale of airpower to "space power," GPS and other satellite systems aided both air and ground forces, enhancing conventional aerial surveillance to offer a network of image-based mapping and navigation.

The twenty-four NAVSTAR GPS satellites and their military and governmental counterparts were not the only orbital technologies that affected the perception and outcome of the war. The conflict in the Persian Gulf in the early 1990s has been characterized as the first televisual war (in contrast to the film-based information broadcast during the war in Vietnam).[41] The speed, immediacy, and accuracy of the real-time images broadcast by twenty-four-hour cable news services such as CNN depended on satellite telecommunications to an unprecedented degree. If during WWII newsreels reached movie theater audiences no less than a month after the occurrence of events depicted, that time lag had been reduced to approximately twenty-four to forty-eight hours between filming on-site and broadcast by TV during the Vietnam war.[42] During the first Persian Gulf war, the seemingly real-time television coverage of Patriot missiles and Scud attacks generated what Robert Stam has called the "pleasures of war spectatorship";[43] CNN's five telecommunications satellites fed simulated "live" accounts twenty-four hours a day, offering greater identification with the military apparatus for many of the viewers glued to their sets during the relatively short-lived conflict.

The truth effect of digital immediacy and the mystique of satellite-aided precision presented a view of the war that built upon the realist documentary tradition.[44] Although most of the visual material that was transmitted was heavily censored by the Pentagon, it was presented as live and unmediated. Many commentators point to the contrast between the Vietnam War, where reporters were able to roam mostly at will (which brought the complexity and atrocities of the conflict into the living rooms of the general U.S. public via TV), and the managed "pool" approach mandated by the Pentagon during the first Persian Gulf war (establishing the approach for succeeding wars). The media image of the conflict in the Gulf, as Stig Nohrstedt explains, was "ruled by restrictions on journalists' freedom to visit front areas, troops, damaged buildings, and so on without military escorts."[45] Reporters, desperate for footage and for any kind of story line, relied on whatever technology could provide. Thus, the "you are there" effect of reporters describing tracer fire from their Baghdad hotel room windows was reinforced by the seeming speed of the transmissions. As many commentators have argued over the years since the war, the media, in general, played technological handmaiden to the U.S. military in its effort to manage the representation of the conflict. As Douglas Kellner relates in his history of the television coverage of the war, "the initial strategy of the war managers was to present an image of the war that was clean, precise, and effective."[46] Since the military was engaged in a public relations campaign as well as a military engagement, the mystique of precision bombing helped to allay concerns about civilian casualties and damage to nonmilitary and religiously significant sites.[47] The airpower doctrine of precision bombing, here aided by GPS and other GIS-related mapping technologies, combined with the seemingly instantaneous media coverage that was enabled by telecommunications satellites to reassure the U.S. population that the heroic project of saving Kuwait from Iraqi invaders was not going to be messy, wasteful in terms of lives and money, or boring. Thus, in the early 1990s, the governmental-industrial-military-complex linked once and for all with the media-entertainment complex—forming new subjects of a militarized visual culture.[48]

If the Norden bombsight captured the imagination of the military and public alike during WWII to assuage concerns about the morality of bombing, engendering discourses of precision and aerial mastery, GPS played a significant role in the public relations war as it was spun in the "military-industrial-media-entertainment network" of the Persian Gulf war. As Daniel Hallin has put it, "overwhelmingly the dominant images of Persian Gulf coverage were the images of triumphant technology."[49] Since the media were barred from battlefield coverage, they resorted to iconographic images that played to the nationalist sentiments of TV watchers at home and kept them tuning in: "the Patriot streaking up to hit a Scud in the night sky; the cruise missiles arching gracefully toward their targets; the jet fighters landing at sunrise or sunset (a favorite TV visual) with soldiers watching and giving the thumbs-up sign; and most characteristically, the smart bomb video."[50]

The visual elements of the "smart" weapons entranced many Persian Gulf war spectators. An editorial published in the *Nation* during the war in February 1991 relates the example of liberal viewers who enthused, "we *hate* the war . . . but we are *into* the planes."[51] Or, as reporter Fred Kaplan recalled in the late 1990s, "seven years have passed since the last time the United States bombed Iraq, but one gripping image lingers—video footage shot on Jan. 17, 1991, the first night of the air war, of a laser-guided bomb plunking straight down the chimney of an Iraqi Air Force building and blowing the place off the map."[52] These "gripping" images were produced by video cameras in the "smart" bombs that were designed to record the strike. In the absence of other visual records, the "smart bomb" footage took on a privileged percentage of the display of technological prowess for which the war is known. The "objective eye" of the smart bomb linked the values of realism, action, and precision that many spectators came to regard as a guilty or not-so-guilty pleasure—watching the U.S. blow stuff to bits in an urban or desert landscape that appeared to be devoid of human beings. The explosions were represented as *precise* strikes "through windows" or "down chimneys" of selected targets. Thus, the guiltless pleasure of viewership was as much due to the belief in the power of precision and the thrill of knowing that the armaments were moving through space and time at enormous speeds to strike a target with exceptional accuracy.

The overwhelming impression conveyed by the military-industrial-media-entertainment network was that the United States and its allies were undertaking precision attacks on military targets, thereby conducting war on a higher moral plane and avoiding unnecessary "collateral damage" and, not incidentally, offering good visual entertainment.[53] However, as numerous commentators have pointed out in the years since the war, although most of the bombs dropped in Iraq (approximately 90 percent) were regular "gravity" or "dumb" bombs without laser or satellite guidance and while a high percentage of those bombs missed their targets (some estimates go as high as 70 percent), what most Americans probably remember about the war is the discourse of precision linked to the imagery produced by the so-called smart bombs. Yet, the precision-guided bombs were also likely to miss their marks. Weather, human error, poor intelligence, and any number of other problems plagued the laser- and GPS-guided missiles and bombs. And, despite the hype, more of the "smart" weapons in the first Gulf war were guided by laser systems than by GPS (which has gained proportionate majority in precision-guided weapons programs in subsequent wars). Significantly, the well-documented *im*precision of the bombing campaign just never gained any traction, since the evidence runs so counter to the discourse of precision and technological mastery that dominated the airwaves during the conflict itself. The most notorious mishap occurred on February 13, 1991, when, based on intelligence identifying the site as a military hard target, a guided missile hit the Ameriya civil-defense shelter at 4:30 a.m., killing between 200 and 300 civilians. In a 402-page published report, Middle East Watch chronicled "needless deaths" during the war due to innumerable violations of the official U.S. military and allied policies, such as daytime bomb and missile attacks on targets in populated areas, lack of warning, strafing attacks on civilian vehicles on highways, attacks on Bedouin tents, and so on. The report concluded that approximately 3,000 civilian Iraqis died from direct attacks, while a "substantially larger" number died or suffered greatly from malnutrition, disease, and lack of medical care caused by "a combination of the U.N.-mandated embargo and the allies' destruction of Iraq's electrical system."[54]

"Space power" and the vast resources of the military-industrial-media-entertainment network generated discourses of precision that obscured information about civilian deaths or rendered them inconsequential. The representation of the war was less embodied than previous representations of wars, with U.S. military casualties going undercover or under the radar, as it were, as well. If the "witnessing" of the war came from the missiles themselves, the point of view was singular, unidirectional, and heavily censored in favor of orchestrated displays of precision. Thus, much of what took place on the ground during the war was never a matter of public record in the globalized televisual experience of the "real." In effect, in the coverage of the Persian Gulf war the U.S. public watched an extended commercial for GPS.

Aftermath: Target Subjects of the Military-Industrial-Media-Entertainment Network

While classical liberal political theory has understood social rights as opposed to market relations, by the late twentieth century and the early twenty-first consumer culture had become central to liberalism and neoliberalism, promoting endlessly the idea of choice

> as central to a liberated subject and enabling the hegemony of both capitalist democracy, American style, and the self-actualizing and identity-producing possibilities of consumption, American style.
>
> —Inderpal Grewal, *Transnational America*[55]

The first Persian Gulf war was not anomalous. If it was the first war to be "driven" by satellite technologies, the logics of those weapons and communications systems built on the practices and problem-solving techniques of previous wars. The amplified opportunities for research, development, and profit making that marked the emergence of the military-industrial complex in the 1960s and its expanded transnational formation, the "military-industrial-media-entertainment network" at the close of the twentieth century, provided fertile ground for the discursive fields of "technoscience," that "world-building" set of alliances that Donna Haraway has identified as "military needs, academic research, commercial development, democracy, access to knowledge, standardization, globalization, and wealth."[56] At the turn of the century, technoscience and its networks produce target subjects through discourses of precise scales and sites of identity. Yet even as these modes of identification promise greater flexibility and pleasure through the proliferation of "choices" among myriad specificities, they also militarize and thus habituate citizen/consumers to a continual state of war understood as virtual engagement. As Jordan Crandall argues, operational media such as GPS-enhanced devices and their ancillary discourses aim to "increase productivity, agility and awareness, yet they vastly increase the tracking capabilities of marketing and management regimes," thereby facilitating the integration of military, corporate, and leisure interests.[57] Lured by "individually tailored enticements," the subjects of technoscience, dedicated to "choice" and to "democracy" as the twin bulwarks of the U.S. "lifestyle," become targets of the information systems they use to satisfy their desires. As Crandall writes, "tracked, the user becomes a target within the operational interfaces of the marketing worlds, into whose technologies state surveillance is outsourced."[58]

Most people who search for driving directions on Web sites or who check out "Google Earth" and other services that offer free satellite photography of specific locations are largely unaware of the military infrastructure that supports such activities. Similarly, most people are not aware of their identification for target marketing through computer databases linked to the use of credit cards, supermarket cards, and driver's licenses. GIS is not a subject of informed discussion among most consumers in the United States even if they may be able to debate the virtues of a GPS-enabled automobile when faced with a choice of rental cars. Thus, as DeLanda and other theorists of militarization have pointed out, the ways in which military institutions, resources, and discourses structure facets of nonmilitary life are mystified in the energetic "forgetting" of the military sources of technologies that many people enjoy or feel required to use in everyday life. Yet, a deterministic approach to military "R&D" can oversimplify the ways in which technoscience and its networks, including media and entertainment, produce hegemonic consent among the citizen/consumers of the present age.

Tim Lenoir has argued that changes in government procurement policies have prompted the military to "spin off" many of their key technologies during the last fifteen to twenty years.[59] The case of GPS is more complicated, since it has been the property of the nonmilitary branches of the government through the Department of Transportation since the mid-1980s. Its partially declassified nature and unusually "open" practices while it was still under development made it particularly suitable for commercial "attention." By the early to mid-1990s, GPS had become ubiquitous in the United States, and its counterparts had become part of the orbital culture of other nations as well. The home page for "GPS World" (a typical site that promotes the technology) currently lists the following industry sectors as relevant to the technology: survey and construction, military and government, avionics and transportation, location-based services, agricultural and natural resources, utilities and communications, and systems design and testing.[60] GPS is used in numerous leisure time pursuits from hunting to yachting, and it has even generated its own sports, such as geocaching (a game that utilizes GIS and GPS to hide "treasure" and log information) and "degree confluence" (a game in which participants compete to "visit" every latitude and longitude "integer degree intersection" in the world). GPS navigational aides can be found in watches, PDAs, phones, and other hand-held units as well as in most forms of transportation, both private and civil. In many ways, GPS has become a powerful metaphor and signifier for consumer culture at the turn of the century.[61] And, of course, it is only more integrated into the warfare in Afghanistan and Iraq as another round of "democratization" is pursued by means of war in the Middle East and South Asia.[62]

If contemporary subjects of technoscience and its military-industrial-media-entertainment networks are constituted as targets, it is imperative that we understand this as a form of mobilization. Peter Miller and Nikolas Rose have argued that the subject of consumption is "mobilized" through the links between "human passions, hopes, and anxieties" and the "specific features of goods."[63] I have tried to show how "precision" has entered the emotional field of subjectivity as the military-industrial complex has grown to encompass more fully the culture industries of media and entertainment. Thinking of consumer subjects as "mobilized" helps us in two regards. First, it allows us to move beyond the model of consumers as feminized, passive targets of unscrupulous advertisers in order to see the ways in which people participate in their construction by "volunteering," if you will, to engage in the products generated by technoscience. Secondly, it allows us to understand how citizens and consumers come together as militarized subjects through target marketing that seeks to identify their tastes, desires, and interests. The ambiguity of subject formation generates the complexities of political and cultural life in an affluent nation. Regardless of whether or not we serve in the military or have the means to afford the latest electronics, residents of the United States are mobilized into militarized ways of being.

The aftermath of the first Persian Gulf war, then, has witnessed not only another war in the same region but also a proliferation of GPS-enhanced consumer goods and civilian applications of the technology. This period has also seen a veritable explosion of data-mining and marketing based on geodemographics.[64] Most recently, the method of identifying consumers by zip code has been challenged by more multileveled cross-referencing. For example, the "old" ACORN Market Segmentation System divided the country into more than 250,000 blocks of neighborhoods. Each block was analyzed and sorted by some forty-nine characteristics, including household income, occupation, age, education, age of the housing stock, and other characteristics of neighborhood purchasing power. Blocks were then recombined under forty-four market segments including, for example, "trendsetting, suburban neighborhoods," "older, depressed rural towns," and "Hispanic and multi-racial neighborhoods."[65] Throughout the 1990s there was mounting evidence of the growing importance of targeting consumers on the basis of "demography and habits rather than on the basis of geographical proximity"; as the maxim from 1980s advertising giant Saatchi & Saatchi had it, there are greater differences between midtown Manhattan and the Bronx than between midtown Manhattan and the seventh arrondissement in Paris.[66]

At the turn of the century, then, it is possible to propose that the citizen/consumer subject in the United States is not so much identifiable in relation to intrinsic territories but mobilized as clusters of identities in and through consumption in the context of militarization.[67] Militarization in the expanded sense in which I have been using it in this essay can be seen as a set of practices at work in sites of war, as well as those of consuming, schooling, worship, and homemaking. Yet, the deterritorializing tendencies of contemporary geodemographics are tempered by the will to locate that subjects of consumption generate and require for identification. GIS- and GPS-linked technologies offer to tell citizen/consumers their precise location, positioning them geographically for any number of reasons. This recourse to terra firma can be seen as a recuperation of geography in the face of digitalized dispersal, but it can also be seen as an articulation of the world that GIS has wrought. The deep meaning of database culture in the age of the Internet is that the less we appear to need geographical information, the more it becomes clear how anchored contemporary power is to geography. That is, the anxiety over security, the call to militarize the borders of the nation, to further police the ports, to conduct satellite surveillance on individuals in their homes and places of work, shows us that the military-industrial-media-entertainment network reworks what geography means in terms of the nation-state under the sign of globalization and in the service of mobilization.

The ways in which the United States conducted war in the Persian Gulf in the early 1990s were made possible by the ways in which many U.S. residents became consumer and citizen subjects—through technoscience and its multiple, national, and transnational networks. Since that time, because power is scattered, unequal, and pervasive, war has become dispersed into many ways of being in the United States.[68] Aiming for precision may be a symptom of that militarized dispersal of power, as fixity of location or identity resists the fluid dissolutions that are claimed to be a by-product of postmodernity. The marketing of location and navigation consumer goods both relies upon and generates discourses and practices of precision. In a time when nationalized empire employs the networks of technoscience and globalization to wage war on cultural, economic, and political fronts, it is necessary to analyze these circuits of power as serving nationalist interests, because the war can be ended only when we recognize our attachments to its subject-making potential.

Notes

I thank Donald Moore as well as Allan Pred, Jake Kosek, Bruce Braun, and Rebecca Stein for their warm encouragement at the very early stages of this project. Versions of this piece have been presented at UC Davis, Berkeley, and Santa Barbara; the American Studies Association; the American Anthropological Association; the International Conference in Critical Geography; and the "Seuils et Traverses III" conference. I have been greatly helped in my research and thinking about militarization and technology by Jennifer Terry's generous sharing of work in progress. Inderpal Grewal's work on security and consumer subjects continues to inspire me. Drafts at various stages have been read and commented upon most helpfully by Eric Smoodin, Minoo Moallem, Angela Harris, Greig Crysler,

Carolyn de la Peña, and the anonymous readers for *American Quarterly*. I am extremely grateful to my research assistants at both UC Berkeley and UC Davis, most recently Mark Plotkin and Renée Vassilos, for their invaluable assistance.

1. Michel Foucault, "Truth and Power," in *Power/Knowledge: Selected Interviews and Other Writings, 1972–1977*, ed. Colin Gordon (New York: Pantheon, 1980), 114.
2. James Der Derian, *Virtuous War: Mapping the Military-Industrial-Media-Entertainment Network* (Boulder, Colo.: Westview, 2001); Tim Lenoir, "Programming Theatres of War: Gamemakers as Soldiers," in *Bombs and Bandwidth: The Emerging Relationship Between Information Technology and Security*, ed. Robert Latham (New York: New Press, 2003), 175–98; Jennifer Terry, *Killer Entertainments: Conditions and Consequences of Remote Intimacy*, work in progress.
3. Susan Mitchell, "Birds of a Feather," *American Demographics*, February 1995, http://www.marketingtools.com/Publications/AD/95_AD/9502_AD/9502AF03.htm (accessed May 26, 2006).
4. John Pickles, "Preface," *Ground Truth: The Social Implications of Geographic Information Systems*, ed. John Pickles (New York: Guilford, 1995), viii.
5. N.A., "Geographic information system," http://en.wikipedia.org/wiki/Geographic_information_system (accessed May 14, 2006).
6. See Denis Cosgrove, *Apollo's Eye: A Cartographic Genealogy of the Earth in the Western Imagination* (Baltimore: Johns Hopkins University Press, 2001); and Susan M. Roberts and Richard H. Schein, "Earth Shattering: Global Imagery and GIS," in *Ground Truth*, ed. Pickles, 171–95.
7. Development of GPS was begun by the U.S. Air Force in 1960. In 1974 the Department of Defense launched the first of twenty-four satellites as the NAVSTAR system.
8. GPS operates through the triangulation of radio signals, fixing position in three dimensions: latitude, longitude, and altitude. Atomic clocks synchronize GPS time, allowing for an unprecedented level of precision in fixing position.
9. Paul Virilio, *War and Cinema: The Logistics of Perception* (London: Verso, 1989).
10. Mark Lewyn, "Where Am I? Ask a Satellite," *Business Week*, October 26, 1992, 116.
11. L. Casey Larijani, *GPS for Everyone: How the Global Positioning System Can Work for You* (New York: American Interface Corporation, 1998), 1–2.
12. Gregory T. French, *Understanding the GPS: An Introduction to the Global Positioning System* (Bethesda, Md.: GeoResearch, 1996), 5.
13. Ibid., 3.
14. Doug Abberley, "The Lure of Mapping: An Introduction," in *Boundaries of Home: Mapping for Local Empowerment* (Philadelphia: New Society, 1993), 1.
15. Navigation and positioning are not the same thing. But most navigation proceeds from a standard of location or reckoning of distance between established locations. Thus, most navigation technologies rely on accurate determinations of positions.
16. Michael Pearce, "An Answer to the Age-Old Cry: Where on Earth Am I?" *The Wall Street Journal*, March 24, 1994, A12.
17. Frank Vizard, "Lost in America—Not!" *Rolling Stone*, May 28, 1992, 67.
18. Steve Ditlea, "Real Men Don't Ask Directions," *Popular Science*, March 1995, 86–89, 120–21.
19. Jeff Bertolucci, "Find Yourself with GPS MapKit SV," *PC World*, July 1993, 84.
20. For recent discussions of "targets" in the humanities and area studies as well as urban studies, see Samuel Weber, *Targets of Opportunity: On the Militarization of Thinking* (New York: Fordham University Press, 2005); Rey Chow, *The Age of the World Target: Self-Referentiality in War, Theory, and Comparative Work* (Durham, N.C.: Duke University Press, 2006); and Ryan Bishop, Gregory Clancy, and John Phillips, "Just Targets," *Cultural Politics* 2.1 (March 2006): 5–28.
21. See Norman J. W. Thrower, *Maps and Civilization: Cartography in Culture and Society* (Chicago: Chicago University Press, 1996); J. B. Harley, *The New Nature of Maps: Essays in the History of Cartography* (Baltimore: Johns Hopkins University Press, 2001); David Turnbull, *Maps Are Territories: Science Is an Atlas* (Chicago: University of Chicago Press, 1989); Walter D. Mignolo, *Local Histories/Global Designs: Coloniality, Subaltern Knowledges, and Border Thinking* (Princeton, N.J.: Princeton University Press, 2000); and Matthew Sparke, "Mapped Bodies and Disembodied Maps: (Dis)placing Cartographic Struggle in Colonial Canada," in *Places through the Body*, ed. Heidi J. Nast and Steve Pile (New York: Routledge, 1998), 305–36.
22. See Virilio, *War and Cinema*; Beaumont Newhall, *Airborne Camera: The World from the Air and Outer Space* (New York: Hastings House, 1969); Caren Kaplan, "Mobility and War: The Cosmic View of U.S.

'Air Power,'" *Environment and Planning A* 38.2 (February 2006): 395–407; Anthony Vidler, "*Terres Inconnues*: Cartographies of a Landscape to be Invented," *October* 115 (Winter 2006): 13–30; Linda Robertson, *The Dream of Civilized Warfare: World War I Flying Aces and the American Imagination* (Minneapolis: University of Minnesota Press, 2003); K. B. Atkinson, ed., *Close Range Photogrammetry and Machine Vision* (Caithness, Scotland: Whittles, 1996); David Buisseret, ed., *From Sea Charts to Satellite Images: Interpreting North American History through Maps* (Chicago: University of Chicago Press, 1990).

23. Albert L. Pardini, *The Legendary Norden Bombsight* (Atglen, Pa.: Schiffer Military History, 1999), 16.
24. I am grateful to one of the anonymous reviewers for this excellent point.
25. Donald MacKenzie, *Inventing Accuracy: A Historical Sociology of Nuclear Missile Guidance* (Cambridge, Mass.: MIT Press, 1990), 3.
26. "In the penultimate draft of the address, Eisenhower initially used the term *military-industrial-congressional complex*, indicating the essential role that the U.S. Congress plays in propagating the military industry. But, it is said that the president chose to strike the word *congressional* in order to avoid offending members of the legislative branch of the federal government." See http://en.wikipedia.org/wiki/Military_industrial_complex (accessed June 1, 2006).
27. Conrad C. Crane, *Bombs, Cities, and Civilians: American Airpower Strategy in World War II* (Lawrence: University Press of Kansas, 1993); and Williamson Murray, *War in the Air: 1914–45* (London: Cassell, 1999).
28. Crane, *Bombs, Cities, and Civilians*, 19.
29. Murray, *War in the Air*, 31.
30. Stephen McFarland, *America's Pursuit of Precision Bombing, 1910–1945* (Washington, D.C.: Smithsonian Books, 1995).
31. Ibid., 5.
32. Crane, *Bombs, Cities, and Civilians*, 20.
33. Pardini, *The Legendary Norden Bombsight*, 16.
34. See for example, autobiographical accounts such as this one at http://www.486th.org/Assn/NL/NL0202.htm#sight (accessed February 24, 2006).
35. Crane, *Bombs, Cities, and Civilians*, 64.
36. Murray, *War in the Air*, 96.
37. Sven Lindqvist, *A History of Bombing* (New York: New Press, 2000).
38. Alastair Finlan, *The Gulf War 1991* (Oxford: Osprey, 2003), 86–87.
39. See Matthew Mowthorpe, *The Militarization and Weaponization of Space* (Oxford: Lexington Books, 2004); and Paul B. Stares, *The Militarization of Space: U.S. Policy, 1945–1984* (Ithaca, N.Y.: Cornell University Press, 1985).
40. Richard Alan Schwartz, *Encyclopedia of the Persian Gulf War* (Jefferson: McFarland, 1998), 171.
41. "The big difference was that for the first time ever, we had the first television war that was live . . . Vietnam was a film war." David Schmerler, NBC news, quoted by Linda Jo Calloway, "High Tech Comes to War Coverage: Uses of Information and Communications Technology for Television Coverage in the Gulf War," in *The 1,000 Hour War: Communication in the Gulf*, ed. Thomas A. McCain and Leonard Shyles (Westport, Conn.: Greenwood, 1994), 65.
42. Ibid., 57.
43. Robert Stam, "Mobilizing Fictions: The Gulf War, the Media, and the Recruitment of the Spectator," *Public Culture* 4.2 (Spring 1992): 101–23.
44. See John Taylor, *War Photography: Realism in the British Press* (New York: Routledge, 1991).
45. Stig A. Nohrstedt, "Ruling by Pooling," in *Triumph of the Image: The Media's War in the Persian Gulf—A Global Perspective*, ed. Hamid Mowlana et al. (Boulder, Colo.: Westview, 1992), 119.
46. Douglas Kellner, *The Persian Gulf TV War* (Boulder, Colo.: Westview, 1992), 139.
47. Journalist John Fialka remembers, "We were escorted away from most of the violence because the bodies of the dead, chopped up by artillery, pulverized by B-52 raids, or lacerated by friendly fire, don't play well, politically." John J. Fialka, *Hotel Warriors: Covering the Gulf War* (Baltimore: Johns Hopkins University Press, 1992), 2.
48. There is an extensive literature on the media and visual politics and practices of the first Persian Gulf War. In addition to works cited elsewhere in this text, and without providing a full bibliography here, the following works have been helpful to me: Judith Butler, "The Imperialist Subject," *The Journal of*

Urban and Cultural Studies 2.1 (1991): 73–78; Victor J. Caldarola, "Time and the Television War," *Public Culture* 4.2 (Spring 1994); Susan Jeffords and Lauren Rabinovitz, eds., *Seeing through the Media: The Persian Gulf War* (New Brunswick, N.J.: Rutgers, 1994); Marita Sturken, *Tangled Memories: The Vietnam War, the AIDS Epidemic, and the Politics of Remembering* (Berkeley: University of California Press, 1997); Ian Walker, "Desert Stories or Faith in Facts?" in *The Photographic Image in Digital Culture*, ed. Martin Lister (New York: Routledge, 1995), 236–52; Georges van den Abbeele, "Lethal Mobilities: Calvin and the Smart Bomb," *Annali d'Italianistica* 21 (2003): 363–78; Elaine Scarry, "Watching and Authorizing the Gulf War," in *Media Spectacles*, ed. Marjorie Garber et al. (New York: Routledge, 1993), 57–73; and Danny Schechter, "The Gulf War and the Death of TV News," *The Independent*, January/February 1992, 28–31.

49. Daniel C. Hallin, "Images of Vietnam and the Persian Gulf Wars in U.S. Television," in Jeffords and Rabinovitz, *Seeing through the Media*, 56.
50. Ibid.
51. Editorial, *Nation*, February 11, 1991, 147, cited in Farrel Corcoran, "War Reporting: Collateral Damage in the European Theater," in Mowlana et al., 109.
52. Fred Kaplan, "U.S. Bombs Not That Much Smarter," *Boston Globe*, February 20, 1998, A1.
53. M. David Arant and Michael L. Warden, "The Military and the Media: A Clash of Views on the Role of the Press in Time of War," in McCain and Shyles, *The 1,000 Hour War*, 36.
54. Middle East Watch, *Needless Deaths in the Gulf War: Civilian Casualties during the Air Campaign and Violations of the Laws of War* (New York: Human Rights Watch, 1991), 19.
55. Inderpal Grewal, *Transnational America: Feminisms, Diasporas, Neoliberalisms* (Durham, N.C.: Duke University Press, 2005), 219–20.
56. Donna J. Haraway, *Modest_Witness@Second_Millennium.FemaleMan©_Meets_OncoMouse™: Feminism and Technoscience* (New York: Routledge, 1996), 51.
57. Jordan Crandall, "Operational Media," www.ctheory.net/articles.aspx?d=441, article a148 (accessed January 18, 2006).
58. Ibid. See also N. Katherine Hayles, *How We Became PostHuman: Virtual Bodies in Cybernetics, Literature, and Informatics* (Chicago: University of Chicago Press, 1999); Paul N. Edwards, *The Closed World: Computers and the Politics of Discourse in Cold War America* (Cambridge, Mass.: MIT Press, 1999); Lev Manovich, *The Language of New Media* (Cambridge, Mass.: MIT Press, 2001); Manuel DeLanda, "Ecologies, Representations, and the Affective Dimension of Image Reception," in *Under Fire 1: The Organization and Representation of Violence*, ed. Jordan Crandall (Rotterdam: Witte De With, Center for Contemporary Art, 2004).
59. Lenoir, "Programming Theatres of War."
60. See http://www.gpsworld.com/gpsworld/.
61. See Carolyn de la Peña, "Ready-to-Wear Globalism: Mediating Materials and Prada's GPS," *Winterthur Portfolio* 38.2–3 (Summer-Autumn 2003): 100–129.
62. Some of the problems that were encountered during the first Persian Gulf War have been purported to be solved by the next generation of GPS guided weapons such as JDAM (Joint Direct Attack Munition) and JSOW (Joint Stand-Off Weapon). Global Security lists 22 bombs guided by GPS in use by the U.S. military in the present conflicts. According to the same site, since the first Persian Gulf War, the Air Force has tripled the number of precision weapons and enhanced accuracy (http://www.globalsecurity.org/military/systems/munitions/smart.htm). However, the "precision" of the strikes in the recent conflicts continues to be debated. Continuing his long-standing attention to the issue of civilian deaths from precision guided weapons, Fred Kaplan reported in 2003 that the accuracy rate of the GPS in use in Iraq remains approximately 10–100 meters and that this combined with any inaccuracy in intelligence results in less than "pinpoint" attacks—with numerous civilians killed in the effort to "hit" high-profile leadership in the opposition. Yet the purported accuracy of the weapons systems encourages the attempt to assassinate individuals via long-range missile strike. Fred Kaplan, "Smart Bombs, Dumb Targets," http://www.slate.com/id/2092759/. (accessed May 10, 2006).
63. Peter Miller and Nikolas Rose, "Mobilizing the Consumer: Assembling the Subject of Consumption," *Theory, Culture, & Society* 14.1 (February 1997): 2.
64. See David J. Grimshaw, *Bringing Geographical Information Systems into Business* (New York: John Wiley, 2000); Richard Harris et al., *Geodemographics, GIS, and Neighborhood Targeting* (New York: John Wiley, 2005); and Irvine Clarke III and Theresa B. Flaherty, *Advances in Electronic Marketing* (Hershey, Pa.: Idea Group, 2005).

65. Donald Lowe, *The Body in Late-Capitalist USA* (Durham, N.C.: Duke University Press, 1996), 64–66.
66. Kevin Robins, "Tradition and Translation: National Culture in Its Global Context," in *Enterprise and Heritage*, ed. John Corner and Sylvia Harvey (New York: Routledge, 1991), 27.
67. See Mark Poster, "Databases as Discourse; or, Electronic Interpellations," in *Computers, Surveillance, and Privacy*, ed. David Lyon and Elia Zureik (Minneapolis: University of Minnesota Press, 1996), 175–92.
68. See Inderpal Grewal and Caren Kaplan, "Introduction: Transnational Feminist Practices and Questions of Postmodernity," in *Scattered Hegemonies: Postmodernity and Transnational Feminist Practices* (Minneapolis: University of Minnesota Press, 1994), 1–36.

The Wire Devils: Pulp Thrillers, the Telephone, and Action at a Distance in the Wiring of a Nation

Robert MacDougall

We can remember the monsters of the Gilded Age, but not the horror they once evoked. The octopus, the spider, the hydra—historians find these images of corporations and the technological networks they built strewn across the culture of the late-nineteenth- and early-twentieth-century United States like the bones of dinosaurs long extinct. Of course, Americans still worry about corporate size and rapid technological change. Today's multinational corporations dwarf the leviathans that alarmed muckrakers and trustbusters a century ago. But that era's most lurid images of corporate and technological networks are somehow alien to twenty-first-century Americans. The political cartoons seem quaint or comical; the oratory of the Populists and the prose of muckraking journalists often read as hysterical or overwrought. In an age of planet-spanning corporations and instantaneous global commerce, it is difficult to apprehend how unnatural—how monstrous, to some—region- and nation-spanning companies and technologies of communication once seemed.

In this article, I reexamine the popular fears of that era by exhuming from obscurity some vernacular literature that directly engaged these anxieties. First, I examine the "wire thrillers," a series of pulp novels published in the 1890s, 1900s, and 1910s that put melodramatic depictions of technological and economic change at center stage. The thrillers were for the most part formulaic potboilers, but they were charged by a simultaneous attraction to, and repulsion from, the technological transformation they described. I then turn to the one advertising campaign that, probably more than any other, pacified fears of America's new techno-industrial order, and drove images of the corporate octopus and wire spider to extinction. This long and influential campaign was the work of the American Telephone and Telegraph Company.

New technologies—in particular the new national networks of telegraph and telephone—were the central subjects of the wire thrillers and the AT&T

campaign. Both the thrillers' authors and the telephone monopoly employed these new technologies as proxies for the nation, and as metaphors for controversial changes to America's political and economic order. While on the surface the wire thrillers celebrated the new technologies of long-distance communication, they also betrayed deep fears of sectional integration and nation-spanning commerce. AT&T publicity worked to answer these anxieties. The telephone company embraced the wire thrillers' metaphorical conflation of technological networks, corporations, and the American nation, but reversed the polarity of their images, celebrating the very sorts of social and economic integration that the authors of the wire thrillers most feared. Ultimately, AT&T succeeded both in constructing a national telephone network and in selling that network as a model representation of the nation itself. The story of how AT&T accomplished that feat is the story of how Americans learned to stop worrying and love the wire, or how technological networks, sectional integration, and nation-spanning corporations were at once conflated and redeemed.

Action at a Distance

Why were Gilded Age Americans so inclined to imagine large industrial organizations, and the technological networks they owned and operated, as monstrous octopi or spiders? In the visual culture of late-nineteenth-century America, tentacles and webs were shorthand for both technological networks—railroad tracks, oil pipelines, telephone and telegraph wires—and for corporate power more broadly (fig. 1). The sinuous tentacles of the octopus are what made it a frightening and powerful symbol—likewise the long limbs of the spider and the ensnaring strands of its web. How often were railroads, telephone and telegraph companies, or oil trusts depicted as octopi stretching their tentacles over a map or even clutching an entire globe? The octopus and the spider were not caricatures of corporate size alone. They were nightmares of *reach*. They were vivid depictions of what was often called "action at a distance."[1]

Figure 1.
A cartoon from 1888 depicts bankers, doctors, and other early telephone users ensnared in the web of the Bell spider. Reproduced from *Judge*, April 7, 1888, 16.

Few features of late-nineteenth-century life seemed more novel or remarkable to observers than the technologies of action at a distance. Chief among these were the railroad, telegraph, and telephone. In the decades after the Civil War, American railroads linked the corners of the continent with thousands of miles of track. The Western Union Telegraph Company, one of the United States' first nation-spanning corporate monopolies, completed a transcontinental

IN THE CLUTCH OF A GRASPING MONOPOLY.

telegraph line in 1861, providing theoretically instantaneous communication from coast to coast. And the telephone, born in the centennial year of 1876, rapidly grew to eclipse the telegraph, promising to connect every home and every life to new national networks of communication and exchange.[2]

New technologies almost inevitably inspire simultaneous enthusiasm and unease. Numerous scholars have documented the anxieties expressed by late-nineteenth- and early-twentieth-century Americans in the face of rapid technological change.[3] Many of these anxieties were provoked by the apparent ability of new technologies to alter spatial categories and boundaries. Railroad cars and streetcars, for example, were new kinds of technological spaces. As Grace Elizabeth Hale shows, these spaces became key battlegrounds in the creation of, and resistance to, systems of racial segregation.[4] The telephone in the home, both Carolyn Marvin and Michèle Martin argue, breached a traditional divide between male public and female private spheres. As a result, the "virtual space" the telephone created appeared to threaten those structures of gender, class, and race that more clear-cut spatial divisions between work and home helped to sustain.[5]

In this article, I am interested in the technological reordering of space on a larger scale—the perceived eclipse of sectional and local economies and identities by a move toward national unity through commerce. One of the era's great clichés held that the rail and wire had "annihilated" space and time.[6] The violence of that phrase is rarely remarked upon. Why were time and space "annihilated," rather than "transcended" or "transformed"? The pace of technological and economic change was indeed violent and wrenching to many Americans. Each advance in the technology of communication and transportation gave new powers to its users, yet also compounded the ability of distant people and events to affect those users' lives.

In the words of Thomas Haskell, "the very constitution of the social universe had changed." As society and economy became more obviously interdependent, Haskell argues, it proved harder and harder to imagine individuals as solitary masters of their fates. Local sources of meaning and order—the family, the sect, the small town—were "drained of causal potency," becoming "merely the final links in long chains of causation that stretched off into a murky distance." America's isolated "island communities," in Robert Wiebe's famous phrase, were absorbed, or feared absorption, into national and even international markets and networks.[7] The local was challenged by the national, the near by the distant, in virtually every area of life. The scale of life changed, and with it changed the social meaning of distance.[8]

Imagined Technologies

This article is more concerned with representations of technology and technological practices than with actual material technologies or the real practices of users.[9] The transcontinental telephone system was at its inception the largest, most complicated machine ever built. Its size and complexity could not be directly perceived or demonstrated in any simple way. "In all the 3,400 miles of line," one AT&T publicist said of the first coast-to-coast telephone call, "there is no one spot where a man may point his finger and say, 'here is the secret of the Transcontinental Line; here is what makes it possible.'"[10] The new networks of national communication could be only viewed in incomplete parts: a telephone here, a tangle of wires there, a line of poles stretching into the distance. "Explanations of it [the telephone network] are futile," wrote journalist Herbert Casson in 1910. "It cannot be shown by photography, not even in moving-pictures."[11] There was no direct way to apprehend the immensity of the system. Electricity in the nineteenth century was, it has been observed, "a force stronger in the imagination than in reality."[12] The first national networks of telephone and telegraph, I argue, were more commonly and powerfully imagined than directly experienced or used.

AT&T executives once flattered themselves in thinking that they alone had taught Americans how and when to use the telephone. The public "had to be educated . . . to the necessity and advantage of the telephone," said AT&T president Theodore Vail in 1909. His company "had to invent the business uses of the telephone and convince people *that they were uses*," declared a company advertisement that same year.[13] That portrait of passive consumers waiting to be instructed by the producers of technology was critiqued in the 1980s and 1990s by historians and sociologists such as Carolyn Marvin, Claude Fischer, and Michèle Martin.[14] Those scholars credited eager and innovative consumers, especially women, with inventing many uses of the telephone neither foreseen nor endorsed by the Bell System.

That case cannot be made, however, for use of the long-distance network. Long-distance calling, especially the very long distance calling made possible by AT&T's transcontinental lines, was a practice in which publicity long preceded consumer demand. AT&T vigorously promoted its coast-to-coast national network as a symbol, but actually using the network was cumbersome and expensive. Few ordinary Americans declared any pressing need to be connected to cities, states, or regions hundreds or thousands of miles away. "No one pretends that the New York–San Francisco line will immediately 'pay'," reported *McClure's* in 1914. "The public will have to acquire the habit of talking transcontinentally, just as it had to learn to use the telephone at all." As late as 1935, AT&T would estimate that less than 1.5 percent of telephone calls crossed even one state line.[15]

Benedict Anderson called the nation an "imagined community" because no citizens of any modern nation can expect to meet or know more than a fraction of their fellows.[16] The long-distance telephone and telegraph were enablers and artifacts of the nationalist project Anderson describes. Like the nations they came to span, the new networks of long-distance communication could be represented—pictured, mapped, performed, described—but not wholly experienced or truly seen. In fact, the new networks of communication were themselves representations of the profound transformation the nation was undergoing. The railroad, telegraph, and telephone offered physical representation of those expanding networks of commerce and exchange that linked section to section, tied livelihoods and fates to distant markets, and seemed to suborn once independent communities to unseen forces and unprecedented conglomerations of wealth and power.

Representations of these networks were not only imaginative, but didactic. Depictions of the wire, even in advertising copy or lowly pulp thrillers, embodied arguments about the nation, and about the ways commerce, power, and information should flow. To understand these technologies and their place in American life, we must examine the metaphors and images that contemporary Americans used to represent them, and the arguments about the nation those metaphors contained.[17]

The Wire Thrillers

"Telephones are only known to me in the kind of novels which a man reads in bed, hoping that they will send him to sleep," reported an English essayist named Andrew Lang in 1906.[18] His remark underlines the power of literary representations to frame early encounters with technology. Many ordinary Americans and others encountered representations of long-distance communication in fiction, advertising, or the press well before experiencing that technology firsthand.

Lang also suggested the sort of literature in which the telephone and telegraph were most likely to be found. Though he did not name the novels by his bedside, he might well have been describing the wire thrillers. Novels such as *The Wire Tappers*, *The Wire Devils*, *Brothers of the Thin Wire*, *Phantom Wires*, and *Fighting Electric Fiends* were the high-tech thrillers of their day.[19] Loaded with technical jargon and slang such as "lightning slingers" and "overhead guerillas," they portrayed the business of telegraphy and telephony as a thrilling demimonde of crime and derring-do—a "secret network of excitement and daring," in one author's words, "which ran like turgid sewers under the asphalted tranquility of the open city."[20] While formulaic in execution, these popular works attested to the excitement and anxiety surrounding the technology of action at a distance when it still seemed dangerous and new.[21]

The formula on which the wire thrillers were constructed is straightforward. They are tales of technology, filled with detailed descriptions of the new communication networks and the endless uses to which they might be put. All manner of masquerades are played out over the telephone and telegraph. Lines are tapped, cut, and redirected. Messages are coded and cracked, intercepted and forged. The wire thrillers are stories of commerce and crime, and they present the line between those activities as blurry indeed. Racetrack gamblers collude with stock market speculators while crooked railroad magnates toy with the lives and livelihood of the nation. Above all, the wire thrillers are breathless depictions of action at a distance—the ability to act in one place and affect the lives of people in another. Electricity, Arthur Stringer rhapsodized in his novel *Phantom Wires*, "hurls your voice half way round the world . . . it creeps as silent as death through a thousand miles of sea . . . it threads empires together with its humming wires; it's the shuttle that's woven all civilization into one compact fabric!"[22]

All this activity requires detailed explanation, and the prose of the wire thrillers is often mired in technical exposition. But exposition was in many ways the point. Long passages of the novels depict corporate machinations and other workings of the new economy in word-pictures that are at once lurid and weirdly loving. The wire thrillers were primers, of a sort, on the operation of the telegraph and telephone, as well as the larger economy of which those devices were a part.

The wire thrillers were enthralled with descriptions of technology and technological mastery. Every novel of the type requires at least one scene in which Morse code, the shared but secret language of the technological elect, is used to communicate under the nose of some unsuspecting quarry. Each novel pays tribute more than once to the power of electricity and those prepared to seize and use it. "Right at the back of this house is a wire . . . a little condensed Niagara of power," one character exults. "I can capture and tame and control that power . . . I can make it my slave, and carry it along with me."[23] The role models of the wire thriller are never bewildered by the technology of action at a distance. They capture and tame it, domesticate it, and make it their tool.

The wire thrillers belong to the "rogue school" of turn-of-the-century fiction. The heroes of the wire thrillers are invariably technologically savvy young men. They are good men, though prone to falling in among bad sorts. The protagonist of *The Wire Tappers* and its sequel, *Phantom Wires*, is Jim Durkin, a former railroad telegraph operator, blacklisted by his employers after failing to prevent a railroad crash. He drifts into the world of wired crime to fund the development of his own personal invention, a form of television. Frank Packard's *The Wire Devils* features as its hero "the Hawk," a gentleman jewel thief who outsmarts both the railway police and a nation-spanning criminal syndicate in a game of cat and mouse played over telephone and telegraph lines. Only in the novel's final pages is the Hawk's true identity, a Secret Service agent conducting an elaborate sting, revealed.

The female protagonists of the wire thrillers are surprisingly similar to the men. They are more passive than the male characters, in the prescribed manner of so much genre fiction, but not exceedingly so. The most typical heroine in the wire thrillers is a young telegraph or telephone operator. These characters tend to be technically proficient in their own right, and no less enmeshed in the shady underworld of the wire. The female operator is an interesting figure in both fiction and in history, a glaring contradiction to her era's nearly automatic equation of technological mastery and masculinity. Young women were a crucial part of the real-life world of the wire from its beginning. Cheaper than men and more reliable than boys, unmarried women were employed nearly everywhere as telephone and telegraph operators. In popular culture, the figure of the female operator united the novelty of the new technology with the novelty of finding so many young women in the worlds of business and industry, and the "hello girl" rapidly became nearly as ubiquitous in prose and song as she was in the Western Union office or Bell exchange. "The young men of the daily press . . . delight to write about that mysterious and nearly always sweet-voiced being, the telephone girl," observed an electrical industry journal in 1889.[24] Many male writers waxed rhapsodic about the female operator and imagined her at the center of a host of romantic scenarios. Mark Twain's Connecticut Yankee is so taken with the operator in his hometown that he names his first child "Hello Central."[25] Other sources, however, portrayed young female operators as giddy and brazen, speaking in slang, garbling messages, and cutting off legitimate callers to flirt with "downtown clerks and dudes."[26] Jeffrey Sconce's analysis of nineteenth-century spiritualism as a discourse on gender and technology notes the common conflation of female operators and female mediums.[27] The complex cultural figure of the female operator demonstrates her era's ambivalence about women's relationship with distance-collapsing technologies. Already powered by a more general ambivalence about action at a distance, the wire thrillers tapped this alternating current of fascination and unease.

The principal female character in *The Wire Tappers* and *Phantom Wires* is Frances Candler, a young telegraph operator who has fallen in with an unsavory bunch. She becomes Jim Durkin's love interest in the first novel and his wife in the second, yet she remains an active participant in his capers. Jim, the novels make clear, is drawn to Frances by her beauty and by her considerable technological skill. These two qualities are united in her deft but delicate touch on the telegraph sounder, a signature "hand" that Jim recognizes over the wires and falls in love with from afar. (To complicate the gender typing of Frances's character, Jim invariably calls her "Frank," which adds a certain homoerotic frisson to the sweet nothings the two lovers transmit over the wires.) If Frances is not quite an equal partner in every action, neither is she portrayed as lost or helpless in the underworld of technology and crime.

Frank and Jim scheme and court by telephone and telegraph. Stringer presents their long-distance romance—"that call of Soul to Soul, across space, along channels less tangible than Hertzian waves themselves"—as the highest achievement of the wire, the one application of the new technology that might be unreservedly embraced as virtuous and good. He describes at length the two lovers' telegraph messages and long-distance phone calls, urging his readers to marvel at the triumph of love and technology over geographic space:

> Not an ohm of . . . soft wistfulness, not a coulomb of . . . quiet significance, had leaked away through . . . hundreds of miles of midnight travel. It almost seemed that [Jim] could feel the intimate warmth of her arms across the million-peopled cities that separated them; and he projected himself, in fancy, to the heart of the far-off turbulence where she stood.[28]

Yet this marvel has a cost. Though "magnetically drawn" to the "excitement and daring" that surrounds the wire, Frances worries constantly about the morality of action at a distance, and its effect on both her femininity and her soul. Through participation in the dangerous world of technological commerce and crime, "the battery of her vital forces"—the novels miss no opportunity for electrical metaphor—is gradually "depleted and depolarized."[29] Frances serves, a little awkwardly, as the moral conscience of both novels, and Stringer's most didactic passages are written in her voice. "What was it that had deadened all that was softer and better and purer within her, that she could thus see slip away from her the last solace and dignity of her womanhood?" she wonders. The answer is the use of the wire for gain without toil. Technology is at once attractive and corrosive. Frances fears she has become "only the empty and corroded shell of a woman, all that once aspired and lived and hoped in her eaten away by the acid currents of that underground world into which she had fallen."[30] Her attraction to and repulsion from the demimonde of the wire parallels the whole genre's conflicting impulses toward the technology of electrical communication and the economic transformation that technology was understood to represent.

Section, Space, and Crime

Frances and Jim and all the other heroes in the wire thrillers are thoroughly upstaged by the villains they confront. The villains of the genre are vividly depicted, and are, even more than the heroes, masters of technology and action at a distance. It is in the figures of these antagonists that the authors of the wire thrillers draw their most direct connections between technology, commerce, space, and crime. In the schemes of those villains, which the thrillers describe at length, the wire's role as a physical representation of economic forces becomes most clear.

The wire thrillers were undoubtedly influenced by *The Octopus* and *The Pit*, Frank Norris's famous novels of the railroad and the stock exchange. Norris and the authors of the wire thrillers looked upon the new networks of technology and commerce with the same mixed feelings. Yet like the authors of the wire thrillers, Norris was also drawn to the challenge of describing commercial systems and illustrating economic change. He conceived *The Octopus* and *The Pit* as the first two parts of an epic trilogy that would capture "the New Movement, the New Finance, the reorganization of capital, the amalgamation of powers, the consolidation of enormous enterprises."[31] And Norris was, like his imitators, particularly concerned with the way new technologies shaped or altered space.[32]

One difference between Norris and his imitators is that in Norris's work, there are few convenient villains. A railroad crushes the California farmers and ruins their lands, but there is no "Master Spider" behind the railroad company to be defeated and unmasked. There is only the complex interdependence of technology, agriculture, and national finance. "You are dealing with forces, young man, when you speak of Wheat and Railroads, not men," says an executive in *The Octopus*. The struggling farmers are left to wonder: "Forces, conditions, laws of supply and demand—were these, then, the enemies after all?"[33] Stephen Kern's recent history of causality in fiction specifically cites *The Octopus* to illustrate a shift toward more complex models of interdependence.

Norris illustrates in fiction the same disorienting recession of causation that Thomas Haskell describes.[34]

In the wire thrillers, by contrast, there is always a man behind the octopus. In *The Wire Devils*, the titular crime ring is controlled by "the Spider," a criminal mastermind in the style of Arthur Conan Doyle's Dr. Moriarty or Sax Rohmer's Fu Manchu. The Spider's master stroke is his capture and use of a railroad telegraph system. He and his Wire Devils use the railroad and its telegraph to direct and cover their criminal activities. They have subverted the system so cunningly, we are told, that the railroad effectively *is* the criminal syndicate, or at least entirely does its bidding. Much is made of this. "Somewhere, hidden away in his web, at the end of a telegraph wire, was the Master Spider," Packard writes. "Where there was a telegraph sounder, that sounder carried the messages, the plans, the secret orders of the brain behind the organization; and the very audaciousness with which they made themselves free use of the railroad's telegraph system . . . was in itself a guarantee of success."[35] Without actually portraying the normal activities of real railroads as criminal, Packard's novel indulges in the fantasy of a criminal railroad and exploits widespread unease with that industry's reach and power.

Arthur Stringer's novels make equally clear connections between the new networks of commerce and power and an imagined world of high-tech crime. Jim and Frances begin their adventures together by conning the crooked operators of a racetrack gambling ring. In Stringer's descriptions of the criminal racing "circuits"—"the huge and complicated and mysteriously half-hidden gambling machinery close beside each great center of American population"—it is easy to read both attraction and repulsion with the workings of commerce and exchange. "Money flashes and passes back and forth, and portly owners sit back and talk of the royal sport," Stringer writes, while "from some lower channel of the dark machine drift the rail-birds and the tipsters . . . the idlers and the criminals." Those criminals, in turn, "infect the rest of the more honest world with their diseased lust for gain without toil."[36]

In the years of the wire thrillers, real-life reformers attacked racetrack gambling in precisely these terms. And because the telegraph and telephone were widely used to collect bets and wire race results to off-track poolrooms, telegraph and telephone companies were often indicted as allies of the gambling industry. The telephone, reformers argued, made it all too easy for seemingly respectable men and women to gamble without having to visit a racetrack or enter a poolroom. "The Western Union Telegraph Company and the telephone interests of the United States are directly responsible for the very existence of the great pool-room game and its aftermath of human wreckage," charged an assault on the industry by *Cosmopolitan* magazine in 1907, the year after *The Wire Tappers* appeared and the year that Arthur Stringer published *Phantom Wires*. "They [the owners of telephone and telegraph systems] tacitly have allowed their companies to become bone and sinew in the body of the pool-room crime."[37]

In both the novel and the magazine, this danger of the new technology is understood as spatial. That is, it involves the way the telephone and telegraph span or reorder social and physical space. The gambling circuits, Stringer tells his readers, connect "New York and Washington, Chicago and St. Louis, Memphis and New Orleans." They bring "idlers and criminals" into close contact with "the more honest world." *Cosmopolitan* is even more explicit: "You want to bet but are afraid to be seen entering a pool-room. We [the telephone company] have arranged with the pool-room criminals so that you can gamble away your money, your employer's money, your husband's money . . . and at the same time never go near a pool-room."[38] The telephone and telegraph, it was feared, broke down spatial boundaries upon which morality—and, not incidentally, gender and class order—seemed to depend.

The attack on racetrack gambling, however, is only a prelude to *The Wire Tappers*' main event. Having swindled the swindlers at the racetrack, Jim and Frances go after bigger prey. Jim learns of a millionaire speculator, a "Napoleon of commerce" who has the power, through access to a private telegraph line from Savannah and New Orleans to the Secretary of Agriculture in Washington, to predict and manipulate the worldwide price of cotton. The character of Curry, the "Cotton King," contains all the ominous connections Americans might imagine between technology and power, crime and commerce, the erosion of section, and action at a distance. "This was the man at whose whisper a hundred thousand spindles had ceased to revolve," Stringer writes, "and at whose nod, in cotton towns half a world away, a thousand families either labored or were idle, had food or went hungry." Curry instigates a panic on the stock exchange, leading to the suicide of a ruined merchant, simply to amuse and impress an actress in the gallery. He uses his control of the telegraph to engineer a bubble in the price of cotton, one that will ruin thousands but make him a fortune. It is, the reader sees, essentially the same swindle perpetrated early in the novel by the gangsters at the racetrack, but on a national scale.[39]

It is fitting that *The Wire Tappers* uses cotton, and a telegraph line from South to North, to illustrate the interdependence of sectional commerce. Cotton was the great global industry of the nineteenth century. "Whoever says Industrial Revolution says cotton," wrote Eric Hobsbawm.[40] For nineteenth-century

Americans, whoever said cotton also said slavery, section, and the Civil War. Cotton tied the economy of the industrial North to that of the slaveholding South. This, abolitionists argued, made Northern mill owners, merchants, and ordinary consumers complicit in the peculiar institution years before the Dred Scott decision or the Fugitive Slave Act.[41] Cotton also linked the fortunes of Southerners to the mills of Manchester and Massachusetts, and to a new industrial order that white Southerners at least neither desired nor admired. The cotton trade wove a "worldwide web," in Sven Beckert's words, that paved the way for global capitalism.[42] A generation after the Civil War, American unease about the economic and moral interdependence of sections could still easily fix on cotton as a symbol.

The issue of race, so crucial and inescapable to America in these years, is conspicuous in the wire thrillers only by its absence. The heroes, villains, and supporting characters of these novels are uniformly white. The wire thrillers have this in common with almost all literature surrounding telegraphy and telephony in this era. Fiction, trade journals, and advertising echo with the silence of whiteness as an unmarked category.[43] The issue of section, however, is everywhere in the wire thrillers. Characters are routinely typed by where they come from; the stories are populated with greedy Yankee capitalists, violent Irish thugs, and delicate Southern belles. More important, the thrillers resonate with anxiety about sectional integration and national commerce.

Section and race in Gilded Age America were of course inextricably intertwined. While the railroad, telegraph, and telephone enabled the incorporation of the South into a unified national economy, segments of white America worked to convince themselves that the nation's culture was or could be similarly united. As the Civil War passed from immediate memory, white Americans in many regions of the country enacted sentimental rituals of sectional "reunion" and "reconciliation"—the terms implying a prior unity that may never have existed. John Hope Franklin, David Blight, and others have shown us that the price of that "re"-unification was a general repudiation of Reconstruction, and national acquiescence to racial segregation and ideas of white supremacy.[44]

It need not trivialize that price to observe that national unification had other costs. Unification, in the sense and service of commercial integration, meant the interconnection of sections and the inescapable unity of the national market. The anxieties of the wire thrillers, like the era's recurring nightmares of octopus and spider, testify to genuine unease about this profound reordering of national space. No specific locations were cast as sinister in *The Wire Tappers* or its fellows. The Cotton King's secret telegraph line from Savannah to Washington to Chicago could be seen as equally ominous by readers in the

South or North or West. Instead, action at a distance is presented as dangerous in and of itself. On nearly every page of the wire thrillers, fascination with new technology confronts a critique, not of technology per se, but of the economic interdependence created by technological networks of wire and rail:

> This Machiavellian operator's private wires were humming with messages, deputies throughout the country were standing at his beck and call, . . .Chicago and St. Louis and Memphis and New Orleans were being thrown into a fever of excitement and foreboding, fortunes were being wrested away in Liverpool, the Lancaster mills were shutting down, and still cotton was going up, point by point.[45]

There is admiration here for the technologies of action at a distance and for those who make them do their bidding. There is a more powerful sense, however, that this sort of commerce must be deeply corrosive. The cotton market is only the racetrack poolroom writ large. "What was criminality from one aspect was legitimate endeavor from another," Stringer writes. "All life . . . was growing more feverish, more competitive, more neuropathic, more potentially and dynamically criminal." With the annihilation of space, the wire thrillers predicted, must come the erosion of morality and the invasion of once virtuous small communities by the imperatives of a national market:

> Timid clerks and messenger boys and widows, even, were pouring their pennies and dollars into the narrowing trench which separated them from twenty cent cotton and fortune. . . . Even warier spirits, suburban toilers, sober-minded mechanics, humble store-traders, who had long regarded [the stock exchange] as a very Golgotha of extortion and disaster, had been tainted with the mysterious psychologic infection, which had raced from city to town and from town to hamlet.[46]

From city to town and from town to hamlet—this is the vision of action at a distance that haunted Gilded Age Americans. The wires are only the means of infection. The deeper fear is a collapse of section and of space, and the corrosive impact of the national corporate economy on the autonomy and morality of America's hometowns.

The networks depicted in the wire thrillers were, therefore, doubly imagined. The novels offered vivid representations of new technologies whose scale and scope could not easily be apprehended in other ways. At the same time, the wire thrillers employed the new networks of telephone and telegraph as metaphors for commercial integration and economic change. Ultimately, these pulp thrillers drew their power less from a critique of technology itself than from frightening visions of the annihilation of space and section, and the assimilation of American communities into one ruthless economy or machine.

AT&T and the Octopus

Nightmare images of space-destroying octopi and monstrous corporate trusts were never confined to fiction, pulp or otherwise. They were ubiquitous in the political discourse of the late nineteenth century. Any large corporation might be caricatured as a monster in this era, but the industries most commonly depicted as octopi and spiders were those in the business of action at a distance—the railroads, telegraph, and telephone (fig. 2). (The only other contender for this honor was Standard Oil, another nation-spanning corporation that played a role in the technological reordering of national space.)

The Western Union Telegraph Company and the great railroad interests of the late nineteenth century often seemed determined to live down to their reputations. There was little in the pages of *The Octopus* or the wire thrillers that could not be matched or topped by the actions of real railroads in the ruthless and loosely regulated competition of the day. Western Union, which secured a national monopoly over the long-distance telegraph in the aftermath of the Civil War, was also famously indifferent to its own public image and to questions of the public good. The American Telephone and Telegraph Company, however, which became the parent company of the Bell telephone interests in 1900, took a different tack.

In the 1880s and 1890s, the Bell telephone companies showed little more interest or aptitude for public relations than their counterparts at Western Union. After 1900, however, and particularly after a change of management in 1907, AT&T became increasingly sensitive to its rather spotty public image. After 1907, the company launched an ambitious public relations offensive. This influential campaign, which continued for many years, did much more than simply polish the telephone company's public face. It took on the idea of the octopus and the fears on display in the wire thrillers, and seems to have quite successfully countered or displaced them. It did so by promoting and redeeming the idea of action at a distance.

Like the wire thrillers, AT&T advertisements in the 1900s and 1910s were efforts to depict what could not be seen—the size and complexity of the new national communications networks. "The Bell System!" enthused Bell promoter Herbert Casson. "Already this Bell System has grown to be so vast, so nearly akin to a national nerve system, that there is nothing else to which we can compare it."[47] Also like the wire thrillers, AT&T advertisements used representations of the telephone network to represent the economy and the nation. The wire represented and embodied a new national network of commerce and information, and it stood for the profound transformation and integration America

Figure 2.
The Bell System's midwestern competitors portrayed their rival as an octopus stretching its tentacles across the plains. Reproduced from *Telephony*, April 1907, 235.

and its economy had undergone. Unlike the wire thrillers, however, AT&T declared that this transformation was nothing to fear.

AT&T's seminal advertising campaign began in 1907, at a time when the company faced fierce competition from thousands of smaller telephone operations known as the independents, and its leaders harbored serious fears of antitrust action or hostile legislation from the government. At the center of the AT&T campaign was a long series of magazine ads created by the advertising agency N. W. Ayer and Son. But the telephone company also planted press releases with friendly editors, subsidized flattering books about the company, and published a flood of "educational" pamphlets, booklets, and films. Roland Marchand called this campaign "the first, most persistent, and most celebrated of the large-scale institutional advertising campaigns of the early twentieth century." It is not too much to say that AT&T blazed the trail that corporate public relations would follow for years to come.[48]

Given public hostility to corporate spiders and octopi, one might have expected AT&T to deemphasize the size and unity of its system. Given also the widespread anxiety about action at a distance, one might expect a retreat from arguments about the way the wire was shrinking and unifying the nation. Yet the telephone company did neither. It offered instead a positive defense—indeed, a celebration—of economic integration and corporate reach. Because it accepted the metaphor that powered the wire thrillers, AT&T had to do more than sell telephone service. It had to sell action at a distance and the new economic unity of the nation.

AT&T publicity described the workings of the telephone and the new national economy in word-pictures that read like the novels of Frank Packard or Arthur Stringer minus the menace or unease. "American business men have been made neighbors through contacts over the wires of a nation-wide telephone system," proclaimed a typical advertisement. "Drawn together by bonds of communication . . . America's industries operate not as individual and isolated enterprises, but as closely coordinated parts of a gigantic mechanism that ministers to the nation's needs."[49] The wire thrillers drew metaphorical connections between new technological and commercial networks; telephone company publicity asserted that the two sorts of networks were one and the same. AT&T executives argued for the essential unity of the telephone network and the corporate network that owned it. "It is not the telephone apparatus, central office equipment, or wires that independently afford or can afford any service," Theodore Vail wrote in 1917. "It is the machine as a whole; all the telephones, all the equipment, all the central offices are vital and necessary parts of that machine. That machine is the Bell System."[50] They went on to equate their national network with the national economy, calling the telephone a "national nervous system," and the "life-blood of commerce." "Intercommunication" is "the basis of all civilization," Vail declared in a 1913 speech, and "prosperity is in direct relation to its completeness and perfection."[51]

Like the wire thrillers, AT&T's portrayal of technology was strictly deterministic. The new communication technologies, from the company's point of view, demanded commercial and political integration. There was no way to resist those technologies and no point in debating the kinds of social and political accommodations they required. "The telephone on the desk must be in contact with . . . every other telephone throughout the continent," said Bell executive James Caldwell in 1911. "This can only be done through one unbroken homogenous system . . . and practically and psychologically that one universal system can only be the American Telephone and Telegraph Company."[52]

AT&T publicity never sought to deny or downplay the scope and power of the new communication networks. Instead, it sought to redeem that power by offering it to users and subscribers. "Your line is connected with the great Bell highways, reaching every state in the union," one advertisement read. "You have the use of switchboards costing upwards of $100,000,000 . . . [and] the benefits of countless inventions." Ads with slogans such as "The Telephone Doors of the Nation," or "You Hold the Key," represented the network as empowering to its users, not powerful in itself. "Every Bell Telephone is the Center of the System," declared a common company slogan.[53] Every man could be a Jim Durkin and every woman a Frances Candler, seizing the technology of action at a distance to "capture and tame and control" the power of the wire. In the publicity for AT&T's national network, as in the wire thrillers, awestruck descriptions of nation-spanning machinery were juxtaposed with seemingly contradictory celebrations of individual empowerment. The wire network was indeed immense, implacable, a leviathan—but men, and perhaps a few women, were bigger still.

As answers to the image of a monstrous corporate octopus, AT&T advertisements presented illustrations of giant telephone operators reaching across the nation, and a series of giant businessmen, looming over a continent the telephone made small (fig. 3). Such images encouraged readers and viewers, in particular the American businessmen whom AT&T deemed the principal market for long-distance service, to imagine themselves as that colossal telephone user. Telephone customers need not see themselves as the prey of the octopus, these illustrations implied. They should see themselves as the octopus, or at least as the square-jawed heroes of the wire thrillers, taking hold of the wire and annihilating space. Readers of the wire thrillers were invited to imagine themselves as men of action at a distance, ready to seize the wire and bend it to their will. Readers of AT&T publicity were encouraged to do the same.

System Over Section

AT&T's boosters and promoters confronted the anxieties the wire thrillers betrayed, addressing many of the subjects the pulps' authors and readers seemed to fear. Novels like Stringer's *Wire Tappers* demonstrated deep unease about the wisdom and the morality of investing and speculating, particularly by those outside the traditional circles of the stock market—"suburban toilers," in Stringer's words, "sober-minded mechanics . . . timid clerks and messenger boys and widows, even," all prone to technological infection by the "diseased lust for gain without toil."[54] AT&T responded to such suspicion by positioning itself as a solid, stable, sensible investment. Decades of advertising depicted AT&T stock as the preferred choice for the very sorts of "suburban toilers," "timid clerks," and "widows, even" that Stringer had described. Thanks to these

Figure 3.
AT&T answered the image of the octopus with visions of giant businessmen empowered by the long distance phone, ca. 1920. Courtesy AT&T Archives and History Center, San Antonio.

images—and to the regular dividends the monopoly paid out—AT&T stock was by the 1930s owned by a greater number of shareholders than that of any other major American corporation.[55]

Novels such as Packard's *Wire Devils* played on popular fears of corporate power, with depictions of "Master Spiders" and "Napoleons of commerce" using the telephone and telegraph to spin webs of influence and greed. AT&T publicity countered such images with the idea that the telephone actually made large companies more egalitarian and democratic. "The telephone arrived in time to prevent big corporations from being unwieldy and aristocratic," company boosters maintained. Because anyone could call anyone else, it was said, the telephone broke down old organizational hierarchies and made rigid chains-of-command obsolete.[56] A pamphlet commemorating the first transcontinental telephone call in 1915 described the transformation of the American economy as a happy *fait accompli*: "The nation became an organized body as it increased its use of the telephone, and there was no loss of the spirit of self-help and democracy that was its birthright."[57]

A remarkable book by Michael Pupin called *Romance of the Machine* took this rhetoric to a millennial extreme. AT&T celebrated Pupin, a physicist and engineer, as one of the fathers of its national network—he had helped to develop the loading coils that made truly long distance telephony possible. In *Romance of the Machine*, Pupin returned the favor, portraying the telephone as an instrument of technocratic utopia. "I wish to describe the romance of the telephone," Pupin wrote. He praised AT&T as "the largest and most perfectly co-ordinated industrial organization in the world." The telephone monopoly ought to be a model, he said, not only for other industries, but for the United States government and the world. AT&T was pioneering a new kind of "economic democracy." The telephone "consolidated" the nation without controlling it, and "harmonized interests" without reducing freedom. "Who can contemplate . . . the industrial democracy inaugurated by our telephone industry," Pupin asked, "without being assured that it is a joyful message of an approaching civilization which will be more just and generous to the worker than any which the world has ever seen?"[58]

Finally, and most strikingly, AT&T worked to counter fears of sectional difference with a rhetoric of national union through technology and commerce. "When the telephone was invented," AT&T proclaimed in 1915, "the United States consisted of 37 commonwealths loosely held together, each filled with energy and enterprise, but lacking in organization and efficiency of action." The arrival of the telephone changed all that. "Loose ends were gathered up. . . . [S]ocial and business methods were put on a broader and more efficient basis, and the passing of sectionalism and race feud began."[59] The telephone had conquered not just "sectionalism" but "race feud"—a bold assertion in 1915. AT&T's long-distance network "put a seal on the fact that there is no longer East and West, North and South," declared another advertisement from that year the first coast-to-coast telephone circuit was completed.[60] This was language well calculated to appeal to the white, upper-class, business-minded Americans whom AT&T saw as its most important consumers in the years of sectional "reunion" and reconciliation.

AT&T was not simply selling telephone service in the 1910s and 1920s. Ultimately, it was selling nothing less than national commercial integration. It was promoting the whole corporate transformation the American economy had undergone. "The waste and folly of competition [have] everywhere driven men to the policy of cooperation," wrote AT&T promoter Herbert Casson in his 1910 *History of the Telephone*. "Mills [are now] linked to mills and factories to factories, in a vast mutualism of industry such as no other age, perhaps, has ever known."[61] This was an audacious and successful campaign that helped to change Americans' understanding of their own economy and their nation.

There were limits to the company's boldness, however, and in the 1910s there were still lines that AT&T publicity was loathe to cross. One way in which the telephone company's representations of its long-distance network were strikingly different from the images of the wire thrillers was in the geographic orientation of that network. In the thrillers, many significant technological connections run from north to south. *The Wire Devils'* Hawk and Master Spider chase each other up and down the eastern seaboard, and the central image of *The Wire Tappers* is the Cotton King's secret telegraph line from Chicago to Washington to Savannah and New Orleans. These were powerful and not unthreatening representations of economic union and integration.

AT&T's representations of sectional interconnection, by contrast, almost invariably ran from east to west. The company vigorously publicized the first telephone call from New York to Chicago in 1892. Connections from New York to Denver were celebrated in 1912. The greatest fanfare, including dozens of commemorative publications, hundreds of demonstrations around the country, and even a lavish song-and-dance number in Florenz Ziegfield's Ziegfield Follies, was reserved for the completion of the transcontinental telephone circuit from New York to San Francisco in 1915.[62] The first telephone calls from New York to Atlanta, by contrast, or from Chicago to New Orleans, are not recorded in AT&T publicity. Though the company constructed long-distance circuits in every direction, they did not trumpet their north-south connections as they did their east-west lines. Those wires, perhaps, were still too live to touch. It was safer for AT&T to illustrate national unity as a matter of east-west communication. This could be done without raising the ghosts of sectional conflict or pressing modern questions around politics and race.

A magazine advertisement from 1913 illustrates the delicacy of AT&T's position. Under the headline "The Merger of East and West," the ad depicts two smiling men speaking on the telephone from either side of the United States (fig. 4). The text of the ad paraphrases Rudyard Kipling's "The Ballad of East and West," a story about an Indian bandit who befriends an English colonel's son. "These men were of different races and represented widely different ideas of life," the ad says. Yet "each found in the other elements of character which made them friends."[63] If that text stood alone, it might be read as remarkably progressive, offering a call for friendship and interconnection across not only sectional but racial lines. But the illustration tells another story. The two men speaking on the telephone are not of different races. Nor do they seem to come from different walks of life. Both appear to be white, business-class Americans. The easterner has a moustache and the westerner wears a hat, but otherwise, they are twins.

The Merger of East and West

"But there is neither East nor West, Border, nor Breed, nor Birth,
When two strong men stand face to face, tho' they come from the ends of the earth!"
—KIPLING.

In the "Ballad of East and West," Kipling tells the story of an Indian border bandit pursued to his hiding place in the hills by an English colonel's son.

These men were of different races and represented widely different ideas of life. But, as they came face to face, each found in the other elements of character which made them friends.

In this country, before the days of the telephone, infrequent and indirect communication tended to keep the people of the various sections separated and apart.

The telephone, by making communication quick and direct, has been a great cementing force. It has broken down the barriers of distance. It has made us a homogeneous people.

The Bell System, with its 7,500,000 telephones connecting the east and the west, the north and the south, makes one great neighborhood of the whole country.

It brings us together 27,000,000 times a day, and thus develops our common interests, facilitates our commercial dealings and promotes the patriotism of the people.

AMERICAN TELEPHONE AND TELEGRAPH COMPANY
AND ASSOCIATED COMPANIES

One Policy *One System* *Universal Service*

The telephone, the ad goes on to say, had "broken down the barriers of distance" and "made [Americans] a homogenous people." This is what AT&T offered, in words and images. This is what was to banish the octopus and take the menace out of action at a distance. Technology and commerce were altering space, connecting Americans to continent-spanning networks of information and exchange. Still, AT&T promised its customers, the telephone would not threaten lines of race and class and outlook—the people on the other end of the line would be people who looked and acted just like them. AT&T's advertising of the national telephone network combined the rhetoric of national integration with subtle assurances that everyone to be so connected was essentially the same. The company's delicate use of these anxieties suggests that it understood the mixture of attraction and repulsion that powered the wire thrillers. Americans were indeed curious and concerned about the new technologies of communication, but what truly attracted them, and repelled them, was each other.

Figure 4.
AT&T worked to counter fears of sectional difference with images of national union through technology, 1913. Courtesy AT&T Archives and History Center, San Antonio.

Conclusion

We routinely credit the railroad, telegraph, and telephone with "shrinking," "unifying," or "consolidating" the United States, but few historians have engaged directly with the wrenching reordering of space and identity this implied.[64] While modern scholars are highly alert to issues surrounding the construction of space on the personal or domestic scale, it requires some imagination to re-create the intensity of Gilded Age anxiety around the alteration of region or section. "It seems impossible," Edward Everett Hale wrote in 1903, to make modern readers understand "how far apart the States were from each other, and how little people knew each other" before the railroad, telegraph, and telephone.[65] This is only more true today. Section, region, and community were profound components of individual identity in nineteenth-century America. They were intertwined, of course, with hierarchies of race and gender, but that made them no less meaningful or evident on their own.

Historians understand that technologies of transportation and communication were essential to constructing America as a modern nation state, but we have tended to describe the "wiring of the nation" in purely functional terms. The new national networks of telephone and telegraph were critical as both tools and metaphors. They not only facilitated a more integrated national economy; they taught Americans how to imagine themselves within that economy, and gave them vivid metaphors with which to do so. "The telephone changes the structure of the brain," proclaimed Gerald Stanley Lee, a pastor turned evangelist for the wire. "Men live in wider distances, and think in larger figures, and become eligible for nobler and wider motives."[66]

The wire thrillers' combination of awe and even subjugation toward technological change with a paradoxical faith in individual agency would become the default rhetoric for talking about communication technology in the twentieth and twenty-first centuries. This was the language of the 1920s radio boom, shared by the wireless hobbyists and their ostensible corporate nemeses at RCA. This was the language of personal computing in the 1970s and 1980s and the language of internet hype in the 1990s. When Microsoft faced antitrust prosecution in the 1990s, the arguments it made and the slogans it offered could have been cribbed from AT&T in the 1910s. Theodore Vail's company said, "The telephone doors of the nation are open to you." Bill Gates's company asked, "Where do you want to go today?"[67]

High-tech adventure novels reminiscent of the wire thrillers are widely read in our own time. Twenty-first-century Americans are no less excited or uneasy than their predecessors about new communication networks and rapid technological change. Yet it is hard to imagine the wire thrillers' criticism of action at a distance, or their profound ambivalence about the legitimacy of large corporations, in the works of Tom Clancy, Michael Crichton, or Dan Brown. Electronic media are the cultural air we breathe; multinational corporations are the dominant features of our economic landscape. In embracing a transcontinental communication system, Americans came to embrace their nation's new political economy.

The first nation-spanning corporations—the great railway conglomerates, Western Union, AT&T, and others—remade the nation in their image, employing the new technologies of transportation and communication to integrate and transform the American economy. The history of the railroad may tell us how a new economic order was born in North America, but the history of the telephone best explains how that order gained wide popular support. AT&T's publicity for the long-distance telephone helped legitimize a new nation-spanning economy dominated by conglomerations of corporate power that earlier generations would have found monstrous. Gilded Age Americans feared the growth of corporate size and reach would fundamentally alter the social and material constitution of their nation. Their fears are foreign to us today, not because they did not come true, but because they did.

Notes

I would like to thank Carolyn Thomas de la Peña, Richard John, and *American Quarterly*'s editors and reviewers for their careful reading and helpful comments.

1. See, for example, "Action at a Distance," *Cassier's*, August 1912, 164–66; "Action at a Distance," *Scientific American*, January 17, 1914, 39. The term comes from the physical sciences, where the source of one object's ability to affect distant objects, as in gravitation, was a mystery going back to Isaac Newton's day. Mary B. Hesse, *Forces and Fields: The Concept of Action at a Distance in the History of Physics* (Westport, Conn.: Greenwood Press, 1970).
2. The literature on these topics is enormous. On the railroads and their impact, begin with Alfred D. Chandler, *The Visible Hand: The Managerial Revolution in American Business* (Cambridge, Mass.: Harvard University Press, 1977). But see also Gerald Berk, *Alternative Tracks: The Constitution of American Industrial Order, 1865–1917* (Baltimore: Johns Hopkins University Press, 1994); William G. Roy, *Socializing Capital: The Rise of the Large Industrial Corporation in America* (Princeton, N.J.: Princeton University Press, 1997). On the telegraph, see Robert L. Thompson, *Wiring a Continent: The History of the Telegraph Industry in the United States, 1832–1866* (Princeton, N.J.: Princeton University Press, 1947); Richard B. DuBoff, "The Telegraph and the Structure of Markets in the United States, 1845–1890," *Research in Economic History* 8 (1983). On the telephone, see John Brooks, *Telephone: The First Hundred Years* (New York: Harper and Row, 1976); Claude S. Fischer, *America Calling: A Social History of the Telephone to 1940* (Berkeley: University of California Press, 1992); and Richard R. John's forthcoming monograph on the history of telecommunications in America.
3. See, for example, Carolyn Marvin, *When Old Technologies Were New: Thinking about Electric Communication in the Late Nineteenth Century* (New York: Oxford University Press, 1988); David Nye, *Electrifying America: Social Meanings of a New Technology* (Cambridge, Mass.: MIT Press, 1990); Linda Simon, *Dark Light: Electricity and Anxiety from the Telegraph to the X-Ray* (Orlando, Fla.: Harcourt, 2004).
4. Grace Elizabeth Hale, *Making Whiteness: The Culture of Segregation in the South, 1890–1940* (New York: Pantheon Books, 1998), 43–51, 121–98.
5. Marvin, *When Old Technologies Were New*, 67–108; Michèle Martin, *Hello, Central? Gender, Technology, and Culture in the Formation of Telephone Systems* (Montreal: McGill-Queen's University Press, 1991), 140–67.
6. For discussion of the phrase see Leo Marx, *The Machine in the Garden: Technology and the Pastoral Idea in America* (New York: Oxford University Press, 1964), 194–96.
7. Thomas L. Haskell, *The Emergence of Professional Social Science: The American Social Science Association and the Nineteenth-Century Crisis of Authority* (Chicago: University of Illinois Press, 1977), 15, 40; Robert H. Wiebe, *The Search for Order, 1877–1920* (New York: Hill and Wang, 1967); Richard R. John, "Recasting the Information Infrastructure for the Industrial Age," in *A Nation Transformed by Information: How Information Has Shaped the United States from Colonial Times to the Present*, ed. Alfred D. Chandler and James W. Cortada (New York: Oxford University Press, 2000).
8. On the social construction of distance and scale, see Henri Lefebvre, *The Production of Space*, trans. Donald Nicholson-Smith (Cambridge, Mass.: Blackwell, 1991); Sallie A. Marston, "The Social Construction of Scale," *Progress in Human Geography* 24.2 (June 2000): 219–43.
9. American studies and the history of technology share some common classics in this vein, beginning with Marx's *Machine in the Garden*. On parallel developments in American studies and the history of technology, and Marx's place in both fields, see Jeffrey L. Meikle, "Reassessing Technology and Culture," *American Quarterly* 38.1 (Spring 1986): 120–26; Jeffrey L. Meikle, "Classics Revisited: Leo Marx's *The Machine in the Garden*," *Technology and Culture* 44.1 (January 2003): 147–59.
10. *The Story of a Great Achievement: Telephone Communication from Coast to Coast* (New York: American Telephone and Telegraph Company, 1915), 10.
11. Herbert N. Casson, *The History of the Telephone* (Chicago: A. C. McClurg & Co., 1910), 141–42.
12. Simon, *Dark Light*, 3.
13. State of New York, *Report of the Joint Committee of the Senate and Assembly Appointed to Investigate Telephone and Telegraph Companies* (Albany, N.Y.: 1910), 398; AT&T advertisement quoted in Fischer, *America Calling*, 62. Emphasis in original.
14. Claude S. Fischer, "'Touch Someone': The Telephone Industry Discovers Sociability," *Technology and Culture* 29.1 (January 1988): 32–61; Fischer, *America Calling*; Marvin, *When Old Technologies Were New*; Martin, *Hello, Central?*

15. "Telephones for the Millions," *McClure's*, November 1914, 45–55; James M. Herring and Gerald C. Gross, *Telecommunications: Economics and Regulation* (New York: McGraw-Hill, 1936), 213.
16. Benedict Anderson, *Imagined Communities: Reflections on the Origin and Spread of Nationalism*, rev. ed. (New York: Verso, 1991), 3–7.
17. Useful studies of machine metaphors in the late nineteenth and early twentieth centuries include Cecelia Tichi, *Shifting Gears: Technology, Literature, and Culture in Modernist America* (Chapel Hill: University of North Carolina Press, 1987); Mark Seltzer, *Bodies and Machines* (New York: Routledge, 1992); Jon Agar, *The Government Machine: A Revolutionary History of the Computer* (Cambridge, Mass.: MIT Press, 2003); Joel Dinerstein, *Swinging the Machine: Modernity, Technology, and African-American Culture Between the World Wars* (Amherst: University of Massachusetts Press, 2003).
18. Andrew Lang, "Telephones and Letter-Writing," *Critic*, June 1906, 507–8.
19. The novels discussed in detail here are Frank L. Packard, *The Wire Devils* (London: Hodder and Stoughton, 1918); Arthur Stringer, *The Wire Tappers* (Boston: Little, Brown, 1906); Arthur Stringer, *Phantom Wires* (Boston: Little, Brown, 1907). Dozens of others follow nearly identical lines.
20. Stringer, *The Wire Tappers*, 263.
21. On reading popular fiction, see Janice Radway, *Reading the Romance: Women, Patriarchy, and Popular Literature* (Chapel Hill, N.C.: University of North Carolina Press, 1984).
22. Stringer, *Phantom Wires*, 291.
23. Stringer, *The Wire Tappers*, 112.
24. "The Telephone Girl Again," *Electrical Review*, August 10, 1889.
25. Mark Twain, *A Connecticut Yankee in King Arthur's Court*, Norton Critical ed. (New York: W. W. Norton, 1982).
26. Examples include "Flirting over a 'Phone," *New York World*, December 27, 1884; "Pete's Pipe Lines," *Telephone Magazine*, December 1902, 274–76; "Cupid Cripples Muncie 'Central,'" *Telephony*, August 1904, 123; "To Marry Telephone Girl," *New York Times*, April 5, 1905, 1; Sylvester Baxter, "The Telephone Girl," *Outlook*, May 26, 1906, 231–39; Wilbur Hall, "A Little Service, Please," *Saturday Evening Post*, April 10, 1920, 18, 155–58.
27. Jeffrey Sconce, *Haunted Media: Electronic Presence from Telegraphy to Television* (Durham, N.C.: Duke University Press, 2000), 21–58.
28. Stringer, *The Wire Tappers*, 141.
29. Stringer, *Phantom Wires*, 242.
30. Ibid., 247.
31. Frank Norris, *The Octopus: A Story of California* (Cambridge, Mass.: Riverside Press, 1958), 72.
32. This observation is made by Richard White in his forthcoming book on North American railroads.
33. Norris, *The Octopus*, 395–96. These passages are also cited by Stephen Kern.
34. Haskell, *The Emergence of Professional Social Science*, 15–40; Stephen Kern, *A Cultural History of Causality: Science, Murder Novels, and Systems of Thought* (Princeton, N.J.: Princeton University Press, 2004), 208–10.
35. Packard, *The Wire Devils*, 122–23.
36. Stringer, *The Wire Tappers*, 43.
37. Josiah Flynt, "The Telegraph and Telephone Companies as Allies of the Criminal Pool-Rooms," *Cosmopolitan*, May 1907, 50–57.
38. Flynt, "The Telegraph and Telephone Companies," 52.
39. Stringer, *The Wire Tappers*, 168, 249.
40. E. J. Hobsbawm, *Industry and Empire: An Economic History of Britain since 1750* (London: Weidenfeld and Nicholson, 1968), 56.
41. Lawrence B. Glickman, "Buy for the Sake of the Slave: Abolitionism and the Origins of American Consumer Activism," *American Quarterly* 56.4 (December 2004): 889–912.
42. Sven Beckert, "Emancipation and Empire: Reconstructing the Worldwide Web of Cotton Production in the Age of the American Civil War," *American Historical Review* 109.5 (December 2004): 1405–38.
43. On whiteness as silence, see Hale, *Making Whiteness*, xi–xii. On race and the telephone, see Venus Green, *Race on the Line: Gender, Labor, and Technology in the Bell System, 1880–1980* (Durham, N.C.: Duke University Press, 2001).
44. John Hope Franklin, *Reconstruction: After the Civil War* (Chicago: University of Chicago Press, 1961); David W. Blight, *Beyond the Battlefield: Race, Memory, and the American Civil War* (Amherst: University of Massachusetts Press, 2002); Hale, *Making Whiteness*.

45. Stringer, *The Wire Tappers*, 237.
46. Ibid., 217, 238.
47. Casson, *The History of the Telephone*, 195.
48. Roland Marchand, *Creating the Corporate Soul: The Rise of Public Relations and Corporate Imagery in American Big Business* (Berkeley: University of California Press, 1998), 48–87.
49. *Telephone Almanac* (New York: American Telephone and Telegraph Company, 1928), n.p.
50. Theodore N. Vail, *Views on Public Questions: A Collection of Papers and Addresses* (New York: privately published, 1917), 344.
51. Vail, *Views on Public Questions*, 16, 99; Casson, *The History of the Telephone*, 233.
52. James E. Caldwell to Cumberland Telephone and Telegraph Company Stockholders, December 27, 1911, Historical Collections, Harvard Business School.
53. AT&T advertisement, *Life*, December 17, 1914, 1137; Herbert N. Casson, "The Future of the Telephone," *World's Work*, May 1910, 12903–18.
54. Stringer, *The Wire Tappers*, 43, 237.
55. Marchand, *Creating the Corporate Soul*, 75–77.
56. Casson, *The History of the Telephone*, 206.
57. "Coordinating the Nation," *Telephone Review* (Supplement), January 1915, 24.
58. Michael Pupin, *Romance of the Machine* (New York: Charles Scribner, 1930), 77–81.
59. "Coordinating the Nation," 24.
60. *The Story of a Great Achievement*, 14–15.
61. Casson, *The History of the Telephone*, 181.
62. *The Story of a Great Achievement*; John Mills et al., "A Quarter-Century of Transcontinental Telephone Service," *Bell Telephone Quarterly*, January 1940, 2–82. On the Ziegfield Follies, see Dinerstein, *Swinging the Machine*, 185–87.
63. AT&T advertisement, *Telephone Review*, August 1913, inside front cover.
64. One exception is Sarah H. Gordon, *Passage to Union: How the Railroads Transformed American Life, 1829–1929* (Chicago: Ivan R. Dee, 1996).
65. Edward Everett Hale, *Memories of a Hundred Years*, vol. 1 (New York: Macmillan, 1903), 230.
66. Gerald Stanley Lee, *Crowds: A Moving-Picture of Democracy* (Garden City, N.Y.: Doubleday, Page, 1913), 19, 65.
67. For analysis of this rhetoric involving several generations of communication technology, see Susan J. Douglas, *Inventing American Broadcasting: 1899–1922* (Baltimore: Johns Hopkins University Press, 1987); Robert W. McChesney, *Telecommunications, Mass Media, and Democracy: The Battle for the Control of U.S. Broadcasting, 1928–1935* (New York: Oxford University Press, 1993); Paulina Borsook, *Cyberselfish: A Critical Romp through the Libertarian Culture of High Tech* (New York: Public Affairs, 2000).

Technology and Below-the-Line Labor in the Copyfight over Intellectual Property

Andrew Ross

Kurt Vonnegut published his first novel, *Player Piano* (1952), at a time of high, dystopian anxiety about the abuse of technology by the state and by industrialists alike. The novel—which depicts an unsavory future in which new technologies make everyone's skills obsolete—dutifully channeled these public concerns. Player pianos do not really figure in the novel, but the title was an explicit allusion to their contribution, historically, to the technological disemployment of musicians. Nor was Vonnegut the only writer of his generation to draw attention to the mother of cultural automation. The threat posed to artists' livelihoods by the mechanical player piano was also shared by William Gaddis, who developed a lifelong obsession with the technology.[1]

It is worth recalling briefly how and why the pianola, which had a short-lived but legally significant career, should have earned such a reputation as the original sinner. Its fin-de-siècle development was arguably the first salient example of an industrial technology designed, in large part, to cut the costs of creative labor. The subsequent pianola boom came at a time when the American Federation of Musicians (AFM) had scored some significant successes in negotiating wage scales and other conditions for its members. Indeed, the union's bristling response to this new technology marked the beginning of the AFM's long struggle against the automation of the jobs of live performers. By 1909, an estimated 330,000 of the pianos produced in the United States were mechanized, and by 1916, 65 percent of the market was still monopolized by player pianos.[2] The roll industry, which serviced the boom, had become one of the chief factors driving the music industries as a whole. While it was promoted as a great equalizer (create your own music in the home!), the pianola met the industry's need to find a less durable consumer product than the standard piano. Aside from the instrument's direct threat to live performers, the production of the player rolls created a low-wage manufacturing industry that offered compensatory factory-style employment to the displaced performers and others who could not find work in vaudeville or in one of the many traveling dance orchestras of the time. As a result, the work of pianists was imperiled and degraded on all sides.

The pianist workforce took further hits with each new commercial technology for recording or broadcasting performances. While the advent of silent movies provided employment for piano accompanists in the theaters, the sound film process introduced by Vitaphone and the use of canned music in motion pictures would put them and thousands more movie and theater pit musicians out of work.[3] Jukeboxes and other uses of phonographs took a further toll. In the space of two decades, pianists who had been the mainstay of virtually all commercial and domestic entertainment were reduced to bit parts in the Fordist assemblies of orchestras and big bands. By midcentury, the piano was more ubiquitous in households as an item of furniture than as an active complement to the hearth. It is fair to say, in keeping with the spirit of Vonnegut's title, that the pianola set in motion a machinery of disemployment that continues to transform the craft of music making to this day.

But the player piano is more likely to be remembered, and cited, today as a key case study in copyright law. Pianists, after all, were not the only group whose livelihoods were threatened by the mercurial rise of these machines. Their use also deprived composers of profits from sheet music sales. Congress was asked to adjudicate whether the pianola companies had to pay copyright holders for permission to play their content. In their landmark decision of 1909, the legislators resolved that whoever wanted to record the music, and make subsequent copies of it, had to pay for the content, though not at a price set by the holder. Instead, the fee paid to the composer or the relevant copyright holder was set by law (at two cents for each copy).

This is a favorite dispute for scholars of intellectual property (IP) to revisit, because its resolution set the precedent not only for ASCAP's licensing and royalty system, but also for the regulation of the radio and cable TV industries, and may yet prove to be a viable model for regulating the use of peer-to-peer file-sharing technologies. For the most part, however, legal scholars' accounts of the case—Lawrence Lessig's treatment, in his book *Free Culture*, is a good example[4]—have nothing to say about the human piano players whose livelihoods were radically affected both by the mechanization of piano playing and by the congressional ruling. The only "musicians" who figure are the composers whose full authorial rights were being compromised by the industrial piracy of the day.

There is much to be learned from this exclusion. Constitutional scholars and First Amendment activists have assumed a natural leading role in the battle against corporate IP monopolies, but the history of creative property and its relationship to technology cannot be left in the hands of law professors to write, nor should it be. Too much is left out, if only because legally-minded coverage of IP disputes tends to revolve exclusively around the interests of claimants: creators, copyright holders, or the general public of users and consumers. The state also figures in these accounts, because its judges and legislators have to decide not only whose interests will prevail in the resolution of disputes, but also how to weigh factors that advance national interests such as high-tech innovation, symbolic prestige, or the IP export trade that garners revenues from other countries.

By contrast, there is little room for those without an immediate legal stake in the disputes. Legal analysts of landmark cases rarely have anything to say about the multitude of jobs and livelihoods affected by the judicial treatment of IP-based assets and new technologies. Not only does this offend our sense of cultural and social history; it also weakens our capacity to understand, and react to, the vast changes occurring today as a result of the technology-driven IP property grab that has resulted in an aggressive expansion of copyright, patent, trademark, or publicity rights. In this essay, I will try to show why labor issues should be a more obligatory component of public debates about new technologies and the so-called corporate enclosure of the "information commons."[5] We ought to acknowledge that efforts to regulate or propertize new technologies have the potential to drastically alter the landscape of work. Yet these consequences tend to go unexamined, whether in case analysis or in the realm of public opinion making, where libertarian concerns about the freedom of consumer choices hold sway to the detriment of attention to labor issues.

It is easy to see why the libertarian response has taken precedence. Since the profits of IP monopolists depend on the creation of information scarcity, corporations such as Time Warner, Microsoft, and MGM have declared all-out war on innovative technologies that can reproduce and disseminate information to users at a cost approaching zero. Consequently, these IP bullies are perceived as blocking our rights to information that "wants to be free." Yet the historical ironies evoked by this assault speak directly to how labor has been discounted in the race to propertize. Consider that the legal vehicle for the new property grab is an expanded version of the limited monopoly rights granted to authors under eighteenth-century copyright laws so that they could pursue an independent living in the marketplace of ideas. Because U.S. law permits corporate entities to be artificial persons, most of the "authors" seizing the copyright and patent claims in the twenty-first century are global firms in multimedia, IT, and biotechnology. Likewise, the technologies under attack—file sharing and other peer-to-peer programs, de-encryption tools for picking digital locks, and each successive generation of reverse-engineering techniques for overriding proprietary measures and "improving" original products—are the brainchildren of the kind of whiz kid innovators that patent laws were initially intended to encourage and assist. The early beneficiaries of patent grants, like their writer peers, were also breaking free of the rigid patronage of monarchs or states to make their own way in the industrializing world.

In the Lockean tradition, property rights have retained a formal, if distant, association with the labor for which such rights are understood to be a reward. In the case of intellectual property, the attachment is ever more tenuous. Legal scholars have explained why entitlement in IP disputes is limited to a relatively small number of economic actors who are close to the claim of being "authors" of the creative property in question.[6] But if the impact of these disputes on the means of production is as profound as some commentators have described, then clearly an infinitely greater slice of the workforce has a legitimate interest in their resolution. Can the claims of those larger constituencies be represented in any adequate way in the current legal wrangling over IP? If the answer is no, then what can scholars and activists do to highlight and remedy this neglect?

The overwhelming evidence from IP law suggests that American courts have little interest in thinking outside of the box of singular authorship. They will not recognize the potentially legitimate IP claims of participants in the kind of collective creative work that is the norm in the culture, IT, and other knowledge-intensive industries, and they have even less interest in hearing the argument that the true source of most creative works is the public domain itself. Instead, judges are increasingly fixed on assigning monopoly rights (and lots of them) to single, indivisible "authors," who are more than likely to be corporate entities. As several scholars have observed, the courts have invested more and more exclusive rights and privileges in the category of proprietary authorship at a time when cultural critics have been doing exactly the opposite—dissolving the Romantic mystique that supports any such notions about the extraordinary rights of creative geniuses.[7] The state has obliged by passing punitive legislation to protect these privileges.

In the court of public opinion, corporate IP warriors can always win points by broadcasting the claim that they are defending the labor rights of vulnerable artists. However, the historical record and the experience of working artists today confirm that the struggling proprietary author has always been more of a convenient fiction for publishers to exploit than a consistent beneficiary of copyright rewards. Culture industry executives who are able to masquerade as the last line of protection for artists when they are systematically stripping them of their copyrights are well set up to wave away claims on their IP assets from broader constituencies.

By contrast, what vision of labor has been put forth by the opposition forces in their public campaigns to raise the alarm about IP monopolies? Liberal advocates of the public domain who argue for a "free culture" (free as in "free speech" not "free beer") have petitioned for the fullest rights of access to information for users and consumers, while continuing to recognize copyright as a valid way of ensuring that individual creators receive their moral dessert.[8] As the foremost public domain proponent, Lessig has compared this campaign for "free culture" to the antebellum free labor movement that fought against chattel and wage slavery alike.[9] It is a loose analogy, and so perhaps it is not entirely fair to observe that this preindustrial ideal of self-reliant artisans—who wanted to sell their products, not their labor—is hardly the most practical response to the broad reality of the hierarchical divisions of labor that knowledge industries command today. On the other hand, it is an ideal that speaks to those whose labor rank puts them closest (but no cigar) to the entitlements due to "authors." For this thwarted class fraction of high-skilled and self-directed individuals, whose entrepreneurial prospects are increasingly blocked by corporate monopolies, the analogy rings true enough.

In loose alliance with the public domain advocates like Lessig are the widely networked ranks of high-tech workers and cognoscenti who rally behind the umbrella term FLOSS, or FOSS (Free/Libre Open Source Software).[10] The production credo of these workers, who are opposed to most proprietary restrictions on the use of information and information technologies, is cooperative in nature, with deep roots in the hacker ethic of communal shareware. The labor ethos of these IT communities leans heavily on volunteerism and mutual support. Because they are generally ill-disposed to state intervention, FLOSS engineers, programmers, and their advocates have not explored ways of providing a sustainable infrastructure for the gift economy they tend to uphold. Nor have they made it a priority to speak to the interests of less-skilled workers who lie outside of their ranks. For the most part, labor-consciousness among FLOSS communities (whether in the relatively distinct "free software" or "open source" subcultures) seems to rest on the confidence of members that their expertise will keep them on the upside of the technology curve that protects the best and brightest from proletarianization. There is little to distinguish this form of consciousness from the guild labor mentality of yore that sought security in the protection of craft knowledge.

Neither the public domain advocates nor the FLOSS evangelists have actively considered the consequences of IP disputes for the mass of workers and employees who do not come close to the legal category of copyright/patent holder. It is odd that such labor concerns have not been more on the agenda. Consider the volume of public anguish expended on the recent "jobless recovery" or on the impact of skilled-labor outsourcing. Ranking politicians have reserved some of their most heated rhetoric, though not their fullest legislative powers, for the purpose of stemming job loss, especially in IP-based industries regarded as strategically important for the national interest. But this backdrop has not insinuated itself very far into the IP wars. The crusade against the IP monopolists continues to be dominated by strains of techno-libertarianism that lie at the doctrinal core of the "information society," obscuring the labor that built and maintains its foundations, highways, and routine production. The result? Voices proclaiming freedom in every direction, but justice in none.[11]

Today's contest over technology-driven copyrights and patents cannot be only about protecting the claims of top-flight knowledge workers, or safeguarding the future of technological innovation, or guaranteeing consumer access to a rich public domain of information. The outcome has far-reaching consequences for the global reorganization of work, and we need to subject these consequences to a serious line of inquiry. Otherwise, it will be safely concluded that the IP wars are simply an elite "copyfight" between capital-owner monopolists and the labor aristocracy of the digitariat (a dominated fraction of the dominant class, as Pierre Bourdieu once described intellectuals) struggling to preserve and extend their high-skill interests.

The Acquisition Race

Though the idea of intellectual property has been around for much longer, IP entered the lexicon of state and corporate bureaucracies only after the 1970 founding of the World Intellectual Property Organization (WIPO). Hitherto a relatively stable niche of U.S. property law, IP legislation began to proliferate after the 1976 revision of the 1790 Copyright Act. No doubt this development reflected the consensus of the nation's economic managers that IP-driven technology growth was becoming the primary industrial asset of the United States.[12] Though it was soon a leading factor in the balance of trade, weighted on the export side by the copyright-based and patent-rich industries of information, media, entertainment, software, and high-value manufacturing, the concept of IP did not fully enter public currency until the 1990s. The 1998 passage of the Sonny Bono Copyright Term Extension Act and the Digital Millennium Copyright Act brought the problem of excessive IP protection to the attention of a wide range of public interest groups. Finally, after 2003, when the recording industry's zero-tolerance crusade against Napster, its clones, and their users hit the courts, the term became all too familiar to the hundreds of millions engaged in online file sharing.

The corporate clampdown on the ubiquitous practice of downloading music and other entertainment products was a sobering initiation for many into the tawdry reality of the IP grab. As a result, everyone has a horror story to tell. There's the one about ASCAP suing the Girl Scouts for singing some of its members' songs around the campfire; George Clinton being sued for singing some of his own songs without permission from the copyright owner of his back catalog; the "Happy Birthday" song, now owned by Time Warner, and restricted to licensed uses until 2030; or the ultimate in class betrayal perpetrated by the litigious copyright owners of Woody Guthrie's "This Land Is Your Land," and the Fourth International's flagship song, "The Internationale." Beyond the music ghetto, things only get more surreal; Donald Trump has trademarked the expression "You're Fired," along with his accompanying hand gesture from the TV reality show *The Apprentice*.[13]

These stories now belong to the demonological archive of consumer folklore. But the truly chilling ones apply to the lifeworld itself, where multinationals like Syngenta, AstraZeneca, DuPont, Monsanto, Merck, and Dow are engaged in a cutthroat race to patent seeds, livestock, plant genes, and other biological raw materials that have been the basis of subsistence farming in the developing world for centuries.[14] The corporate privatization of biodiversity is a colossal act of plunder, infinitely more damaging to the basic income and health of mass populations than the petty piracy of movies in developing countries is to those who work in the Hollywood entertainment system. Neoliberal pillage of nature and indigenous knowledge is an imminent threat to food security and livelihoods across the global South.

From the perspective of countries with few IP assets, the demand, on the part of rich nations, to respect and protect the IP rights of foreign multinationals is little different from the traditional imperial call on a vassal to pay tribute. Nor, as an economic arrangement, is it much of a departure from the colonial pattern by which the periphery supplied raw materials to be processed and branded in the core. Today, the materials come in the form of traditional knowledge—seeds, folklore, healing remedies—and are converted into IP by the likes of Monsanto, Disney, and Pfizer. Resistance to this arrangement surfaced at a stormy April 2005 meeting of WIPO in Geneva. A bloc of global South nations, including key players such as Brazil, Argentina, and India, flatly challenged the efforts of the rich countries to continue imposing IP rights protection through global trade treaties. According to the statement submitted by India, "no longer are developing countries prepared to accept this approach, or continuation of the status quo." The fighting words continued, albeit in the jargon of progressive policy wonk diplomacy: "Given the huge North-South asymmetry, absence of mandatory cross-border resource transfers or welfare payment, and absence of domestic recycling of monopoly profits of foreign IP rights holders, the case for strong IP protection in developing countries is without any economic basis. Harmonization of IP laws across countries with asymmetric distribution of IP assets is clearly intended to serve the interest of rent seekers in developed countries rather than that of the public in developing countries."[15]

The rationale behind the uprising was plain enough. Why should poor countries spend their scarce resources on IP policing operations for foreign multinationals? They see none of the benefits of Sony, Bertelsman, Microsoft, or Aventis's profits, nothing in the way of technology transfers, and precious little that could be viewed as a development asset.[16] By contrast, their underground pirate economies do a passable job of providing much-needed drugs, software, consumer technologies, seeds, and all manner of cultural products at affordable prices, and at cost margins that filter into the pockets of local producers and distributors. Piracy, from this viewpoint, is just another name for catering to community needs and staving off predatory outsiders.

The 1960s saw a similar revolt against the Western copyright laws embodied in the multilateral Berne Convention of 1884 and largely written to benefit the major IP exporters. The break, conceived by African nations at a Brazzaville meeting in 1963, resulted in the Stockholm Protocol Regarding Developing Countries. Like the Development Agenda put forth in Geneva this year, it was vigorously opposed by Washington. The world's leading pirate nation for two centuries, the United States had lately become a net exporter, and though it would not become a full Berne signatory until 1989, it was beginning to flex its muscles as a global IP bully. The outcome of the wrangling—the Paris revisions of 1971—preserved intact the broad international membership of Berne but relaxed restrictions on IP uses for scholarship, teaching, and research in developing nations.[17] The outcome of the 2005 insurgency remains to be seen, but copyright powers in the North will be less likely to agree to concessions on educational materials than they were thirty-five years ago. In the intervening years, higher education has become a key site of capital accumulation in the knowledge economy.

Academics don't have to hail from Africa or India to see the evidence in their own workplace. The chilling effects of the IP clampdown extend into every corner of campus. Institutions increasingly claim ownership of traditional academic works—from syllabi and courseware to published research—that had hitherto been assigned to the independent copyright jurisdiction of their faculty creators. Now these materials are increasingly regarded as "works for hire," prepared by employees in the course of fulfilling their contracts, in much the same way as an industrial corporation asserts ownership of its employees' ideas and research.[18] Well-established trends confirm that the research university is behaving more and more like an adjunct to private industry; the steady concentration of power upward into managerial bureaucracies, the abdication of research and productivity assessment to external assessors and funders, the pursuit of intimate partnerships with industrial corporations, the pressure to adopt an entrepreneurial career mentality, and the erosion of tenure through the galloping casualization of the workforce.[19] From the perspective of increasingly managed academic employees, the result is systematic de-professionalization; the value of a doctoral degree has been degraded, while new divisions of labor have emerged that are corrosive to any notion of job security or peer loyalty.[20]

As Clark Kerr once prophesied, academics are now more like "tenants" than "owners" of their university institutions, but today's university is not quite the high-tech "knowledge factory" that he, and his critics, described.[21] The research academy—with its own bulging portfolio of patents, copyrights, trademarks, and corporate funding contracts—is undoubtedly a conduit for capitalizing and transmitting knowledge to the marketplace, but it is also an all-important guardian of the public domain. As Corynne McSherry points out in *Who Owns Academic Work?*, the academic workplace is characterized by a tension that lies at the heart of knowledge capitalism. As the academy increasingly hosts property formation and incorporates the customs of the marketplace, ever greater care must go into maintaining its function as a guarantor of truth and unreservable knowledge.[22] This is not just window dressing, or money laundering. Without an information commons to freely exploit, knowledge capitalism would lose its primary long-term means of reducing transaction costs. Nor, if all knowledge were propertized, could faculty entrepreneurs poach on the community model of academic exchange to advance their own autonomy and status as knowledge owners. Consequently, the traditional academic ethos of disinterested inquiry is all the more necessary not just to preserve the symbolic prestige of the institution but also to safeguard commonly available resources as free economic inputs, in much the same way as manufacturing, extractive, and biomedical industries all depend on the common ecological storehouse for free sources of new product.

High-Tech IP and Outsourcing

Though the academy is the natural home of this tension, its side effects are familiar to all knowledge professionals who enjoy a degree of autonomy in their workplace. This is because the collegiate model of the self-directed thinker has steadily migrated to knowledge-intensive industries, where no-collar employees emulate the work mentality and flexible schedules of disinterested research academics on corporate "campuses" or in surveillance-free work environments. Arguably, the diffusion of this temperament is much better evidence of the character of knowledge capitalism than are departmental water-cooler tales about the corporatization of universities.

As the knowledge and work customs of the academy infiltrate the high-tech corporate world, they are employed to extract IP-rich value from employees in ways that were impossible in more traditional, physically bounded workplaces. In return for ceding freedom of movement to workers along with control over their schedules and work initiatives, employers can claim ownership of ideas that germinate in the most free times and locations of their employees' lives.[23] New mobile technologies aimed at ubiquitous computing and telecommunications have directly facilitated employers' annexation of that free time. In principle, employers can now harvest IP returns from their employees anytime, anywhere. With the advent of globally networked technologies, the value collecting has extended its reach even further.

This new geographical scope has opened the way to a wave of outsourcing of skilled work, which cuts costs drastically, and, just as important, imposes labor discipline on each end of the transfers. Under pressure to hold on to their hard-earned skills, onshore employees struggle to keep their jobs above the red line, while their offshore counterparts are warned that their new jobs could move to a cheaper location at any time. The process of outsourcing, moreover, depends on an implicit understanding that the skills and every other facet of the work being migrated are the intellectual property of the employer. IP, in this context, is much more than technology-driven legal entities such as patents, copyrights, and trademarks. It is the whole range of assets—processes, techniques, methodology, and talent—that are required to operate and make use of technologies and that business analysts often refer to as "intellectual capital."

In the course of researching my recent book *Fast Boat to China*, I tried to trace some of the operations involved in "knowledge transfer," the favorite corporate euphemism for the outsourcing of skilled work.[24] This process involves extracting knowledge and skills from the heads of onshore employees and moving these assets to replacements in a cheaper part of the world. Global work-flow platforms and other business process technologies have been developed to gather, coordinate, and reintegrate the results from far afield. These technologies, and the modular units of information that they move around, are designed to minimize IP leakage or theft. Naturally, this process of knowledge diffusion and recomposition is a much more complex and fraught logistical operation than shipping out machinery to set up a plant offshore. But with the requisite technologies, global firms are now in a position to shift around bits and pieces of ever more skilled occupations at will. Perhaps only science fiction can help us imagine the kinds of technologies that might be developed to ensure the truly efficient extraction of knowledge. But the basic steps in this process are already considered routine in many multinational corporations.

In high-tech industries, where job-hopping is endemic among valued employees, managers have learned to build into the cost structure the risk of workers walking off with the company's IP. Yet the perception that they are the true, "dispossessed" authors of corporate IP helps to explain why engineers in these sectors are often the most zealous participants in FLOSS projects during their downtime.[25] In FLOSS's cooperative nonproprietary mode of production, they do not see the product of their labor as alienated. More to the point, free or open source software is a product that reflects their class consciousness; it is a flattering tribute to their collective labor, and the philosophical zeal for it to be used by everyone, with only minimal restrictions, endows the claim to universality to which any rising class must aspire.

For that reason alone, the much-lionized history of shareware and its maturation into the dual-track ethos of free software and open source can be seen as the narrative of a distinctive class fraction—a thwarted technocratic elite whose libertarian worldview butts up against the established proprietary interests of capital-owners.[26] While they see their knowledge and expertise generating wealth, they chafe at their lack of control over the property assets. Their willingness to work against the proprietary IP regime is directly linked to their entrepreneurial-artisanal instincts, but, more important, it is a power-test of their capacity to act upon the world. The class traitors in their midst are engineer innovators who go over to the dark Gatesian side of IP monopoly enforcement.

But what about those further down the entitlement hierarchy, who are not direct participants in this power struggle, and whose prospects in the chain of production do not extend to the profile of the master craftsman straining at the corporate leash? They are much more distant from the rewards of authorship, and are less likely to feel personally disrespected when IP rights are expropriated from above. When their jobs are outsourced, they are simply told to retrain, or seek occupational niches that are secure from flight. (Middle-class parents may soon hope their children grow up to be plumbers.) Alternatively, if they belong to unions, say, in the copyright industries (one of the few union strongholds in the private sector), their affiliates may find it strategic, for the purposes of job protection, to side with employers engaged in the punitive clampdown against IP infringement.

In any event, their interests do not coincide with the highly skilled auteurs-manques or with the general public, the other claimant to partial ownership of IP rights. Adjunct teachers in the academic workforce are a good example. Full-time, tenured faculty, whose claim to authorial status is relatively strong, barely regard them as colleagues, rarely speak on their behalf, and are disinclined to oppose initiatives that clearly involve institutional expropriation of faculty IP rights but that affect adjuncts disproportionately. This is surely one of the reasons for the largely unobstructed growth of remote learning programs and private, for-profit, online institutions (made possible by Internet technologies) such as the University of Phoenix, Walden University, Kaplan University, Westwood College, and DeVry University.[27] Lack of full-timer opposition also explains the steady march to outsource writing, or other "remedial," programs from four-year institutions to the underpaid staff at extensions or in two-year community colleges.[28] It is also driving the overseas expansion of American collegiate brands, in the form of a "global campus" system, which, ultimately, will adopt some of the fiscal logic of global corporations, balancing the onshore budget against a network of offshore units.

Such initiatives are aimed at cutting teaching labor costs and establishing control over curricular materials and rights. Except at the height of the dot-com boom, when digital technology fever penetrated even the fantasies of Ivy League administrators, full-timers have generally viewed such developments as a threat only to those who do not share their own guild privileges. Even so, contingent faculty constitutes strong, articulate voices, and they have sought to unionize in great numbers to protect their interests.[29] Compared to the marginalized in other knowledge industries, they are taking steps to clarify their relationship to IP rights in their workplace.[30]

Their counterparts in the technology industries have a harder time making claims on IP. Consider the landmark, decade-long court case (*Vizcaino v. Microsoft*, first filed in 1993) brought against Microsoft by its "permatemps." Thousands of longtime employees who had worked alongside full-timers but who were denied benefits because they were classified as "independent contractors," sued the corporation for undercompensation. One of the biggest claims in the case revolved around their exclusion from the Microsoft Employee Stock Purchase Plan, which would have brought indirect benefits from IP assets the permatemps helped to create. Faced with several rulings that established the workers as "common-law" employees, Microsoft settled out of court in 2002 and immediately established hiring rules designed to restore the status quo ante by circumventing the new legal and tax regulations that applied to long-term serial temporary assignments. This revised policy has been widely copied throughout corporate America. Temps are now more carefully segregated within corporate culture, further distancing whatever IP-related claims they might have on the products of their labor. In addition, the permatemp case helped to spur corporate flight. Jobs hitherto assigned to pools of temporary workers were added to those of regular employees slated for overseas "knowledge transfer."

But it is the entertainment industry and its hierarchy of craft unions that offers the clearest, single example of the stratification of creative labor. Indeed, performers, writers, and directors are commonly referred to as above-the-line employees. Their unions—the American Federation of Television and Radio Artists, the Writers Guild of America, the Directors Guild, the Screen Actors Guild, and the American Guild of Variety Artists—have negotiated successfully for residuals payments, basically "royalties" from rebroadcasts or reuse of film, TV, or commercials. That these talent unions can extract such fees from the Alliance of Motion Picture and Television Producers, which represents most studios and independent producers, is the source of their strength and their relative health. By contrast, below-the-line technician employees have been hit hard by a combination of de-skilling from new technologies and runaway production to non-union locations.[31]

Here we see two sides of the impact of globalization. As the entertainment industry has expanded its ability to distribute overseas through each new technological generation of media formats, the additional residuals have brought handsome benefits to those above the line. Below the line, however, the capacity to produce overseas or in right-to-work states has decimated the livelihoods of technicians, set designers, sound engineers, cinematographers, and grips. While no one, either above or below the line, enjoys full authorial IP rights, the ability of talent to piggyback on copyright for its claim on royalties has made all the difference between the two classes of employees. Clearly, the development of the new technologies has only accentuated the uneven distribution of income that is governed by the line.

Union Resistance

In the 1930s, the American Federation of Musicians took a militant last stand against technological automation, establishing a Music Defense League to combat the use of canned music in movie theaters. But the union soon made its peace with the motion picture and other entertainment industries in the form of collective bargaining contracts. The advent of "business unionism" in these industries ensured a new intimacy between the interests of owners and their employees. Accordingly, the resistance of unionized musicians to new technologies that reduced their employment prospects now ran in tandem with the resistance of corporate owners to new technologies that undermined their control over IP.

In the annals of IP scholarship, unions, when they appear at all, are almost always portrayed in the role of antimodernizers, instinctively set against the march of progress, rather like the fuddy-duddy folkies who famously booed and pulled the plug on Bob Dylan's electrified set at the Newport Folk Festival. Let me recount just one example. In *Copyright's Highway*, Paul Goldstein's generous history of copyright, the printer and bookbinder unions are fingered as the chief lobbyists behind the blocking of efforts, in the early decades of the twentieth century, to conform the U.S. Copyright Law to the international standards of the 1884 Berne Convention.[32] For their part, these unions were defending the favorable position their members had enjoyed as the chief beneficiaries of the trade in reproducing foreign titles during the golden age of American piracy. But Goldstein's account sees the protectionism of organized labor as standing in the way of international cooperation, perceived as more enlightened. As a result of what we are then invited to interpret as union intransigency, Washington was forced to maintain its outcast status in the international copyright community for several decades.

Yet there is another way to interpret the anachronistic feel of the unions' response. It reminds us that a variety of hands were once understood to have an equivalent stake in the production process. Martha Woodmansee has shown that, in Germany as late as the 1750s, the author was still regarded as "just one of the numerous craftsmen involved in the production of a book—not superior to, but on a par with[,] other craftsmen." Under its definition of "book," a dictionary of the time lists writers alongside papermakers, type founders, typesetters, printers, proofreaders, publishers, bookbinders, gilders, and brass-workers as equal beneficiaries of "this branch of manufacture."[33] The subsequent crusade to elevate the authors' labor from that of craftsman to originator of special value was the heady product of Romantic ideology about the singularity of artistic creation. A multifaceted response to the onset of industrialization and commerce in culture, this ideology was expediently taken up to justify the generous rights extended to authors under copyright law. In the European legal tradition, these inalienable natural or "moral" rights are limited to flesh-and-blood authors and cannot be assigned to corporations. By contrast, in the American legal tradition, which seeks to balance the interests of copyright holders against the needs of consumers, real authors have no such moral standing.

It is unlikely that the printers and bookbinders who opposed the U.S. move to join Berne were acting out of some high-minded principle about the maldistribution of copyrights benefits. They were simply holding on to a good racket. But their formal claim on the trade, and the considerable influence that it carried for so long, demonstrates how union power can be used to effectively represent workers whom copyright law does not recognize as author-worthy contributors to cultural production.

In other contexts, union resistance can be a useful and persistent reminder that industrial technologies, especially those served up with a supersized helping of utopian modernization, are developed and programmed to control the labor process in every way possible. For the harried employee, new technologies in the workplace are invariably the bearers of speed-up and ever more sophisticated forms of managerial surveillance. They are packaged and introduced with the warm promise of job enrichment, but are more likely to be deployed as a way of de-skilling or disciplining a workforce.

In the case of creative technologies, this dark side is more difficult to distinguish, especially when the results are vaunted as important advances in the arts. Simon Reynolds, the pop music critic, once told me that the British musicians' union launched a campaign against the synthesizer in the 1980s, at a time when electronic dance music had begun its mercurial ascent to the status of a mass cult in European pop. The threat was clear enough, and, indeed, the subsequent reign of dance music proved to be a long, cruel season for performers of live musical instruments. In many quarters, they were as uncool and obsolescent as an eight-track tape; no one wanted to book them, especially if they came in the form of guitar bands.

As a devotee of the genre, I could certainly count myself among those who believed that its inventive use of drum machines, samplers, and sequencers ushered in a quantum leap in musical progress. Indeed, it took some convincing to persuade me that the result had anything remotely in common with the worker layoffs that came with automated factories. Yet whenever I asked no-name working musicians who depended on live club and bar bookings what they thought of DJ music, I was guaranteed to get an earful. There was no question in their minds that owners of live venues welcomed and encouraged a DJ-based economy of prerecordings or musical acts because it cut their overheads and labor costs by eliminating drummers, keyboard players, guitarists, and vocalists. Killing off live music may have been sold to fans as a worthy crusade against the pretensions to "authenticity" of the rock aristocracy, but it was also a serious labor problem.

Labor concerns were also an issue in the early hip-hop scene. High-minded advocates of vinyl-based sampling argued that it was a way of paying homage to the ancestral archive. In the black musical tradition, according to this view, ideas, phrasing, and melodies were more likely to be seen as common property than as a matter of personal ownership; a version of someone else's people's music was a tribute, not an act of plagiarism. Even after the commercial introduction of the MIDI digital interface in 1982, which transformed hip-hop into a reliable, recording industry product, the ancestor-worship theory endured as a worthy rebuttal to accusations that sampling was just a virtuoso form of theft. But sampling was just as likely to be viewed as disrespectful by the very elders who were supposed to be recipients of the homage. In addition, for every layperson's casual dismissal—"it's not real music"—there was a musician who saw DJ-based hip-hop as a threat to his or her livelihood as a contract performer. Nor did it help when leading rap producers declared war on musicians. Listen to Hank Shocklee, from Public Enemy's Bomb Squad, the most formidable crew of sound engineers working in the 1980s: "We don't like musicians, we don't respect musicians. . . . We have a better sense of music, a better concept of music, of where it's going, of what it can do."[34]

The Labor Theory of Value

If those who labor are routinely neglected in the field of IP law, the concept of "labor" has hardly been absent from it. Indeed, one of the fundamental philosophical precepts informing the field is the labor theory of value that took its most canonical form in John Locke's influential views on property. For Locke, property rights accrued to individuals as the fruits of their labor upon resources that were held in common, or that were unclaimed. For example, these property rights could be earned by improving an object of nature. This was the doctrine notoriously applied to justify the legal appropriation of commonly used native land by colonist settlers; Native Americans who had a view of land-use rights more akin to usufruct were stripped of these rights when they signed agreements bound by European laws that honored exclusive individual property rights.[35] Locke also argued that labor was a property of personhood, and that individuals had a right to own whatever they "mixed" with that labor. If the resources worked upon were held in common, and "if there is enough and as good left in common for others" after the appropriation, then the result could justly be regarded as a natural property right.

This latter proviso was a serious consideration for any attribution of physical property rights (about which Locke himself was writing), but appeared to be less consequential in the field of creative property, where ideas and facts are "nonrivalrous" goods whose value is not diminished, in principle, by their shared use by others. It was this acknowledgment, by Jefferson and others, that creative property was not like real property that led to the special provisions made for IP in the Constitution. Though it was a relatively uncontroversial observation, it was destined to be abused in a market civilization wherein monopolists depend on artificially imposed scarcity to generate wealth. No would-be monopolist will pass up the opportunity to exploit public confusion about the difference between physical and intellectual property or relinquish the invaluable, moral stigma attached to theft and piracy in order to do so. The development of IP-driven technologies has directly strengthened the hand of those in a position to profit from the outcome.

Industry gatekeepers like the MPAA's Jack Valenti, the voice of Hollywood's vested interests for several decades, could be depended on to never concede an inch on this distinction—VCR owners, for example, who copied a TV show in the home were no less felons than the stick-up artists at the local drugstore. Prone to grandstanding, Valenti often went much further. Indeed, in the 1984 case (*Sony v. Universal Studios*) that established the legality of the VCR, he testified to Congress that "the VCR is to the American film producer and the American public as the Boston strangler is to the woman home alone."

It was also Valenti who declared that the duration of copyright terms should be "forever minus a day." By perfect contrast, in a British parliamentary debate in 1841 about a proposed extension of the British copyright term, Thomas Macauley argued that any form of monopoly is an evil, and "the evil ought not to last a day longer than is necessary for the purpose of securing the good." Copyright, in his view, was "the least objectionable way of remunerating" writers.[36] It is difficult to imagine what possible labor theory of value could bridge the positions of Valenti and Macauley, and yet the indeterminate labor proposition that underpins copyright law has been able to do so.

On the one hand, Valenti and other hired corporate PR guns shamelessly cite the labor of poor struggling artists in their efforts to expand the bundle of rights assigned by IP law. Celebrity actors and musicians are needed to front the corporation's cause, though, like striking baseball players on multimillion-dollar salaries, the stars run the risk of exhausting public credibility by claiming that their livelihoods are sufficiently harmed by unlicensed access to their performances. But the risk is easily borne because the Romantic concept of the artist as a neglected genius still holds sway over a sector of the public imagination. Its staying power overrides what people know about the distinction between the creative labor of authorship and the copyrighted ownership of the product—which is primarily in the hands of a corporate entity. Indeed, invocations of the underpaid proprietary author were a charade in the eighteenth century—since most authors signed over copyright to publishers before publication—and are even more so today, when for legal purposes, the entity that paid to have the work created is regarded as the author, rather than the real author. Still, copyright's reward is a highly visible formal expression of the Lockean principle that individuals are not only naturally entitled to the fruits of their labor, but also that property is an appropriate, if not inviolable, part of the reward.

On the other hand, the moderate view of Macauley leans toward a much more limited definition of just desserts. Creative work is not a special category of labor, deserving of extensive rights to collect rents. On occasions when authors have no other means to eke out a living, then limited copyright monopolies are justified to ensure some modest reward. Though the industrial economy of his day was characterized by wage slavery, Macauley's position resonates with the liberal ethos of a society of possessive individuals. Property rights in such a society are easier to reconcile with another favorite rationale for copyright—the utilitarian injunction to maximize net social welfare. Indeed, this harmonious balancing act is the default position today of liberal opposition to the corporate IP fundamentalists. For pragmatic advocates of the public interest, the goal is to ensure just compensation for the honest labor of individuals but not at the cost of the broadest public benefit from that labor. In their view, careful copyright observance in statutes and case law was the key to preserving that balance for two centuries, and it must be restored.

Yet any demotion of creative work to an ordinary social status opens the door to other kinds of challenges. For one thing, the worth of that toil is now subject to a wider pool of critiques of the labor theory of value. Some of these are based on a more frank assessment of the relative worth of creativity. Many critics, for example, see no reason for retaining a system of individual ownership that enshrines notions of originality that are increasingly implausible and unworkable in an age of ubiquitously networked information. The abundant availability of information, ideas, and data makes it more and more transparent that "creators" really only add something to what their predecessors have thought or done. At times, what they add can be said to be truly original, but most of it is staggeringly mundane and almost entirely derivative of the public domain of ideas. In any event, this value hardly justifies granting an exclusive monopolistic property right, which can then be assigned to a multinational corporation, for up to ninety-five years, on average.

Other critiques are less ethnocentric. In non-Western societies where Anglo common law and the continental legal systems were colonial impositions, the property traditions that these codes honor are not always the best fit. Individual IP rights do not resonate well in cultures in which creativity and knowledge are more likely to be considered a collective characteristic or expression. Western efforts to impose IP regimes on China for example, have repeatedly foundered on a combination of Confucian legacies and pastoral rule on the part of the state.[37] In India, analysts like those in the Sarai new media group are striving to conceive how new forms of information networking can resonate with communal traditions to form a working basis for economic policy in developing countries.[38] Relatedly, the assumption of usufruct, rather than property, rights has been proposed as a more suitable way of handling creative works.[39]

A third source of opposition stems from the acute embarrassment generated by laws that are quite simply untenable. Mass use of peer-to-peer technology is crumbling the ground beneath existing IP laws. No legal arrangement can subsist for long when it makes outlaws of most citizens. Legislators feel uncomfortable when their laws are so out of synch with customary practice, and so, increasingly, those appointed to legitimize the social order will feel the need to find a practical substitute.

The General Intellect

Notwithstanding the travails of IP law in the age of digitization, one might fairly ask of the entire Lockean tradition why private property should be the reward of labor, of any kind. Why would we expect to own what we had mixed our labor with? Most workers, in any case, have not come anywhere near to exercising that privilege. As James Boyle notes, copyright presents us with the blatantly unfair proposition that "property is only for the workers of the word and the image, not the workers of the world."[40] Nor is there any reason why work, in and of itself, should have to be ennobled, except, of course, to legitimize it to those already burdened with loathsome toil.

Those who helped to idolize work in the nineteenth century were, by and large, middle-class intellectuals, such as Thomas Carlyle, Horace Greeley, William Ellery Channing, and, of course, Marx himself.[41] Yet it was Marx who saw most clearly that the labor theory of value, pioneered by Locke, Smith, and Ricardo, should be viewed in the context of social division of labor as a whole rather than as an explanation of individual acts of exchange, as classic liberalism prefers. Accordingly, he had a system-level response to the notion that labor earned the right to property. Only collective forms of ownership could dissolve the exploitative inequalities that private property promotes.

Of necessity, Marx had much more to say about the direct labor of producers than about forms of input we would recognize as intellectual in nature. Nonetheless, his thoughts, in the Grundrisse, on what he called "the general intellect," have been a stimulant to recent debates about property formation in a knowledge economy.[42] As the capitalist use of science developed apace, Marx saw that the generation of profit would depend less on direct labor time and more on the harnessing of mental powers and knowledge resources—"the general productive forces of the social brain." Technology, in the form of fixed capital, would be the most efficient way for owners to coordinate and absorb mental labor. Yet to the degree to which the general intellect is a collective entity, production would become more and more social in nature. Ever alert to evidence of the bourgeoisie digging its own grave, Marx imagined that this latter development might lead to the dissolution of wage labor and private ownership along with capital itself.

That moment is not yet upon us, but it is plausible to conclude that the conflicts manifest in the IP wars are, in large part, a consequence of the potential harbored within the general intellect. Efforts to administer an effective division of labor within the knowledge industries increasingly depend upon control over the IP inside employees' heads and the capacity to affect "knowledge transfers" without too much friction. Information monopolists have undertaken a massive property grab to prevent leaks in the system. But the leaks are being sprung nonetheless, and the Internet, which teems with unauthorized content, is the most porous of all public entities. Corporate managers bent on disciplining rogue users, through the use of electronic locks or other forms of Digital Rights Management (DRM), are now in a running battle with the ever-proficient hackers of the technocratic fraternity. Punitive policing among the general population runs the risk of adding to the record of case law that supports fair use and casts doubt on the legitimacy of all-out privatization.

No one can doubt that these coercive efforts will continue apace. IP-driven industries—from microchips and biomedicine to multimedia entertainment—stand at the commanding heights of a rapidly globalizing economy, and their owners are bent on hammering out a property regime that will keep them there for decades to come. As always, their ability to shape and create their own opposition makes it easier to recruit their enemies. At least two generations of hackers have agonized over accepting lucrative offers of employment within corporate or government IP security. That is hardly the model of job creation or income security that we need. But it is clear that we do need one.

The cooperative labor ethos of the FLOSS initiatives has not yet become a practical inspiration for those without the resources to survive in a gift economy. Moreover, the burgeoning corporate interest in the open source alternatives to proprietary software is testing the clarity of FLOSS idealism daily. Open source software is no longer a fringe option for corporate America. The outstanding technical performance of Linux (built with the volunteer labor of 3,000 engineers in over ninety countries) has attracted the patronage of multinationals. Yet few can doubt that this interest in nonproprietary standards is not driven by the opportunity to take advantage of unpaid, highly innovative labor.[43] As for the free software movement, for all its admirable political advances, it has done little to address the suspicion that a predominantly volunteer labor model poses a threat to the livelihoods of future engineers. Nor have Free/Libre Culture contestants in the IP wars made a priority of thinking about the bread-and-butter interests of lower cohort employees in the knowledge industries, let alone those of workers whose service labor supports the knowledge economy. Neither the reformists—petitioning to rescue the IP system from the monopolists—nor the abolitionists—dedicated to alternative forms of licensing—have so far been able to link their issues to the needs of those further back in the technology race.

In part, this has been due to the need to keep eyes on the prize. But if the IP wars are not a single-issue skirmish—if they are about altering the relations of production rather than just restoring the status quo ante—then it is time to ask questions about how the prize is to be distributed. How can we ensure that the interests of those who fall "below the line" are more fully represented in the resolution of disputes? How can the campaign for a free information domain take up the challenge of conceiving a sustainable income model? What kind of state action is required to ensure that inequalities in the private sphere are minimized by the establishment of a public sphere that is knowledge rich and monopoly free? Which new technologies and policies are best suited to furthering these goals? These are not easy questions, but the ability to answer them should not be beyond the conceptual limits of a technologically advanced people.

Notes

1. Though his lifelong research did not surface in any complete published form, *Agape Agape* (New York: Viking Penguin, 2002) illustrates some of Gaddis's obsessive interest.
2. Cyril Ehrlich, *The Piano: A History* (New York: Oxford University Press, 1990), 134. William Gaddis, "Stop Player. Joke No. 4," *The Rush for Second Place: Essays and Occasional Writings* (New York: Penguin, 2002).
3. James Kraft estimates that by 1934, 20,000 theater musicians—"perhaps a quarter of the nation's professional instrumentalists and half of those who were fully employed"—lost their jobs as a result of the talkies. Exhibitors "saved as much as $3,000 a week by displacing musicians and vaudeville actors." *Stage to Studio: Musicians and the Sound Revolution, 1890–1950* (Baltimore: Johns Hopkins University Press, 1996), 33, 49.
4. Lawrence Lessig, *Free Culture: How Big Media Uses Technology and the Law to Lock Down Culture and Control Creativity* (New York: Penguin Press, 2004), 55–56.
5. For broad coverage of the debates generated by the IP wars, see David Bollier, *Silent Theft: The Private Plunder of our Common Wealth* (New York: Routledge, 2002): *Brand Name Bullies: The Quest to Own and Control Culture* (New York: John Wiley, 2004); Lawrence Lessig, *The Future of Ideas: The Fate of the Commons in a Connected World* (New York: Random House, 2001); David Lange, "Recognizing the Public Domain," *Law and Contemporary Problems* 44.4 (1981); Michael Perelman, *Steal This Idea: Intellectual Property Rights and the Corporate Confiscation of Creativity* (New York: Palgrave Macmillan, 2002); Seth Shulman, *Owning the Future: Inside the Battles to Control the New Assets That Make up the Lifeblood of the New Economy* (New York: Houghton Mifflin, 1999); Siva Vaidhyanathan, *Copyrights and Copywrongs: The Rise of Intellectual Property and How It Threatens Creativity* (New York: New York University Press, 2002), and *The Anarchist in the Library; How the Clash Between Freedom and Control Is Hacking the Real World and Crashing the System* (New York: Basic Books, 2004); Ronald Bettig, *Copyrighting Culture: The Political Economy of Intellectual Property* (Boulder, Colo.: Westview, 1996); Peter Drahos, with John Braithwaite, *Information Feudalism: Who Owns the Knowledge Economy?* (London: Earthscan, 2002); Adam Thierer, ed., *Copy Fights: The Future of Intellectual Property in the Information Age* (Washington, D.C.: Cato Institute, 2001).
6. James Boyle, *Shamans, Software, and Spleens: Law and the Construction of the Information Society* (Cambridge, Mass.: Harvard University Press, 1996); Jessica Litman, *Digital Copyright Protecting Intellectual Property on the Internet* (New York: Prometheus Books, 2001) William Fisher, *Promises to Keep: Technology, Law, and the Future of Entertainment* (Stanford, Calif.: Stanford University Press, 2004).
7. Some of the fruits of the interdisciplinary "critique of authorship" project launched in the early 1990s can be found in Peter Jaszi and Martha Woodmansee, eds., *The Construction of Authorship: Textual Appropriation in Law and Literature* (Durham, N.C.: Duke University Press, 1994).
8. Fueling the movement for an information commons, a bevy of legal blogs provide daily input on the rapidly changing IP landscape: Copyfight, Importance OF (both on Corante), Berkman, Furdlog, GrepLaw, Law Meme, Tech Law Advisor, CopyFutures, bIPlog, CIS Blog, Academic Copyright, The Trademark Blog, Lessig Blog, Copyright Readings, and others. Public interest organizations have formed to lobby around IP disputes: Public Knowledge, Future of Music Coalition, Digital Future Coalition, Electronic Freedom Frontier, Center for the Public Domain, Union for the Public Domain. The most high-profile cases in recent years in the IP wars have been *A&M v. Napster* (2001), *Eldred v. Ashcroft* (2003), *Tasini v. New York Times* (2001), and *MGM v. Grokster* (2004).
9. Russell Roberts, "An Interview with Lawrence Lessig on Copyrights," *Library of Economics and Liberty*, April 7, 2003, http://www.econlib.org/library/Columns/y2003/Lessigcopyright.html (accessed June 19, 2006).
10. Richard Stallman, Lawrence Lessig, Joshua Gay, eds., *Free Software, Free Society: Selected Essays of Richard M. Stallman* (Boston: Free Software Foundation, 2002); Eric Raymond, *The Cathedral & the Bazaar* (Sebastopol, Calif.: O'Reilly, 2001); Steven Weber, *The Success of Open Source* (Cambridge, Mass.: Harvard University Press, 2004); Sam Williams, *Free as in Freedom: Richard Stallman's Crusade for Free Software* (Sebastopol, Calif.: O'Reilly, 2002); Chris DiBona, Mark Stone, and Sam Ockman, eds., *Open Sources: Voices from the Open Source Revolution* (Sebastopol, Calif.: O'Reilly, 1999).
11. Paula Forsook, *Cyberselfish: A Critical Romp through the Terribly Libertarian Culture of High Tech* (New York: Public Affairs, 2000).

12. William Landes and Richard Posner note that the increasing regulation of IP from 1976 was out of synch with the general movement toward economic deregulation. *The Economic Structure of Intellectual Property Law* (Cambridge, Mass.: Harvard University Press, 2003).
13. Examples are culled from Kembrew McLeod, *Freedom of Expression (R): Overzealous Copyright Bozos and Other Enemies of Creativity* (New York: Doubleday, 2005).
14. Vandana Shiva, *Biopiracy: The Plunder of Nature and Knowledge* (Boston: South End Press, 1997), and *Protect or Plunder? Understanding Intellectual Property Rights* (London: Zed Books 2001); Jeremy Rifkin, *The Biotech Century: Harnessing the Gene and Remaking the World* (New York: Jeremy Tarcher, 1998); Silke von Lewinski, *Indigenous Heritage and Intellectual Property: Genetic Resources, Traditional Knowledge and Folklore* (The Hague: Kluwer Law International); Michael Brown, *Who Owns Native Culture?* (Cambridge, Mass.: Harvard University Press, 2004).
15. Statement by India at Inter-Sessional Intergovernmental Meeting on a Developmental Agenda for WIPO, April 11–13, 2005, available at http://weblog.ipcentral.info/archives/2005/04/more_wipo_india.html (accessed June 19, 2006).
16. Carlos Correa, *Intellectual Property Rights, the WTO, and Developing Countries: The TRIPS Agreement and Policy Options* (London: Zed Books, 2000); Peter Drahos and Ruth Mayne, *Global Intellectual Property Rights: Knowledge, Access and Development* (New York: Palgrave Macmillan, 2002); Christopher May, *Global Political Economy of Intellectual Property Rights* (New York: Routledge, 2000).
17. Paul Goldstein, *Copyright's Highway: From Gutenberg to the Celestial Jukebox* (Stanford, Calif.: Stanford University Press, 2003), 153–54.
18. Robert A. Gorman, "Intellectual Property: The Rights of Faculty as Creators and Users," *Academe* 3.14 (May–June 1998): 84.
19. Sheila Slaughter and Larry Leslie, *Academic Capitalism: Politics, Policies, and the Entrepreneurial University* (Baltimore: Johns Hopkins University Press, 1997); Sheila Slaughter and Gary Rhoades, *Academic Capitalism and the New Economy: Markets, State, and Higher Education* (Baltimore: Johns Hopkins University Press, 2004); Jennifer Washburn, *University, Inc.: The Corporate Corruption of American Higher Education* (New York: Basic Books, 2005); Stanley Aronowitz, *The Knowledge Factory: Dismantling the Corporate University and Creating True Higher Learning* (Boston: Beacon Press, 2001); Dorothy Nelkin, *Science as Intellectual Property: Who Controls Research?* (New York: Macmillan, 1984); David Kirp, *Shakespeare, Einstein, and the Bottom Line: The Marketing of Higher Education* (Cambridge, Mass.: Harvard University Press, 2004); Randy Martin, ed., *Chalk Lines: The Politics of Work in the Managed University* (Durham, N.C.: Duke University Press, 1999); Sheldon Krimsky, *Science in the Private Interest: Has the Lure of Profits Corrupted Biomedical Research?* (New York: Rowman & Littlefield, 2003); Benjamin Johnson, Patrick Kavanagh, and Kevin Mattson, *Steal This University: The Rise of the Corporate University and the Academic Labor Movement* (New York: Routledge, 2003); Cary Nelson and Stephen Watt, *Academic Keywords: A Devil's Dictionary for Higher Education* (New York: Routledge, 1999).
20. *Workplace: A Journal of Academic Labor* (online at *http://www.cust.educ.ubc.ca/workplace*) is an indispensable source of commentary.
21. Quoted in Jeff Lustig, "The Mixed Legacy of Clark Kerr: A Personal View," *Academe* 90.4 (April 2004): 51–53.
22. Corynne McSherry, *Who Owns Academic Work? Battling for Control of Intellectual Property* (Cambridge, Mass.: Harvard University Press, 2001). Christopher Newfield offers a longer historical view on this paradox in *Ivy and Industry: Business and the Making of the American University, 1880–1980* (Durham, N.C.: Duke University Press, 2004).
23. Andrew Ross, *No-Collar: The Humane Workplace and Its Hidden Costs* (New York: Basic Books, 2002).
24. Andrew Ross, *The Fast Boat to China: Corporate Flight and the Human Consequences of Free Trade—Lessons from Shanghai* (New York: Pantheon, 2006).
25. For example, it was a group of AOL employees who created Gnutella. The program "escaped" into the public domain in the course of the few hours it was posted on the AOL Web site before managers took it down.
26. McKenzie Wark describes this battle over property between what he calls the "vectoralist" and the "hacker" classes in *A Hacker Manifesto* (Cambridge, Mass.: Harvard University Press, 2004).
27. George Keller, *Higher Ed, Inc.: The Rise of the For-Profit University* (Baltimore: Johns Hopkins University Press, 2003); David Noble, *Digital Diploma Mills: The Automation of Higher Education* (New York: Monthly Review Press, 2002).

28. Sandra Baringer, "Repositioning the Ladder: A Cautionary Tale about Outsourcing," *Minnesota Review* 63–64 (Spring–Summer 2005): 195–202.
29. Tony Scott, Leo Parascondola, and Marc Bousquet, *Tenured Bosses and Disposable Teachers: Writing Instruction in the Managed University* (Carbondale: Southern Illinois University Press, 2003).
30. See the AAUP position papers on contingent faculty and intellectual property (www.aaup.org).
31. Lois Gray and Ronald Seeber, *Under the Stars: Essays on Labor Relations in Arts and Entertainment* (Ithaca, N.Y.: ILR Press, 1996).
32. Goldstein, *Copyright's Highway,* 151.
33. Martha Woodmansee, "The Genius and the Copyright: Economic and Legal Conditions of the Emergence of the 'Author,'" *Eighteenth-Century Studies* 17.4 (Summer 1984): 425–448; and "On the Author Effect: Recovering Collectivity," in *The Construction of Authorship*, ed. Woodmansee and Jaszi, 15–16.
34. Quoted in Tricia Rose, *Black Noise: Rap Music and Black Culture in Contemporary America* (Middletown, Conn.: Wesleyan University Press, 1994), 81.
35. William Cronon, from *Changes in the Land: Indians, Colonialists and the Ecology of New England* (New York: Hill and Wang, 1983).
36. Thomas Macauley, *Speeches on Copyright* (London, 1914), 23.
37. William Alford, *To Steal a Book Is an Elegant Offense: Intellectual Property Law in Chinese Civilization* (Stanford, Calif.: Stanford University Press, 1995).
38. The activities and research programs of Sarai are archived at www.sarai.net.
39. See the manifesto by Marieke van Schijndel and Joost Smiers, "Imagining a World Without Copyright," summarized in *International Herald Tribune*, October 8, 2005. For a more general, up-to-date overview of the debates in information studies, see Siva Vaidhyanathan, "Critical Information Studies: A Bibliographic Manifesto," *Cultural Studies* 20.2–3 (March–May 2006): 292–315.
40. James Boyle, *Shamans, Software, and Spleens*, 57.
41. Jonathan Glickstein, *Concepts of Free Labor in Antebellum America* (New Haven, Conn.: Yale University Press, 1991), 17.
42. Nick Dyer-Witheford, *Cyber-Marx: Cycles and Circuits of Struggle in High-Technology Capitalism* (Urbana: University of Illinois Press, 1999), 219–38; Paolo Virno and Michael Hardt, *Radical Thought in Italy: A Potential Politics* (Minneapolis: University of Minnesota Press, 1996); Johan Soderberg, "Copyleft vs. Copyright: A Marxist Critique," *First Monday* 7.3 (March 2002), http://www.firstmonday.org/issues/issue7_3/soderberg/ (accessed June 19, 2006).
43. Tiziana Terranova, "Free Labor: Producing Culture for the Global Economy" *Social Text* 18.2 (2000): 33–57; D. M. Berry and G. Moss, "On the "Creative Commons": A Critique of the Commons Without Commonalty," *Free Software Magazine* 5 (2005), http://www.freesoftwaremagazine.com/free_issues/issue_05/commons_without_commonality (accessed June 19, 2006).

Failing Narratives, Initiating Technologies: Hurricane Katrina and the Production of a Weather Media Event

Nicole R. Fleetwood

> The preparation for and response to Hurricane Katrina show we are still an analog government in a digital age. We must recognize that we are woefully incapable of storing, moving, and accessing information—especially in times of crisis.
>
> Many of the problems we have identified can be categorized as "information gaps"—or at least problems with information-related implications, or failures to act decisively because information was sketchy at best.
>
> —*A Failure of Initiative*

The *Final Report of the Select Bipartisan Committee to Investigate the Preparation for and Response to Hurricane Katrina* is a bold statement about the centrality of technology to securing the nation's population and its future. From the first page, the report frames the crisis resulting from the hurricane as a failure of individuals to access accurate information and to actualize proper technologies both prior to the hurricane's arrival and after New Orleans flooded. By staging the disaster in these terms at the outset, the Republican-led congressional committee displaces the privilege given in public culture to the media coverage of the disaster, in particular the media's focus on abandoned black residents in the city's floodwaters. In fact, the committee asserts that part of the disastrous consequences of Katrina resulted from the media's spreading unfounded information about the extent of the disaster and civil unrest, as opposed to the scientific accuracy of reports by the National Hurricane Center and the Army Corps of Engineers.[1] In assigning accountability, the document states: "The media must share some of the blame here . . . its clear accurate reporting was among Katrina's many victims. If anyone rioted, it was the media."[2] The *Final Report* is the federal government's attempt to be the authoritative voice through which the public understands the impact of Katrina. This is crucial given how the media's coverage has shaped the nation's understanding of what happened prior to and after the arrival of the hurricane. The congressional committee's report has the effect of shoring up the grand narrative evoked in the epigraphic quote. This narrative depends on technological determinism and scientific truth; and it reduces the Katrina event to the misdeeds of individuals, misinformation, and outdated technologies.

The term the "Katrina event" is one I will use throughout this essay to describe a host of activities and processes surrounding the hurricane and its aftermath, including the reportage during and afterward, the ensuing flooding of New Orleans, the displacement and evacuation of thousands of people stranded in the city, and the ongoing and contested reconstruction process. The Katrina event refers to the material and social impact of the storm, as well as the complex set of social, technological, and economic narratives and processes reported by the news media and through governmental reports. The committee's assertion that Katrina revealed the need for institutions to embrace the digital age relies on a belief in technology to decrease, if not overcome, human failure. As evidenced in this report and other post-Katrina efforts, the government invokes technological determinism—or technology as the central force behind progress and social development—as a response to the public's belief that the Katrina event represents a massive failure in governmental accountability and social and technological progress. At the same time, the Katrina event is the result of the success of other forms of technologies, that is, the power of meteorological technologies to predict disaster and that of news media technologies, as the privileged producers and disseminators of knowledge and as consumable goods for the national public. These two sets of technology were instrumental in producing Hurricane Katrina as a weather media event, what Marita Sturken describes as a convergence of news and meteorology, and as a national crisis.[3] The application of these technologies constructed Hurricane Katrina as the national disaster that has been imprinted on the nation's psyche, especially through the images of suffering, emoting, and abandoned black bodies in the floodwaters of New Orleans. Turning attention to the importance of technological mediation and narratives in the production of the crisis and in devising its solution reveals a great deal about the operations of race, class, and risk in the United States. Technology here should be understood as a media process of production and as a discursive tool by which particular narratives are naturalized and certain bodies made vulnerable. In this context, Hurricane Katrina reveals a different kind of determinism—the stark operations of technology in determining who lives and who dies.

This essay focuses on the complex relation between concepts of technological determinism, the definition of a crisis within technological discourses as a failure of specific technologies and operators, and the role played by media technologies in framing thousands of blacks as disposable beings. In tracing the role of technology in the production of Katrina, my analysis engages with two of the most widely distributed sources by which Katrina has been recorded and which simultaneously reveal the significance of technological determinism as a national motif: (1) the news media archive of the storm and its aftermath (a diffuse set of stories and images that dominated television for much of September 2005), and (2) *A Failure of Initiative*, the report of the Bipartisan Committee to Investigate the Preparation for and Response to Hurricane Katrina. These sources, on many levels, compete with each other to be the authority through which this event gets marked and understood. Working with these two sets of documents, I attempt to track the technological production of the storm as it developed in public discourse, specifically through news media coverage.

In focusing on the unfolding production of Hurricane Katrina as a weather media event, this analysis exposes how a particular narrative of technology revealed itself in the moments after the hurricane arrived on land. For a brief period, the Katrina event showed the instability the dominant technological narrative (in particular, the notion that technology and scientific truth trump racial and social barriers). Katrina exposed how systems of power invest in material and infrastructural technologies to maintain certain imbalances. Take, for instance, the notoriety of the convention center in New Orleans, a place where thousands fled for safety. The masses of black bodies in despair housed in a site meant for industry, public conventions, and "state of the art" expositions defeated dominant notions of technological advancement and progress as colorblind processes and objectives. Analyzing the significance of the abandoned subjects in the convention center, Michael Ignatieff writes: "What has been noticed is that the people with the most articulate understanding of what the contract of American citizenship entails were the poor, the abandoned, hungry people huddled in the stinking darkness of the New Orleans convention center . . . Having been abandoned, the people in the convention center were reduced to reminding their fellow citizens, through the medium of television, that they were not refugees in a foreign country."[4] The convention center no longer was a site of innovation and progress but a dumping ground for housing a group of displaced subjects, or "body objects."

While the Katrina event challenged the dominant U.S. technological narrative and made it temporarily vulnerable, the government's attempt to mend this failing narrative was to initiate more technology as a material solution to the disaster of Katrina. Technology as public, governmental, and media discourse follows a positivistic model that festishizes specific commodities, frames the nation-state at the center of technological advances, and posits the future of labor, citizenry, national security, and public life through a deterministic model of progress. More important, technological determinism is complicit in producing certain populations as disposable for the state, or what Achille Mbembe theorizes as necropolitics. Necropolitics refers to the state's ultimate expression of sovereignty as "the power and the capacity to dictate who may live and who must die."[5] States may sanction the annihilation of specific populations through various means, including "technologies of murder" and by producing the conditions in which many poor and racialized groups found themselves before Katrina arrived, the state of living dead.[6] In the days following Hurricane Katrina, as the national audience sat glued to the television, the imagistic power of the living dead on the streets of an American city shook the sovereignty of the nation-state and its foundational narratives.

Technologies of Forecasting: Imaging the Weather, Predicting Disaster

> So tragically, so many of these people, almost all of them that we see, are so poor and they are so black, and this is going to raise lots of questions for people who are watching this story unfold.[7]
>
> —Wolf Blitzer, CNN anchor

Wolf Blitzer's candid and perplexed statement during his CNN news show exposes precisely the moment in the immediate aftermath of Hurricane Katrina in which dominant narratives of technology failed to account for the overwhelming images broadcast by news media organizations and the lack of organized governmental response to the disaster. At the same time, Blitzer's confusion highlights the flexibility and power of media technologies to make visible the failure of these narratives as camera crews arrived long before first responders and governmental agencies. Blitzer, like many others, was caught off guard by the disarray that resulted after Hurricane Katrina hit the Gulf Coast. Much of the media's tracking of Katrina was an impressive and orchestrated exercise in meteorological technologies to build anticipation over a period of days. While the Weather Channel and the major news media organizations accurately predicted the hurricane and activated a range of technologies to forecast, simulate, and record the disaster, these technologies and their application could not provide an adequate narrative to explain the flood of "so poor" and "so black" bodies in despair as a result of the storm.

During the late summer of 2005, disaster reportage dominated much of television news, a time when viewership tends to be low. News shows were dedicated to predicting and witnessing weather media events, including a record number of hurricanes. This coverage was bolstered by sophisticated hurricane tracking technologies and on-the-ground reporting by well-known anchors and newspeople. When the Weather Channel, the major networks, and cable news sent crews to the U.S. Gulf Coast to track Hurricane Katrina, a massive storm that had reached Category 5 status, news reporters transformed into thrill seekers. After days of following the movement of the storm across Florida and the Gulf Coast, they were there when the hurricane finally landed on Monday, August 29, 2005. Because Katrina had weakened and made landfall as a Category 3 storm, the general assessment on the morning of its arrival was that New Orleans had been spared the destruction predicted by meteorologists across the nation.[8] Yet, as journalists began to assess (visually) the storm's damage across Alabama, Louisiana, and Mississippi, and public officials learned that the levees of New Orleans had failed, news outlets, government agencies, and emergency response organizations all scrambled to gather information even while vivid images, with often scattered and contradictory commentary, aired live. The hurricane was precisely the storm that the National Hurricane Center had predicted and yet its consequences and aftermath came as a shock to both government officials and the weather-focused media. In spite of meteorological accuracy, Katrina became the terrorist attack that the American public had feared would hit at some unsuspecting moment, and years after 9/11, the nation-state proved unprepared to handle such a crisis.

While Hurricane Katrina is widely understood as a spectacular moment of failure in the national imaginary, the technologies available through the National Hurricane Center (NHC) and National Weather Service (NWS) allowed the storm to be predicted with uncanny accuracy. On the day before Katrina made landfall in the Gulf Coast, the National Weather Service put out the following alert: "Most of the area will be uninhabitable for weeks . . . perhaps longer . . . human suffering incredible by modern standards."[9] Partly in coordination with updates from NHC and NWS, weather media reports began to develop a narrative of the hurricane. In producing the Katrina event, news media organizations straddled a complex nexus of imaging technologies, entertainment, information, and communication technologies. According to media scholar Katherine Fry, such dramatization of weather events and the primacy of their coverage, known as the "CNN Syndrome," have grown more prominent since the early 1990s.[10] The CNN Syndrome uses both technological instruments (weather maps, Doppler radar, interactive diagrams) and narrative strategies (anticipation, climax, denouement) to make weather must-see TV news.

On some level, meteorological technologies give the power to predict the future. We are able to witness the storm as weather media event from its inception as tropical storm to its naming to its levels of categorization to its arrival. The detailed sophistication afforded the viewer promises to lower our risk by increasing preparedness. Marita Sturken argues that computer visualization technologies, such as Doppler radar, "are used to convey the sense that weather-tracking technologies can actually help to control the weather itself."[11] Further, Sturken argues that the production of weather media events and the illusion of controlling the future have little impact on preparedness: "Weather prediction is in fact a very limited kind of knowledge that promises protection and reassurance yet which bears no relationship to the social infrastructures that would ensure preparedness. Indeed, it could be argued that prediction not only has little impact on people's daily lives but serves to screen out the politics of disaster."[12] Taking Sturken's argument further, one could say that prediction becomes an instrument of necropolitical operations by effacing "the politics of disaster" and assigning responsibility for protection to individual consumption.

If anything, the ability to track and monitor weather has evolved into a particular form of spectator pleasure that writer Mark Svenvold terms catastrophilia.[13] Weather becomes the big media event, similar to the anticipation around the Super Bowl or other risk-involved live events. Further complicating the type of participatory spectatorship and consumption involved in watching weather media events, Sturken argues, "the weather is the site of a production of knowledge that functions as a means to erase political agency and to substitute the activity of witnessing in its place. Watching it becomes the central experience; indeed it subsumes all other experiences."[14] In fact, Sturken says, the consolidation of media and news technologies has created the "weather citizen," a social position that combines civic responsibility and consumerism with spectatorship and technological fetishism.[15] Assuming the role of the weather citizen, connected to the disaster through news media production, was the dominant method by which the public responded to the unfolding events in late August and early September 2005. Yet, those stranded on the Gulf Coast were not afforded this position of civic participation through spectatorship and consumption. Instead, they were denied their more fundamental and substantive rights as citizens and residents to be protected and sheltered by the government during times of catastrophic disaster, or what the federal government labels an Incident of National Significance (INS).

Quite certainly, Hurricane Katrina made for excellent television programming, as it also unified the nation of weather citizens in a shared sense of shock and horror at the evolving footage of the Gulf Coast. The combination of the storm's cinematic destruction of a major metropolitan area, the unfettered images of human suffering, the emotional outburst of media personalities and celebrities, and the portrayal of governmental ineptitude fixated viewers across the country on their television screens. Part of the power of Katrina as weather media event had to do with the news media's juxtaposition of, on the one hand, human desperation and need and, on the other, neglect on various levels. For example, it has been widely reported that President George W. Bush was on vacation when the National Hurricane Center predicted that Katrina would be a catastrophic weather event, and he remained on vacation for two days after the hurricane hit. (Even more troubling, several months after Hurricane Katrina, the Associated Press released video footage of top FEMA and National Hurricane Center officials briefing Bush on the potential catastrophic damage that could result from the hurricane and Bush responding that the federal government was fully prepared to handle the situation.[16]) *Newsweek*, examining the president's initial lack of response, reports that Bush was not aware of the severity of the storm until one of his aides made a DVD of news coverage and showed it to him on Thursday, three days after the storm's arrival.[17] This account foregrounds the centrality of news media in making the hurricane a weather media event worthy of watching, as well as the role of technology in providing access to the nation, even the president, as he watched the "highlights" of Katrina news coverage. In essence, Bush and his administration chose not to heed the National Weather Service's warnings and responded only after technological interfaces produced images that validated the warnings.

The Katrina event showcased the power of imagistic news technologies and the ineffectiveness of rescue/security/disaster technologies, as they have become even more closely connected in the era of George W. Bush's War on Terror. Under Bush's leadership, domestic emergency and rescue operations, most notably FEMA, have become housed in the Department of Homeland Security. With Katrina, the bureaucracies and technologies of crisis that Bush has spent several years developing in the post-9/11 era were revealed as impotent when faced with immobilized bodies, rendered as such by news media and consumer technologies (camera phones, digital cameras, consumer camcorders). Furthering this point is the fact that governmental officials were reliant on the visual narrative unfolding through media coverage to execute technologies of war and rescue. In *A Failure of Initiative*, one top-ranking military official reported to Congress that military and first responders relied on media coverage for situational updates: "We focused assets and resources based on situational awareness provided to us by the media, frankly. And the media failed in their responsibility to get it right . . . we sent forces and capabilities to places that didn't need to go there in numbers that were far in excess of what was required, because they kept running the same B roll over and over . . . and the impression to us that were watching it was that the condition did not change. But the conditions were continually changing."[18] The jarring contrast of technologies of circulation and mobility (such as portable news cameras, Coast Guard rescue helicopters, and later, military convoys) functioned in stark contrast to the stasis of thousands of black bodies framed in congested, filthy, dire conditions. News helicopters hovered above flooded homes, with cameras zooming in on people stranded on their roofs with signs such as "Help Us." Searching for narratives of heroism, news crews would occasionally wait until a Coast Guard or other rescue force appeared and then showed live coverage of the rescue mission. At other times, the reporter on duty would explain that the news media helicopter was not equipped to rescue but that the reporter would notify authorities of the person's location: "Oh gosh, there's nothing we can do to help these people but notify the Coast Guard."[19] The rooftop rescues made for exciting, dramatic television, as opposed to the bureaucratic wrangling and the structural repairing that were taking place at the same time.[20] Portable technologies—handheld digital video cameras, still cameras, videophones, and camera phones—captured the corporeality of the body as the social machines necessary to support it broke down. Visual media exposed bodies emoting, bodies suffering, bodies bloated and decaying, bodies—live and dead—as obstacles to be removed so that "disaster capitalism" could begin its work of rebuilding what has been described as a dead city.[21]

On the Ground Scramble: Infrastructural Failure and Technologies of Survival

> We saw buses, helicopters, and FEMA trucks, but no one stopped to help us. We never felt so cut off in all our lives. When you feel like this you do one of two things; you either give up or go into survival mode. We chose the latter. This is how we made it. We slept next to dead bodies; we slept on streets at least four times next to human feces and urine. There was garbage everywhere in the city. Panic and fear had taken over.
>
> —Patricia Thompson, New Orleans resident and evacuee[22]

The level to which Katrina destroyed, interrupted, and severely damaged already weakened technological infrastructure and capacities in New Orleans is immense. The lack of technological access after the storm arrived affected the operations of public and private sectors in countless ways. Downed power lines littered the landscape; cell phone coverage was absent for days; and batteries were in short supply. News footage showed thousands of cars in ruin, often mangled into barely recognizable forms. Images of parking lots filled with flooded buses, potentially available to evacuate many before the storm had arrived, further underscored the neglect of those stranded in the floodwaters. Hospital officials were forced to abandon some of their most ill patients and to euthanize others because of loss of power and failing generators, which cut off critical medical interventions. Governmental computers with vital court records were destroyed. Oil rigs were left slammed against the coast, and production platforms were destabilized. Tide gauges failed in the New Orleans area, making a precise measure of the storm surge in the urban region almost impossible.[23]

News reportage showed stranded people, mainly black residents, making use of materials and objects for survival. Stranded individuals turned refrigerators and blow-up mattresses into flotation devices. Some trapped residents "chopped their way to their roofs with hatchets and sledge hammers, which residents had been urged to keep in their attics in case of such events."[24] Reports surfaced of carjackings becoming a means of escaping what for many was an apocalyptic environment. One of the more dramatic stories was of a young teenage boy who commandeered a yellow school bus. He filled the bus with stranded Orleanians and drove it to Houston for shelter. When reporters asked him if he had ever driven a school bus, he stared into the camera blankly and said no. After days of nonstop coverage of flooded New Orleans, television news organizations turned to recycling several familiar images of black suffering and survival. News media selected specific images and clips from their disaster coverage to stand in for a range of black pathos. For example, there was the repeated image of a dead elderly woman at the convention center, left in a wheelchair, covered by a plaid blanket. Another commonly replayed image was that of the young woman who held two listless babies outside of the convention center. Out of these images emerged stories of individualized hardship and struggle.

One story that became the symbol of black tragedy and hope was that of Hardy Jackson of Biloxi, Mississippi. Speaking to a reporter at WKRG, a local CNN affiliate station, Jackson, confused, disheveled, and tearful after narrowly escaping drowning, describes having to let go of his wife's hand as she is swept away by the waters. The reporter asks of the whereabouts of his wife now; Jackson responds, "Can't find her body. She gone." He then cries: "She told me, 'You can't hold me, . . . take care of the kids and the grandkids.'" The news camera focuses on Jackson's emoting face but with occasional wider shots that include his two young grandchildren and the young, white, blond reporter. The story ends with Jackson looking aimlessly at the devastation around him, as his grandchildren stand next to him silently looking in other directions.

CNN titled the news clip "Hurricane Heartache" and replayed it numerous times in the immediate aftermath of the storm. News organization produced several follow-up stories on Jackson's plight. In one brief segment on *Good Morning America*'s Web site titled "Story of Man Whose Wife Was Swept Away in Flood Touched Nation," Jackson's story is framed in moral terms: "It's his devotion to his family that has captured Americans' hearts. Each day since Katrina, Hardy has returned to what is left of his home—and to thoughts of what he will do about his wife's last words."[25] A week later, CNN followed up with a news story that began: "His face is etched into our collective memory." The story traced Hardy Jackson to Georgia, where he, his children, and his grandchildren were housed temporarily with a relative. In another update attempting to resolve Jackson's loss, we learn that an anonymous benefactor donated a house to him.[26] The emphasis on individual profiles continued for several months after the hurricane. In part, this appeared to be an attempt by media organizations not to forget the faces of Katrina or a response to the criticism that dominant news media have only short-term memory, that is, one sensational story supplants another. Yet, the process of individuating the disaster through news media representations plays into grander narratives of technological determinism and personal responsibility. This is in part because the focus on personal suffering and survival has sidetracked the public away from the larger structural causes that produced the Katrina event. Instead, the major news corporations and the federal government promote narratives of individual survival and technological reinforcements, focusing on rebuilding efforts, such as strengthening houses and making more durable levees. The proliferation of individual Katrina narratives supports the quick fix strategy of resolving the disaster, an issue I explore further in the following section.

While the news media have focused on the individual tales of people like Hardy Jackson, Katrina's impact was so far-reaching in large part because of the level to which poor and black communities in New Orleans had been systemically neglected and rendered as obstacles to the city's and ultimately the nation's progress. The racial and environmental history of New Orleans is one of structural inequality, particularly in housing, infrastructural development, and protection against weather events. Since Reconstruction, blacks have occupied the swampier lowlands of New Orleans.[27] Even more critically, Jim Crow–era housing segregation led to blacks, poor whites, and immigrants occupying the bottom of "the bowl" and other flood-prone areas. In discussing the racial implications of early-twentieth-century public works and civil engineering projects in New Orleans, Craig Colten writes: "Although the improved drainage system opened new areas to black residences, at the same time it contributed to segregation, reflecting established patterns of turning low-value land associated with environmental problems to minority populations."[28] In recent years, the Ninth Ward has flooded repeatedly, most notably during Hurricane Betsy in 1965 and during the winter of 1983, leading residents of the district to file a lawsuit against the levee board for inaction and neglect due to the 1983 flood.[29]

In addition to flood vulnerability, the environmental toxicity of these districts is another major condition that contributes to the structural, environmental, and social neglect of poor and black (pre-Katrina) inhabitants. Julie Sze argues, "The 'toxic soup' that has received much public attention is filled with the effluence from the oil and petrochemical industry, and stands as a visible manifestation of the everyday environmental and human disaster that poor, African-American Gulf Coast communities faced before Katrina on an everyday basis."[30] With many industrial and manufacturing plants, from petrochemical to vinyl chloride, located there, southern Louisiana is one of the most polluted regions in the country. Three Superfund toxic waste sites exist within or in close proximity to New Orleans, due to years of toxic accumulations, all of which were flooded by Katrina.[31] Theodore Steinberg argues that poor development policies manufacture "natural disaster" risk.[32] As Katrina made obvious, as one's risk of living in an area prone to natural disaster increases based on being raced as black and classed as poor, so too does one's risk of governmental neglect.[33] These analyses make evident that the notion of risk is central to the positioning of certain subjects as the living dead. Given that nation-states rarely take responsibility for creating vulnerable populations, dominant narratives typically frame risk as caused by "acts of God" or by the pathological behavior of marginalized groups, which reproduces their own vulnerability. New Orleans's environmental toxicity, infrastructural neglect, compiled with its high poverty, illiteracy, and crime rates contribute to the city as an inescapable "death-world," a site occupied by those who have been rendered the living dead.

Considering the persistence of the living dead in the nation, Sharon Holland asks: "What if some subjects *never* achieve, in the eyes of others, the status of the 'living'? What if these subjects merely haunt the periphery of the encountering person's vision, remaining, like the past and the ancestors who inhabit it, at one with the dead—seldom recognized and, because of the circum-Atlantic traffic in human cargo or because of removal, often unnamed?"[34] Holland's questions then provide a context to understand Blitzer's confusion, that is, how do we make sense of all those poor and dark bodies on the nation's television screen in light of our overinvestment in technologies as the nation's future (as tools

Figure 1.
Deborah Willis, *Where the Levee Broke,* February 2006. Digital C-Print 11x14. Courtesy of the artist and Bernice Steinbaum Gallery, Miami.

Figure 2.
Wyatt Gallery, *Casino Boat and Chair,* Biloxi, Mississippi, October 2005. Digital C-Print Edition #1/5 30x38. Courtesy of the artist, Watermark Fine Arts, and www.wyattgallery.com.

Figure 3.
Cheryl Finley, *Sorting Debris: Intersection of Royal and Dauphine*, 9th Ward, 2006. 16x24. Courtesy of the artist.

Figure 4.
Wyatt Gallery, *Personal Files*, St. Bernard Parish, LA. October 2005. Digital C-Print Edition #1/5 30x38. Courtesy of the artist, Watermark Fine Arts, and www.wyattgallery.com.

Figure 5.
Deborah Willis, *Lower 9th*, February 2006. Digital C-Print 11x14. Courtesy of the artist and Bernice Steinbaum Gallery, Miami. Forida.

Figure 6.
Cheryl Finley, *"I'm Back! R U?"*, 2006. 16x24. Courtesy of the artist.

to predict disaster with uncanny precision, machines for capitalistic growth, systems to offer longer life expectancy, and operations that enforce national security). In other words, those stranded in the floodwaters had been deemed dead, even prior to Katrina's arrival. A state-supported narrative of decline and death, one that seems to contradict the narrative of technological progress but in fact colludes with it, underlines the reality that most working-class and poor blacks in the United States have already been given the status of the "living dead." The visual images and material conditions of thousands of stranded black subjects provides evidence of how death and decline and technological progress work together; the "so black" and "so poor" were rendered as object bodies clogging the social machines that worked to contain or erase their presence. Those desperate figures on rooftops, streets, and huddled in the Superdome and convention center were the subject(s) of death. Their morbidity and expendability circulated through the ultra-flexible portable imaging and informational technologies available to media outlets. Appropriating Susan Sontag's analysis, we—the audience—were given uncanny access "to look at people who know they have been condemned to die."[35]

While risk of disaster is increased by structural and governmental neglect, the discourse of personal responsibility often places the burden of preparedness on individuals, personal wealth accumulation, and private technologies. Such is the case in the emphasis placed on automobiles during and after Katrina. The primary mode of evacuating before the storm hit, and in most urban disaster plans in the United States, is the private automobile. Until the levees failed, this plan had been considered a success in the case of New Orleans, then undergoing the largest evacuation in the region's history. The *New York Times* reported that "at one point during the evacuation of New Orleans on Sunday, more than 18,000 cars an hour were leaving the city."[36] In response to the early public questioning about why so many remained in New Orleans after the government called for a mandatory evacuation, writer Anne Rice argued in an editorial that "thousands didn't leave New Orleans because they couldn't leave. They didn't have the money. They didn't have the vehicles. They didn't have any place to go. They are the poor, black and white, who dwell in any city in great numbers; and they did what they felt they could do—they huddled together in the strongest houses they could find. There was no way to up and leave and check into the nearest Ramada Inn."[37] Because of the centrality placed on the consumer vehicle in times of crisis, the lack of automobile ownership among the New Orleans poor became the default explanation as to why so many were left behind before the hurricane hit. Yet, I find this a wholly unsatisfying explanation. For one, it does not account for the more than

250,000 vehicles left in New Orleans during Hurricane Katrina.[38] Nor does this address the role of automobiles in helping to create the Katrina disaster. Scientists have noted how greenhouse gases contributing to global warming might affect the strength and severity of tropical storms. Further, the evacuation of Houston during Hurricane Rita, predicted to hit dangerously near the city a few weeks after Katrina, demonstrated how disastrous this strategy can be: more people died in the Houston area during attempted evacuation by automobiles than from the damage of the hurricane. Furthermore, the irony of FEMA's long-term relief plan for those affected by Katrina was to house them in hotels and mobile homes, often on the outskirts of existing communities. These mobile homes, which are even more vulnerable to natural disaster, will be the residences for thousands indefinitely, thus rooting them in another type of flexible, transitional housing without the social mobility to flee.[39]

The focus on the lack of personal automobiles is a myopic attempt to grasp all of those dark bodies in the Katrina aftermath, when there are no easy answers to make sense of racialized and classed human suffering. Explanations offered by journalists and pundits about the lack of automobiles in poor communities, the lack of news and emergency information and resources circulating in these communities, the culture of poverty argument about irresponsibility and neglect in these communities, as well as the fact that Katrina hit at the end of the month (when many on government assistance are *anxiously* without money until the first of the month) reproduce black subjects of New Orleans as already static beings prior to the arrival of Katrina.

The reasons so many residents stayed in New Orleans after the mandatory evacuations are more complex than lack of access to vehicles. They return us to structural neglect, narrative violence, and discursive erasure. More tangibly, many Orleanians had to remain in the local region for medical reasons or to care for sick and elderly relatives. In fact, the Superdome, before being designated a shelter of last resort, had been established as the primary shelter for "special needs" populations. Prior to the storm, the city used buses and paratransit vehicles to bring "'special needs' citizens" there.[40] After the city flooded, it was revealed that many deaths occurred when ill patients ran out of oxygen or could not get access to their dialysis center. For others, especially—but not exclusively—those with limited social networks and financial means, the most secure place to be in times of crisis is home. The safety associated with one's domestic environment takes on heightened relevance in postindustrial urban areas with neglected infrastructural systems. In examining why the majority of heat-related deaths during the 1995 Chicago heat wave were elderly blacks, Eric Klinenberg considers the social ecology of the neighborhoods where the death toll was the highest. Arguing that the high mortality was connected to the failing infrastructure in these neighborhoods, Klinenberg writes that "the dangerous ecology of abandoned buildings, open spaces, commercial depletion, violent crime, degraded infrastructure, low population density, and family dispersion undermines the viability of public life and the strength of local support systems."[41] During the Katrina event, rumors and reports about happenings, especially violent attacks on other stranded people, in and around the Superdome and the convention center simply reinforced this belief for many that home was the safest place to be during the crisis. The infrastructural neglect and the overreliance on private technology (i.e., the automobile) are discourses of abandonment and neglect presented as individual failures that obscure the perpetuation of "death-worlds."

Technological Solutions and Narratives of Reconstruction

In his infamous speech given in New Orleans's French Quarter two weeks after the city flooded, President Bush, responding to the criticism that had been launched at his administration, raised the specter of an unjust historical past without locating the blame on the nation-state. Bush connected racism to narratives of individualism and capitalism, arguing that attempts at redress must operate through national progress, business development, and technological determinism.

> Within the Gulf region are some of the most beautiful and historic places in America. As all of us saw on television, there's also some deep, persistent poverty in this region, as well. That poverty has roots in a history of racial discrimination, which cut off generations from the opportunity of America. We have a duty to confront this poverty with bold action. So let us restore all that we have cherished from yesterday, and let us rise above the legacy of inequality. When the streets are rebuilt, there should be many new businesses, including minority-owned businesses, along those streets. When the houses are rebuilt, more families should own, not rent, those houses. When the regional economy revives, local people should be prepared for the jobs being created.[42]

In the same speech, Bush described Katrina as wiping clean the region ("Along this coast, for mile after mile, the wind and water swept the land clean. In Mississippi, many thousands of houses were damaged or destroyed. In New Orleans and surrounding parishes, more than a quarter-million houses are no longer safe to live in"). While lamenting the history of racial injustice, Bush locates these issues in a bygone past. His speech makes no reference to the conditions that make certain populations vulnerable to "acts of God." Bush instead sees economic and technological opportunities in the aftermath of Katrina. According to him and the business and political interests that support him, Katrina created a blank slate for developers and businesses to reconstruct without the burden of displacing the already displaced, thousands of poor, black, and undocumented residents now dispersed throughout the country.

The displacement process was the result of a haphazard evacuation that relied primarily on charter and school buses. It has led to large populations of Orleanians rebuilding their lives in other regions of the country and a whitening of the city. The *New York Times* referred to the movement as "a diaspora of historic proportions. Not since the Dust Bowl of the 1930s or the end of the Civil War in the 1860's have so many Americans been on the move from a single event. Federal officials who are guiding the evacuation say 400,000 to upwards of one million people have been displaced from ruined homes, mainly in the New Orleans metropolitan area."[43] The rescue strategy in New Orleans evidenced a complete disregard for social relations among black and poor residents, while prison populations and the undocumented remained off the media and government's radar, which was tracking those who "should" be saved. One evacuee was quoted in the *New York Times* as stating: "In the middle of the flight they told us they were taking us to New Mexico. . . 'New Mexico,' everyone said. 'My God, they're taking us to another country.'"[44] The separation of family members and disruptions of community networks were possible because the thousands of subjects stranded were considered impediments to both the process of draining and the process of rebuilding the urban environment.

What is more, developers and federal public officials have portrayed the City of New Orleans as an obstacle to its own development and progress. Through reports of looting and disorder, background stories on the pre-Katrina conditions of the city, and comparisons between the city and less developed regions of the world, New Orleans was portrayed as both premodern and decaying. While the national audience was encouraged to sympathize with individualized narratives through the consumption of images and through charitable donations, simultaneously New Orleans was portrayed as deserving its demise. Its high crime rate, failing school system, and bureaucratic inefficiency were emphasized, along with its racial demographics: a city in which nearly two-thirds of its pre-Katrina population and most of its public officials were black. To solidify this point, *A Failure of Initiative* rests much of the blame for the high number of deaths in New Orleans on Mayor Nagin's shoulder, making him (and Governor Blanco) largely accountable for evacuating those who remained behind after a mandatory evacuation was ordered: "Finding: The failure to order timely mandatory evacuations, Mayor Nagin's decision to shelter but not to evacuate the remaining population, and decisions of individuals led to incomplete evacuation."[45] The weight of this statement grows heavier when compared with the report's assessment of President Bush and Michael Chertoff, Secretary of Homeland Security: "Finding: It does not appear the president received adequate advice and counsel from a senior disaster professional."[46] On policy and urban development levels, Hurricane Katrina was the panacea that wiped the city clean of its disease, that is, lawless, lazy, and premodern blacks in large numbers. The framing of displaced black residents, instead of Katrina or the years of social and structural neglect, as the cause of the city's destruction allows for developers to begin the process of reshaping the city without regard for the majority of the city's population prior to the hurricane. Urban renewal projects of recent years have focused on repopulating U.S. cities with middle-class white singles and families. In most places, these initiatives have met with much opposition as the movement of whites into these areas is instrumental to the removal of blacks, Latinos, and poor, disenfranchised communities from areas where many have lived for generations. New Orleans, however, one of the blackest cities in the United States, one with a much reported failing infrastructure, provides the opportunity for capital to accelerate and expand now that the traditional obstacles have been removed: disenfranchised subjects, public housing, and locally based community networks.

Thus, the military's late arrival to provide rescue operations for those stranded in New Orleans was not a failed operation, but rather an instrument in facilitating the city's redevelopment by protecting and guarding contractors and the interest of entrepreneurs over private citizens. Given that the technologies of disaster recovery, war, and reconstruction overlap, military forces were right on time and well positioned for the work of creating New Orleans anew. In part, the use of amphibious vehicles, armored tanks, and helicopters to re-establish order and to create paths for recovery and reconstruction highlights these technologies' effectiveness as the tools of war and diffusing social unrest but not the tasks of humanitarian and rescue missions. The links that Bush attempted to make between the Katrina event and the War on Terror solidified in the activities and plans following the evacuation of those stagnant bodies literally and figuratively impeding progress. Investigative journalists pointed to the fact that the same companies receiving large security and rebuilding contracts in Iraq were put to work in New Orleans a week after the hurricane arrived.

> As business leaders and government officials talk openly of changing the demographics of what was one of the most culturally vibrant of America's cities, mercenaries from compa-

> nies like DynCorp, Intercon, American Security Group, Blackhawk, Wackenhut, and an Israeli company called Instinctive Shooting International (ISI) are fanning out to guard private businesses and homes, as well as government projects and institutions . . . Some, like Blackwater, are under federal contract. Others have been hired by the wealthy elite, like F. Patrick Quinn III, who brought in private security to guard his $3 million private estate and his luxury hotels, which are under consideration for a lucrative federal contract to house FEMA workers.[47]

These security forces, along with the U.S. military, operate as "war machines," which Mbembe defines as "diffuse organizations" that are highly mobile and "are characterized by their capacity for metamorphosis."[48] These loose structural organizations and their flexible technologies stand in contrast to older forms of militarization and occupation. Their organizational structures and tools can be used for multifarious purposes, from destroying infrastructure and landmarks to reconstituting social order to reconstruction through the erasure of certain populations.

Two of the most important lessons of Katrina, according to the federal government's investigation, are individual accountability and better access to technology. The emphasis on individual accountability furthers the emergence of the weather citizen, who consumes weather as media spectacle and consumes preparedness goods for comfort and as civic duty. In terms of governmental responsibility, *A Failure of Initiative* argues for the need of all levels of government to strengthen their technological infrastructure and to invest in more sophisticated machines. The report emphasizes efficiency, protocol, and technological consumption. The privileged site of technology in this dominant account of Katrina further mitigates the suffering and disposability of displaced subjects. Much of the media coverage of reconstruction has focused on mending the levees to survive the next hurricane season. The general interest technology magazine *Popular Mechanics* ran a cover article, "Now What? The Lessons of Katrina," in which the writers debunked a series of "myths" about Katrina by presenting the scientific and technological "reality" of the Katrina event. In privileging scientific truth and technological progress, the writers present a series of suggestions for the next-time scenario, focusing on better technological investment: "In disasters, the right tools are everything. PM [*Popular Mechanics*] chose three Katrina-tested technologies that should be part of every emergency manager's arsenal."[49]

While one might expect a publication such as *Popular Mechanics* to focus on equipment-based solutions, this same framework defines government and private industry's post-Katrina rebuilding efforts. The government has focused on adding technologies to its arsenal of disaster rescue and recovery, while thousands of Orleanians remain in temporary housing, often isolated from pre-Katrina family and community networks. The inability of many black, poor, and undocumented Orleanians to return to the city aids in the redevelopment initiatives of the government and private contractors. The technological fix to Katrina necessitates a clean slate, that is, the erasure of historical struggles and the removal of disposable bodies. The Katrina event was a result of a narrative of progress and disposability that technology makes visible and can be used to enforce, but technology does not create the narrative. Through the Katrina event, we see both the vulnerability and recalcitrance of the nation-state's investment in a deterministic narrative of progress, one in which the central belief in technology is essentially integrated with (and dependent on) the marginalization and disposability of certain populations. The Katrina event is not a revelation of American racial and class stratification, as those social and discursive processes were not hidden but instead have always hovered at the spectral margins of public discourse and media culture. Instead, the Katrina event reveals the operations of American necropolitics and how dominant discourse about the nation's stability and progress necessitate the decline and death of certain black subjects as the embodiment of national anxiety and a historical past that must be overcome. This revelation was facilitated both by flexible technologies and by the participation of the weather citizen as national spectator. Even as we, as weather citizens, witnessed the state produce living dead subjects, we see historical erasure and technological determinism subsume the state's violence into a narrative of reconstruction in which Katrina "swept the land clean" to begin anew.[50] The technological fix to Katrina then provides dangerous comfort in the post-9/11 nation, where we watch as disaster hovers on the horizon and frames public discourse.

Notes

I am grateful to Carolyn de la Peña, Marita Sturken, Julie Livingston, Elisabeth Soep, Kyla Tompkins, and Vilna Bashi for their comments on earlier versions of this essay. Also thanks to Christa Collins, Deb Willis, Ann Fabian, Nancy Hewitt, Benton Greene, and Jasbir Puar for suggestions and resources. This essay benefited from discussions at the Institute for Research on Women, Rutgers, New Brunswick; the Hurricane Katrina Teach-In at Rutgers University, New Brunswick; and the "Feminist Responses to Disaster" panel at Montclair State University, March 2006.

1. "The lack of a government public communications strategy and media hype of violence exacerbated public concerns and further delayed relief." U.S. House of Representatives, "Executive Summary," *A Failure of Initiative: Final Report of the Select Bipartisan Committee to Investigate the Preparation for and Response to Hurricane Katrina* (Washington, D.C.: U.S. Government Printing Office, 2006), 4. In other sections, the committee spends a great deal of time attempting to correct media reports, in particular on how contracts were divvied out and services dispersed.

2. "Conclusion," *A Failure of Initiative*, 360.
3. Marita Sturken, "Weather Media and Homeland Security: Selling Preparedness in a Volatile World," *Understanding Katrina*, Social Science Research Council, October 5, 2005, online at http://understandingkatrina.ssrc.org/Sturken/ (accessed October 26, 2005).
4. Michael Ignatieff, "The Broken Contract," *New York Times Magazine*, September 25, 2005, 15–16.
5. Achille Mbembe, "Necropolitics," trans. Libby Meintjes, *Public Culture* 15.1 (2003): 11. Although Mbembe's analysis focuses on the colonial and postcolonial state, I find his theory relevant for discussing how the United States assigns disposability to populations of blacks historically and in the present.
6. Mbembe, "Necropolitics," 19.
7. "Aftermath of Hurricane Katrina: New Orleans Mayor Pleads for Help; Race and Class Affecting the Crisis?" *The Situation Room*, CNN, aired September 1, 2005; transcript at http://transcripts.cnn.com/TRANSCRIPTS/0509/01/sitroom.02.html (accessed October 24, 2005).
8. Hurricane Katrina made landfall at Buras, Louisiana, 65 miles south-southeast of New Orleans on August 29, 2005, at 6:10 a.m. CDT. "Pre-Landfall Preparation and Katrina's Impact," *A Failure of Initiative*, 71.
9. As quoted in "Investigative Overview," *A Failure of Initiative*, 12.
10. Katherine Fry, *Constructing the Heartland: Television News and Natural Disaster* (Cresskill, N.J.: Hampton Press, 2003).
11. Sturken, "Weather Media and Homeland Security."
12. Ibid.
13. Mark Svenvold, "Look, Dear—More Catastrophes!" *Washington Post*, November 6, 2005, B1, B5. Also see Svenvold, *Big Weather: Chasing Tornados in the Heart of America* (New York: Henry Holt, 2005).
14. Marita Sturken, "Desiring the Weather: El Niño, the Media, and California Identity," *Public Culture* 13.2 (2001): 187.
15. Ibid., 172.
16. Scott Shane and David D. Kirkpatrick, "Unaware as Levees Fell, Officials Expressed Relief," *New York Times*, March 2, 2006, A16. Also see "Video Shows Bush Katrina Warning," BBCNews, http://news.bbc.co.uk/1/hi/world/americas/4765058.stm (accessed March 15, 2006).
17. "The reality, say several aides who did not wish to be quoted because it might displease the president, did not really sink in until Thursday night. Some White House staffers were watching the evening news and thought the president needed to see the horrific reports coming out of New Orleans. Counselor Bartlett made up a DVD of the newscasts so Bush could see them in their entirety as he flew down to the Gulf Coast the next morning on Air Force One." Evan Thomas, "After Katrina: How the Response Became a Disaster Within a Disaster," *Newsweek*, September 19, 2005, 32.
18. Lt. Gen H. Steven Blum, as reported to the Select Committee on October 27, 2005. "Conclusion," *A Failure of Initiative*, 361.
19. According to the report of the Select Committee, the U.S. Coast Guard rescued more than 33,000 and the Louisiana National Guard more than 25,000. "Evacuation," *A Failure of Initiative*, 116.
20. A few days after Katrina hit, CNN showed much less exciting footage of the levees being plugged. For minutes the audience could hear the whir of helicopters and intermittent commentary from a local reporter or anchorperson as engineers attempted to drop massive sandbags into the gaps left by the breakage in the levees.
21. Here I refer to Naomi Klein's analysis of profiteers of disaster: "The Rise of Disaster Capitalism," *The Nation*, May 2, 2005, posted April 14, 2005, at http://www.thenation.com/doc/20050502/klein (accessed September 22, 2005). Also see "Needed: A People's Reconstruction," *The Nation*, September 26, 2005, 12.
22. As quoted in "Investigation Overview," *A Failure of Initiative*, 6. Thompson was interviewed by the committee on December 6, 2005.
23. "Levees," *A Failure of Initiative*, 93.
24. "Pre-Landfall Preparation and Katrina's Impact," *A Failure of Initiative*, 73–74.
25. From abcnews.com, September 2, 2005, http://abcnews.go.com/GMA/HurricaneKatrina/story?id=1090173&CMP=OTC-RSSFeeds0312 (accessed November 10, 2005).
26. "Anonymous Benefactor Gives Jackson a Home in Atlanta," abcnews.com, September 3, 2005, http://abcnews.go.com/GMA/HurricaneKatrina/story?id=1093853 (accessed November 10, 2005).
27. See chapter 3, "Inequity and the Environment," of Craig E. Colten's *An Unnatural Metropolis: Wresting New Orleans from Nature* (Baton Rouge: Louisiana State University Press, 2005).

28. Colten, *An Unnatural Metropolis*, 106–07.
29. Ibid., 154.
30. Julie Sze, "Toxic Soup Redux: Why Environmental Racism and Environmental Justice Matter after Katrina," *Understanding Katrina: Perspectives from the Social Sciences*, Social Science Research Council, October 24, 2005, online at http://understandingkatrina.ssrc.org/Sze/ (accessed October 26, 2005).
31. Juliet Eilperin, "Flooded Toxic Waste Sites Are Potential Health Threat," *Washington Post*, September 10, 2005, A15. Also see U.S. Army Corps of Engineers' reports on the sites: "Bayou Bonfouca Superfund Site," last updated April 26, 1999, at http://www.mvn.usace.army.mil/pd/iis/BayouBonfouca.HTM (accessed November 2, 2005); "Agriculture Street Landfill Superfund Site New Orleans, Louisiana," last updated July 6,1999, at http://www.mvn.usace.army.mil/pd/iis/agriculture.htm (accessed November 2, 2005).
32. Theodore Steinberg, *Acts of God: The Unnatural History of "Natural" Disaster in America* (New York: Oxford University Press, 2000), 98.
33. Journalist Jon Ellison wrote that the year previous to Katrina, FEMA had denied three requests from Jefferson Parish, one of the hardest hit areas of southeastern Louisiana, for grants to decrease its flood vulnerability. Jon Elliston, "Confederacy of Dunces," *The Nation*, September 26, 2005, 5.
34. Sharon Holland, *Raising the Dead: Readings of Death and (Black) Subjectivity* (Durham, N.C.: Duke University Press, 2000), 15.
35. Susan Sontag, *Regarding the Pain of Others* (New York: Picador, 2003), 60.
36. Joseph B. Treaster and Abby Goodnough, "Powerful Storm Threatens Havoc along Gulf Coast," *New York Times*, August 28, 2005, A12.
37. Anne Rice, "Do You Know What It Means to Lose New Orleans?" *New York Times*, Editorial, September 4, 2005, Sec. 4, 11.
38. "Evacuation," *A Failure of Initiative*, 116.
39. Keith Naughton and Mark Hosenball, "Cash and 'Cat 5' Chaos," *Newsweek*, September 26, 2005, 34–36.
40. "Pre-Landfall Preparation and Katrina's Impact," *A Failure of Initiative*, 65.
41. Eric Klinenberg, *Heat Wave: A Social Autopsy of Disaster in Chicago* (Chicago: University of Chicago Press, 2003), 91.
42. Presidential Address, "President Discusses Hurricane Relief in Address to the Nation," Jackson Square, New Orleans, La., September 15, 2005. Transcript accessed at http://www.whitehouse.gov/news/releases/2005/09/20050915-8.html (accessed September 28, 2005).
43. Timothy Egan, "Uprooted and Scattered Far from the Familiar," *New York Times*, September 11, 2001, Sec. 1, 1. What has been seriously underrepresented in coverage of Katrina is the number of migrant workers and undocumented people living in the region.
44. Ibid., 1.
45. See "Evacuation," *A Failure of Initiative*, 111. In more severe terms, this section concludes: "New Orleans' decision to shelter instead of evacuate the population, as well as individuals' reluctance to leave, further resulted in an incomplete evacuation. The thousands of people left in New Orleans suffered death or had to be rescued to await an evacuation that should have already occurred before landfall" (123).
46. "National Framework," *A Failure of Initiative*, 132.
47. Jeremy Scahill, "Blackwater Down," *The Nation*, October 10, 2005, 18.
48. Mbembe, "Necropolitics," 32.
49. Camas Davis et al., "Now What? The Lessons of Katrina," *Popular Mechanics* (March 2006), 66.
50. Presidential Address, "President Discusses Hurricane Relief in Address to the Nation."

Boundaries and Border Wars: DES, Technology, and Environmental Justice

Julie Sze

According to a recent study, newborn babies in the United States absorb an average of 200 industrial chemicals and pollutants into their bodies via their mothers' wombs. The umbilical cord blood contains pesticides, consumer product ingredients (Teflon, stain and oil repellants from fast food packaging, clothes, and textiles), and wastes from coal, gasoline, and garbage.[1] Even in a highly polluted society, the extent to which newborns come into the world with a "human body burden" comes as disturbing revelation for environmental activists and health researchers. From a cultural perspective, these findings raise questions about the impact of industrial production on the human body, especially in pregnancy. That newborns enter the world marked by pollution highlights the contradiction between the idealized notion that babies are innocent and pure and the reality that they are born in and of the toxic soup that comprises the post–World War II landscape of pesticides, chemicals, and plastics. In other words, polluted babies are troubling creatures because they collapse the boundaries of the bodily and the natural with the technological, the man-made, and the synthetic.

Arguably, the first well-known example of the problems of synthetic intervention during pregnancy and childbirth arose with the widespread use of diethylstilbestrol (DES), a man-made estrogen created in 1938 by Englishman Charles Dodds. It was "a novel and a daring product" that produced the same feminizing effects as estrogens derived from plants and animals but was three times more powerful.[2] DES was a man-made version of a "natural" hormone presumed, because it was derived "from nature," to be beneficial. In reality, DES was toxic for the main populations to whom it was given: women and livestock animals. Between 4 and 6 million women in the United States from 1948 to 1971 were prescribed DES to prevent miscarriages.[3] For women, it was promoted by pharmaceutical companies and prescribed by doctors as a "miracle drug," also used to treat menopause, to dry up breast milk in non-nursing mothers, as a "morning after" contraceptive, to prevent girls from growing "too tall," and to aid male-to-female transsexuals before sex change operations.[4] It was also standard practice for farmers to give DES to chickens, cows, and other livestock because they grew fat, their meat became succulent, and males were chemically castrated.[5] Mothers given DES during pregnancy were later suspected of having higher risk of breast cancer, and DES-exposed children proved to have higher risks of various cancers and genital abnormalities, representing the first known human occurrence of transplacental carcinogenesis (cancer-causing effects) due to in utero exposure.[6]

How do human and animal bodies interact with technology, and how do these interactions illuminate the contested terrain between technology and environmental studies? The traditional opposition of technology/machine or environment/nature is explicitly rejected by Donna Haraway in her influential account of the cyborg.[7] Building upon Haraway's articulation of the cyborg as a cybernetic hybrid of organism and machine, this essay argues for an American studies analysis of DES and a revision of the cyborg concept through the framework of "technologically polluted bodies." This analysis complicates hybridity, purity, and nature as cultural concepts in technological and environmental studies.[8] If, as Jacques Ellul argued in 1964 in his influential *The Technological Society* that technology is the "stake of the century," then the forms of the technologically polluted bodies (both women and animals) that I discuss are the stakes of this new century, and need serious cultural analysis from the perspectives of technology and environmental studies.[9]

DES offers a unique prism to understand nature's social construction through technology and the human body, as opposed to the wilderness as the idealized site of "nature" in environmental studies. Environmental historians have questioned the centrality of the idea of "nature" as pristine green space absent of people, and the racial and gendered implications of this construction.[10] The emerging social movement and academic field of environmental justice offers another challenge to the "nature of nature" as a category unmarked by race or class. At the core of the term *environmental justice* is a redefinition of the "environment" to mean not only "wild" places, but the environment of human bodies, especially in racialized communities, in cities, and through labor (exemplified by the movement slogan that the environment is where people "live, work, play, and pray"). Defining and redefining "nature" and "environment" are thus cultural questions, of which narratives and stories are a central part.[11] The explosion of research on environmental justice has taken place primarily in the social sciences, although it is increasingly being seen as an important area of humanities and American studies research.[12] Using the environment and environmental justice to understand DES deepens the projects of denaturalizing nature and understanding the body's interactions with technologies in American political, cultural, and corporate contexts.

Using DES as the case study, I advance an understanding of technologically polluted bodies—not as Haraway's hybrid of organism and machine, but hybrids of bodies—*animals and human, and particularly female*, with nonmachine-based forms of technological intervention, such as the pharmaceutical, petrochemical, and livestock industries and the products they create and normalize through their production processes. DES is a rich example of how human and animal bodies interact with a variety of technological and environmental systems. Understanding DES bodies as technologically polluted is to argue, as Haraway suggests, against the purity and integrity of social, natural, and bodily categories, and in favor of what she calls "boundary breakdowns." American studies offers an interdisciplinary focus on cultures of consumption, the politics of corporate practices, and their impacts on bodies and communities. My analysis also involves components that have been rarely used to understand DES, specifically literary analysis, and critical questions of race, class, and pollution. This approach to DES enables different epistemological and political questions about the nature of nature, the relationship between the technological and the natural, and between the environmental and human harms that result from particular technologies. Analyzing DES through the framework of technologically polluted bodies highlights how categories of race, gender, human/animal, and nature are unstable, shaped and contested by ideas and cultures, and through corporate industries, which actively shape these categories through their products and processes.

What do new forms of technology mean to our understanding of what it means to be human (or animal)? Crucially, how do these technologies emanate from particularly American values and views of nature, progress, control, and optimism? As a case study in polluted women and livestock (animal bodies), DES illustrates changes in the human relationship to nature and what these changing relationships might mean for the possibility of justice and ethics in a hyperpolluted, highly technological world of corporate concentration. If we are what we eat in food and medicine (DES in animal bodies consumed as food for humans, and given to women as medicine), then what we eat alters our body in a feedback loop that calls into question any idea of the body or nature that is pure or unadulterated. DES bodies, like the newborns, illuminate how we are already hybrid and that there is no nature or body that is not shaped by culture, technology, or medicine, no purity that we can stand upon to define concepts of nature, race, gender, or humanity itself. Those ubiquities of border crossings, hybridity and cyborgian alterations, mean that the need for politics or philosophy of justice and/in technology becomes even more urgent.[13]

Female Bodies and Scientific Bodies of Evidence

DES, a man-made and nonsteroidal estrogen, was synthesized in 1938 by chemist-physician Edward Charles Dodds in London.[14] Derived from coal tar, it differs from steroidal estrogen in that it lacks the four interlocking carbon rings that characterize natural steroid hormones and their derivatives. As Cynthia Orenberg explains, DES lacks the chemically distinct four-ring structure. DES is a "ringer" and "*fundamentally different*" in its chemical structure. Although it behaves empirically "like estrogen," it isn't. It is three times more powerful and not destroyed or affected by gastric secretions as is natural estrogen. For Orenberg, the central paradox is that DES was "accepted as the real thing" by physicians who "never thought that they might be tampering dangerously with nature."[15] Dodds "resolved to diverge radically from nature in order to mimic it."[16] He was also an inherently "conservative" physician, who rejected the idea of DES as a "miracle" cure. He expressed "humble" respect and awe for the female reproductive body.[17] He saw its impacts when male workers in his laboratory who handled DES grew female breasts and became impotent (eventually DES was produced by women to avoid this problem).[18] He also suspected other problems, specifically, that in being highly estrogenic, it was also carcinogenic.[19] By 1939, forty articles demonstrated the carcinogenic effects of synthetic and natural estrogens in animals, including several that focused specifically on DES. By 1941, 257 papers demonstrated the value of DES for menopausal symptoms and other uses.[20]

Although DES was first synthesized in London, its widespread popularity can be considered an American phenomenon stemming from the crucial role that U.S. doctors associated with elite medical institutions and pharmaceutical companies played. Although he conducted his research as part of the British system and was prohibited from taking a patent for DES, Dodds endorsed the idea that his invention stay within the public domain so that it could be used for the "greater good of humanity" (especially since he created DES in a wartime race with the Nazis).[21] No restraints held back pharmaceutical companies who distributed samples widely and encouraged research on DES's "miraculous effects." The widespread use of DES during pregnancy began in 1947 because of a husband-and-wife research team at Harvard Medical School, George and Olive Smith. Her 1948 ground-breaking paper and their 1949 follow-up in the *American Journal of Obstetrics and Gynecology* encouraged the use of DES to prevent miscarriages.[22] According to their theory, elevated estrogen levels during pregnancy stimulate progesterone, which is essential for the uterus to receive and sustain the egg. Inadequate levels of either would lead to complications or failure of the pregnancy.[23] At a 1949 medical meeting, the Smiths announced that DES benefited *all* first-time pregnancies. In their words, DES seemed to "render normal gestation 'more normal.'"[24] DES was considered benign because it was making a "natural," "biological," and "normal" process more effective.[25] How DES was relentlessly promoted exemplifies a utopian belief that technologies could harness and "improve" on nature itself. While this belief system was not unique to the United States, certain factors made it particularly dangerous, specifically the tactics of the pharmaceutical sectors in ensuring that the Food and Drug Administration (FDA) was a weak regulatory agency and in their marketing power.

The Smiths' theory was refuted by a 1953 University of Chicago study by William Dieckmann, which definitively found that DES was *ineffective* at preventing miscarriages. However, doctors continued to widely prescribe DES in normal pregnancies like "vitamins" or "a little extra insurance."[26] They continued because of the highly aggressive sales tactics of pharmaceutical representatives and because it was widely advertised in medical journals and the popular press. Influenced by the advertisements, and in their desperate desire to avoid miscarriages, women demanded DES. One such 1957 advertisement featuring a happy baby read, DES is "recommended in ALL pregnancies . . . desPLEX tablets also contain vitamin C and certain members of the vitamin B complex to aid detoxification in pregnancy and the effectuation of estrogen."[27] Its use was unabated in pregnant women until a 1971 study that confirmed the link between in utero DES exposure and a rare vaginal cancer, clear cell adenocarcinoma, was published in the *New England Journal of Medicine*. Only then did the FDA issue an alert about DES use in pregnancy.[28] According to DES Cancer Network, none of the 267 pharmaceutical companies who produced and distributed DES have accepted any responsibility for DES's health effects.

Reading DES and Gender

Existing histories of DES analyze it through the lenses of medical history and the history of regulation/ industry in the livestock or pharmaceutical sectors, and ignore certain cultural questions. There is, in other words, no parallel to Rachel Carson's incisive critique of DDT and pesticides in postwar American culture as captured in her historic 1962 *Silent Spring*.[29] In contrast, what we have is what science and technology scholar Joseph Dumit describes as a set of histories that detail "its incredibly tragic history within a kind of enlightenment narrative. They state that DES was not studied carefully enough at first, and those studies which showed problems were ignored by the medical community at large. When irrefutable proof of DES's harm was provided in 1971 (the narrative goes) the medical community responded, the public was outraged, and more research was conducted."[30] These accounts treat DES in animals and women alongside parallel tracks, making its time line difficult to decipher (DES was banned in chickens in 1958, while its use in cattle continued until 1979, and its use in women continued through the 1980s).[31] They generally portray DES as a tragic or peculiar historical episode that tells a particular tale about medical knowledge or about the history of government regulation.[32] For others, DES is "a modern meat production milestone, perhaps the most important single occurrence in the chain of events that culminated in the current methods of production."[33]

The particularly *American* aspects of DES tell a story about the relationship of production and dissemination that, as Alan Marcus suggests, emerge from a complicated dance between corporate and academic scientists and government regulators. The use of DES in beef came from a researcher at Iowa State, its use in chickens from the University of California at Davis.[34] The corporate-academic-government nexus in which DES emerged is specifically a U.S. model of research that flourished in the postwar era. How can the power of the pharmaceutical and agricultural lobbies and the inefficacies of the FDA and the U.S. Department of Agriculture (USDA) as regulatory agencies be compared against each other? These kinds of questions cannot be asked or answered using existing analytic frames. And one cultural question that is almost never asked is: how can we think about the relationship between women and animals through DES?

When DES histories focus on women, they tend to insufficiently consider gender with complexity. Popular and historical accounts on DES's use in women deploy an early feminist framework that emphasizes women's victimization.[35] Orenberg describes DES as "her story" as well that of her daughter exposed in utero. DES is "only one example of the consequences of thinking that modern medicine is infallible, that the physician is sacrosanct, and that the patient (particularly the *woman* patient) is an object to be 'done to.'"[36] DES represents prevailing medical practices and the community's attitude toward research on women without their consent, as shown by the "chilling number of trial-and-error medical experiments using DES on women."[37] Personal narratives are gendered insofar as they focus on the perspectives of the women who took it or DES daughters were exposed in utero, whose genitalia were arguably more altered than those of DES sons, and whose cancer risk was higher.[38] Lastly, the FDA's continued inaction on DES was successfully challenged by the women's health movement, particularly by DES-exposed mothers and their children.[39] Women's magazines, which had promoted DES earlier, later became effective venues for communicating its risks.[40]

What these narratives and histories lack are a sustained consideration of how gender and sexual development as constructed categories can illuminate the cultural significance of DES. For example, in 1958, DES use was suspended in poultry when cases of early sexual development in young children in Puerto Rico and Italy were correlated with high chicken consumption.[41] Also, male farm workers who fed DES to chickens and men who ate large amounts of chicken developed breasts, reported sterility, lost facial hair, and developed high-pitched voices.[42] What made DES's use in chickens unpalatable (literally), was the way in which it visibly made men "women," and made children sexually mature. The "unnatural" sexual and bodily developments that DES triggered were visible, embodied, and therefore grotesque. When these categories of gender and sexual development were made manifest, DES was banned. Although the long-term harms of DES in other livestock and in women were as harmful, they were not visible in quite the same way, and not acted upon until there was a greater body of scientific evidence of DES's harms and changing political and cultural climates.

DES highlights how *female identity* itself was defined medically and socially through hormones. If being female and male can be defined through hormones, then a whole host of female and male "problems" could be solved. Throughout the twentieth century, female hormones were given to categories of women who were considered insufficiently female, such as menopausal women and lesbians.[43] The larger epistemological and cultural questions—why these "conditions" were considered a medical problem in the first place requiring medical intervention—have been explored by scholars in queer studies and intersex studies. This same line of critical interrogation has analyzed the parallel "problem" of the menopausal woman, since the first popular and regulated use for DES's "miraculous" effects was on menopausal women. The search for a remedy for menopause has been replete with cultural stigmas about female identity in postreproductive years.[44] As one popular book in the 1960s by Robert Wilson titled *Feminine Forever* stated, "a woman is not 'complete'" unless she takes hormone replacement pills, and will be "condemned to witness the death of her own womanhood."[45] National advertisements in the 1950s and 1960s for hormone replacement therapy were often openly sexist, depicting menopausal women as "repulsive, witchlike . . . angry or depressed, menacing" and prone to violence once their reserve of estrogen was gone.[46]

Lastly, the most obvious way in which gender shapes DES is through the critique of the gender normative definitions of a woman's identity through pregnancy and childbirth. Miscarriage was to be avoided at all costs, at least in part because women defined their female identity with successful fertility, childbirth, and ultimately motherhood. Thus, with the goal of having a successful pregnancy and birth, millions of women took a drug derived from their natural hormones that ultimately led to their daughters having higher cancer risks and drug-altered uteruses that would make their own pregnancies difficult and dangerous.

Cultures of DES: Technology, Cyborg, and Hybrid Stories

In addition to rereading DES through critical frames of American values and corporate culture, gender construction and sexual identity, we need to consider it from the perspective of technology studies. Is DES a reproductive technology?[47] As one feminist critic notes: "technological interventions in the womb are extraneous parties (*objects or people*) that hinder, modify, or enhance female reproduction."[48] Although most descriptions of reproductive technology do not include DES, it *does* fall under a broad definition of technology in general and reproductive technology specifically as an application of scientific knowledge to assist in making babies (or in this case, by supposedly preventing miscarriages).

Ellul's definition of "technique" is illuminating in reframing DES vis-à-vis technology studies. He writes that although we automatically link technology with machines, "it is a radical error to think of technique and machines as interchangeable."[49] Rather, "technique does not mean machines, technology, or this or that procedure for attaining an end. In our technological society, technique is the totality of methods rationally arrived at and having *absolute efficiency*."[50] Broad definitions of reproductive technology that include attempts to "improve" pregnancy thus include DES under their umbrella. Expanding the definition of reproductive technology to include DES also draws the pharmaceutical sector into a history that predates in vitro fertilization and other more easily recognized reproductive technologies.[51] DES is neither object nor person, but its function was indeed to "enhance" female reproduction. It was used, in the spirit of the Smiths and a generation of doctors treating pregnant women, as a tool to achieve "better," more natural and normal, pregnancy—in Ellul's words, the goal of *absolute efficiency* in pregnancy and childbirth. DES—as technology and technique—complicates definitions of technologies as object or machine-based.

Reconsidering DES vis-à-vis technology studies also situates it within the framework of Haraway's influential cyborg theory. As a reconsideration of the cyborg, DES provides an ideal case study for understanding how to integrate cultural theory, pedagogy, and activism. I use Haraway's cyborg theory to suggest how to teach DES in a classroom using literary analysis. My chosen text is Ruth Ozeki's narrative *My Year of Meats* (1998).[52] The novel is narrated by Jane Takagi-Little, a "DES daughter." Jane is a mixed race (Asian/white, Japanese/American) aspiring documentary filmmaker who begins her story as corporate tool for a Japanese TV show, *My American Wife!*, sponsored by an American meat export lobby to increase its sales in Japan. I focus on this novel because it resurrects the DES story long buried from popular consciousness. Like Upton Sinclair's *The Jungle*, it is a novel with muckraking intentions. It also brings up complex themes of hybridity and of the cyborg, which can illuminate larger questions about culture, technology, and the body, in an accessible text that is easily read and taught in an undergraduate classroom.

Haraway writes that cyborgs are a "cybernetic organism, a hybrid of machine and organism, a creature of social reality as well as a creature of fiction."[53] Her focus on the cultural production of the cyborg as a "creature of fiction," and our collective complicity in their existence, is to place agency, pleasure, and politics of their construction and existence into our own hands. As a creature of "fiction," the cyborg is located where the appropriation of nature in the production of culture meet and mesh, in which "the stakes in the *border war* have been the territories of *production, reproduction, and imagination*."[54] Her focus on these intertwined realms is echoed by Dumit in his call for an investigation of the epistemology of the "facts" of DES.[55] A fact is "a word used to describe the situation where (our) culture and nature agree. To call something a fact is to represent a cultural consensus on the nature of nature."[56] Dumit argues that drug-altered bodies, in particular DES bodies, are cyborg and that more research needs to be done on how and why original cyborgian alteration with drugs extends to further alterations with technologies.[57] If DES bodies are cyborgian, what are the politics of that identity? Haraway suggests that the cyborg myth is about "transgressed boundaries, potent fusions and dangerous possibilities which progressive people might explore as one part of needed political work."[58] Three boundary breakdowns are central to understanding cyborgs and the boundaries they transgress: the breaching of the boundary between human and animal, between animal-human (organism) and machines, and between the physical and the nonphysical.[59] Transgression enables freedom from epistemological and historical constraints and dominations. Cyborgs complicate long-standing dualisms that have functioned to dominate women, people of color, nature, workers, and animals.[60] For Haraway, "cyborg politics is the struggle for language and the struggle against perfect communication, against the one code that translates all meaning perfectly . . . .That is why *cyborg politics insists on noise and advocates pollution, rejoicing in the illegitimate fusions of animal and machine.*"[61]

In thinking about how Haraway's cyborg analysis can be used to understand DES, several questions emerge that confirm and possibly trouble her analysis. Are "illegitimate fusions" of animal and machines necessarily something to "rejoice"? Is the pleasure inherent in the transgression, in the border crossing, and in collapsing categories? Why do cyborg politics necessarily "advocate pollution"? What does this "pollution advocacy" mean, vis-à-vis the *actual* case of DES? Are technologically polluted bodies just a naive wish to return to a precyborgian, prehybrid state of the unpolluted body? What are the "dangerous possibilities" that cyborgian transgressions, boundary breakdowns, and rejections of dualisms represent, if we are to take up Haraway's call as "progressive people" exploring "political work"? These are questions I now turn to through a close analysis of DES in *My Year of Meats.*

DES Narratives: Gender, Race, and Hybridity in *My Year of Meats*

As Haraway suggests, the "border war" between nature/culture and environment/technology is contested in the terrains of production, reproduction, and imagination. Thus, production, reproduction, and imagination are useful frames for analyzing cyborgs and, by extension, DES as cultural narratives. In photography collections, documentaries, and plays, DES has been a topic of cultural production.[62] This focus is unsurprising, because the personal stories that these cultural productions reveal serve as important counternarratives to the overwhelming statistical and medical tone of dominant DES narratives.[63] What sets *My Year of Meats* apart from most DES histories is its dual focus on both women and animals. Takagi-Little is a "DES daughter" who works for *My American Wife!*, which highlights a variety of American women, their families, and their meat dishes from particular regional and ethnic subcultures in the United States. The aim of the show is to increase American meat sales in Japan by teaching Japanese women how to cook unfamiliar kinds of meat.

As a racially mixed DES daughter, Jane embodies Haraway's critique of static categories of identity, specifically of racial and national identities. The author ties Jane's mixed-race status to two key terms in the novel: hybridity and sterility. As a person of mixed race and binational heritage, Jane acts as a cultural broker between two nations and cultures. As she explains, "being racially 'half'—neither here nor there—I was uniquely suited to the niche I was to occupy in the television industry. . . . Although my heart was set on being a documentarian, it seems that I was more useful as a go-between, a cultural pimp, selling off the vast illusion of America" (9). While Jane's mixed-race status gives her authority as a cultural broker between nations and races, Ozeki also ties her status to nonhuman (specifically plant) hybrids. Hybrids are clear rejections of "nature." But this rejection is complex for Jane. Neither unproblematic nor idealized, hybrids stand as both a *warning sign* and an *opportunity* to escape a cultural past obsessed with notions of purity. Thus, race and culture are likened to native species that are increasingly moving to nonnative locales. For Jane, the cautionary tale of this crossing is in the story of kudzu, an invasive plant species that represents the dangers of careless botanic transplantation and "biological invasions."[64] Kudzu is a Japanese plant that was touted as a "miracle plant" in the United States and brought to the South to "rescue" the depleted southern soil. But it soon overran more than 500,000 acres in the South, due to its "predaceous and opportunistic" and fast-growing nature (it often grows up to a foot a day). It echoes DES itself and thus represents the unforeseen consequences when miracles (whether technological or botanical) go awry. Jane explains that kudzu is used as a disparaging metaphor by American nativists for the economic "invasion" of the Japanese in the South (77). At the same time, the movement of plants and people represents an outcome that Jane welcomes. As she describes, "all over the world, native species are migrating, if not disappearing, and in the next millennium, the idea of an indigenous person or plant or culture will just seem quaint. Being half, I am evidence that race, too, will become a relic. . . . Some days, when I'm feeling grand, I feel brand-new—like a prototype. . . . Now, oddly, I straddle this blessed, ever-shrinking world" (15). Kudzu is an ambiguous metaphor for both Jane and the author, representing both freedom and danger in the flows and movements of peoples, plants, and cultures around and within the world.

Jane's racial hybridity is also linked to her fertility, both actual and perceived, and like the discussions of human and plant hybrids in the novel, complex and ambivalent. On the one hand, as Joichi Ueno, the Japanese advertising executive in charge of the Beef-Ex account and the show says in a moment of drunken flirtation: "You, Takagi, are a good example of hybrid vigor . . . . We Japanese get weak genes through many centuries' process of straight breeding. Like old-fashioned cows. Make weak stock. But you are good and strong and modern girl from crossbreeding" (43). Part of her perceived "strength" comes from her height, as she is taller than most Japanese women; and she is more direct and a social nonconformist. "Hybrid vigor" is a well-known term used by breeders and farmers that applies to the "exceptional sturdiness" of the first generation of cross-breeds in animals and plants.[65] Thus, the author uses Jane's racial hybridity to stand for agricultural and nonhuman "natural" ideas of fertility and strength. Jane's "hybrid vigor" is contrasted with her narrative double, Ueno's wife, Akiko. Their marriage is plagued with problems, in part because of Akiko's infertility, as well as his "stony rage." The irony is that Jane has more difficulties getting pregnant and carrying a pregnancy to term because of the structural changes in her uterus as a result of her DES exposure, which she discovers through the course of the narrative. Her infertility caused the fractures that led to her first divorce and is linked to her mixed-race status, because "like many hybrids . . . I was destined to be nonreproductive" (152), referring to the tendency of certain hybrids to be sterile, like mules.[66] Jane's infertility is dually ironic: the internal tragedy and transformations of DES daughters (whose mothers took DES to help themselves be successfully fertile) and because of the external perception that her strength comes from her racial hybridity. Jane expresses the freedoms that come with hybridity that Haraway suggests for cyborgs. But with geographic, biological, racial, and national freedoms come dangers, ambivalence, and complexity. With hybridity, perceptions of hyperfertility and strength work alongside themes of infertility and sterility.

Technologically Polluted Bodies: We Are What We (M)eat[67]

While Jane embodies Haraway's notion of a racial and technological hybrid and cyborg, a close reading of *My Year of Meats* suggests that critiques of fixed categories of identity that cyborgs implicitly embody need not lead to a celebration of technologically polluted bodies. Haraway's cyborg theory provides a crucial and important critique of fixed identities, but in doing so, may make it difficult to critique pollution, a stance that the novel explicitly challenges. The key to this challenge is how Jane is further shaped and altered through what she eats, specifically meat, and what she learns about the industry that uses hormones like DES to "improve" their product. In the novel, Ozeki connects women and livestock in a complicated stance that simultaneously embraces hybridity and rejects pollution.

Jane starts the book as the coordinator of the production team for *My American Wife!* and in that capacity interprets the directives from the Tokyo office on behalf of the show's sponsor (Beef-Ex), to remember that "Beef Is Best." As she writes in a memo describing the show,

> Meat is the Message. Each weekly half-hour episode of *My American Wife!* must culminate in the celebration of a featured meat, climaxing in its glorious consumption. It's the meat (not the Mrs.) who's the star of our show! Of course, the "Wife of the Week" is important too. She must be attractive, appetizing, and all-American. She is the Meat Made Manifest: ample, robust, yet never tough or hard to digest.

Meat, in other words, is how ultimately one can make sense of DES's rise in the United States, both as *a symbol* connecting women and animals, and as a *technological process* to *control nature* and *maximize efficiency* through technology. Anthropologists have long analyzed meat's cultural and symbolic importance, arguing that it represents the domination of humans over nature and nonhumans.[68] Thus, it comes as no surprise that meat is so central to the novel because it acts as a rich signpost of cultural and moral values. As Jane, Ozeki describes the American wives on the show: "through her, Japanese housewives will feel the hearty sense of warmth, of comfort, of hearth and home—the traditional family values symbolized by red meat in rural America" (8).

But the darker side of the "values" of meat is also represented by DES. Thus, women and animals are linked in DES, not accidentally or incidentally, but through an American technological and medical culture that saw the improvement of nature through technology and increased efficiency as central to the larger cultural project of improvement and progress. Further, the links between reproduction and production and between animals and women perform and create *new kinds* of technological violence in the bodies they inhabit. These links between animals and women are not surprising, because, as Haraway notes, one of the key boundary breakdowns that characterizes the cyborg is through the human/animal breakdown. Further, the boundary breakdown between *woman* and *animal* has been a central feature of the history of hormone development, since animals have been central to the development of the birth control pill and reproductive technologies such as IVF and hormone replacement therapy, and as the sources of popular estrogen therapies.[69] DES was first used as a treatment for menopause, but it soon was supplanted by Premarin, the most commonly prescribed drug used to treat menopause introduced in 1942.[70] Because it is derived from a pregnant mare's urine during the third through tenth month of the equine gestational period (hence, the name PREgnant MAre uRINe), an estimated 35,000 mares stand in barns throughout Canada and parts of the midwestern United States for about six months out of every year with urine collection devices strapped onto them, and most of their foals sold for slaughter.[71]

In making animals, meat, and female reproduction increasingly efficient, DES stands not only as a symbol for pollution in the processes of meat and baby-making, but for changing systems of production more generally. Meat functions as a larger symbol of forms of racialized American violence and violence against animals in meat production. Jane describes the well-publicized murder of a Japanese exchange student, Yoshihiro Hattori, shot to death by Rodney Dwayne Peairs when he rang a doorbell to ask for directions to a Halloween party. Jane notes that Peairs worked as a meat packer in Louisiana:

> Hattori was killed because Peairs had a gun, and because Hattori looked different. Peairs had a gun because . . . we fancy that ours is still a frontier culture, where our homes must be defended by deadly force from people who look different. And while I'm not saying that Peairs pulled the trigger because he was a butcher, his occupation didn't surprise me. Guns, race, meat, and Manifest Destiny all collided in a single explosion of violent, dehumanized activity. (89)

Although numerous scholars have described how notions of the frontier, Manifest Destiny, and American imperialism were racialized, Jane adds meat and systems of its production and consumption to this cultural history with contemporary "real-world" implications for how people (especially disenfranchised people, such as people of color and immigrants) live, work, and consume meat.[72] In this sense, the novel expands on Roger Horowitz's examination of race, gender, ethnicity, and technology in the American meat-packing industry. Horowitz asks, but does not answer, the question: is there a role for the household or the consumer in this story? *My Year of Meats* offers precisely that view, of race, gender, and ethnicity in the household and consumption aspects of meat production in the United States.[73]

Individual violent acts like the Hattori-Peairs murder act as microcosms of larger cultural forms of systemic and technological violence enacted through meat production. The health effects of the hormones, drugs, and chemicals are countless, and the cruel conditions that the animals (cramming them into pens, cutting off chicken beaks) live in are well-documented. The conditions in which the animals live in the livestock industry are a logical outcome of the story that began with DES, in other words, the changing technologies of food production and consumption. As Jane describes: "DES changed the face of meat in America. Using DES and other drugs, like antibiotics, farmers could process animals on an assembly line like cars or computer chips. Open-field grazing for cattle became inefficient and soon gave way to confinement feedlot operations or factory farms, where thousands upon thousands of penned cattle could be fattened at troughs. This was an economy of scale. It was happening everywhere, the wave of the future, the marriage of science and big business" (125). Meat, in other words, became increasingly mechanized, animals became things, economies of scale grew, and older forms of food production and consumption were abandoned.

Thus, the key cultural question that *My Year of Meats* asks is not "whether eating meat is natural/right/ethical," but rather, how has meat been made *different* technologically, what kinds of food and social systems have developed in the last fifty years that are significantly different in scale and scope from older systems of production and consumption, and what do race and gender have to do with these changes?[74] What emerged was a production system that completely mechanized its product, in other words, taking animals away from nature and into technology. As Horowitz suggests, "convenience (in meat production) . . . rested on ever greater intervention into *nature* . . . .Implicit in the very notion of convenience was using *technology* to help mankind claim victory over the organic subduing of animals and their parts to the imperative of the human race. Altering animal biology and growth patterns, tinkering with forms of processed meat, adding chemicals to feeds, creating more automated production methods—all were elements of relentless efforts to turn *nature's bounty into products that fit with the modern lifestyles of our civilization*."[75] DES is just one particularly salient (salacious?) example of this tension between nature and technology, reflecting cultural desires of control and domination that shape the particular contours of production.

Yet another logical outcome of these changes in production and its utter moral bankruptcy is represented in *My Year of Meats* by Gale Dunn, the son-in-law of Bunny Dunn, a featured "Mrs." on *My American Wife!* and wife of Colorado rancher-patriarch, John Dunn. Gale describes the cattle feed as "recycled" to Jane: "We even got by-products from the slaughterhouse—recycling cattle right back into cattle. Instant protein . . . the formulate feed we use is real expensive, and the cattle shit out about two-thirds before they can even digest it. Now there's no reason this manure can't be recycled into perfectly good feed . . . you should be really happy, 'cause this pretty much takes care of the 'organic waste' problem . . . .Feed the animals shit and it gets rid of the waste at the same time" (258). The process Gale describes is linked to mad cow disease. Although cows are naturally herbivores, the industry turns them into cannibals by feeding them meal ground from beef and beef bones.[76]

At the same time that systems of meat production became increasingly mechanized, systems of consumption altered to create new markets for the global food industry. Beef-Ex, the trade group sponsoring *My American Wife!* does so explicitly to familiarize the Japanese housewives with meat and how to cook it, in a culture based historically on nonmeat food sources. The strategy was "'to develop a powerful synergy between the commercials and the documentary vehicles, to stimulate consumer purchase motivation.' In other words, the commercials were to bleed into the documentaries and the documentaries were to function as commercials" (41). The novel thus connects meat production with global consumption, including advertising, that functions to create and shape the needs and desires of individual consumers and in national and global markets. Whether that consumer is a housewife buying meat for her family's table (in Japan or the United States), or a pregnant woman pressing her doctor to prescribe DES as a "magic" pill to prevent miscarriages, individual/group consumption and production are inextricably linked through cultural ideas and images, linked through the discourse of the wonders of technological progress and improvement.

DES and the Search for Environmental Justice

Reframing DES and meat as problems of technological production and consumption updates Haraway's cyborg. She argues that "high tech culture's challenge of existing categories and dualisms" is liberatory because these dualisms are challenged and because "it is not clear who makes and who is made in the relation between human and machine."[77] But in using DES as a case study in technological culture, the maker and the made *are* clear. The pharmaceutical and livestock industries made DES, used and promoted it widely, made sure that regulatory agencies were ineffective at protecting the public interest, and altered women's and livestock's bodies in terribly troubling and culturally complex ways. To embrace the cyborg and the hybrid as emblematic cultural figures and to reject notions of bodily and environmental purity does not mean that we can't have a *politics* and *ethics* of technologically polluted bodies. This accountability lies squarely with corporate polluters, weak regulatory agencies, and the consumers who depend on existing structures of production and consumption. Thus, I turn to the field of environmental justice to consider DES as a contemporary parable.

The connections between DES, *My Year of Meats*, and environmental justice can be seen in the topical links between the issues portrayed in the book and examples of current environmental justice activism, and in the expansion of definitions of environmentalism to include race, class, gender, and injustice frames—cultural questions in which narratives and stories play a central part. One link is in the struggles of Native women in the Arctic organizing against their exposure to environmental pollutants from persistent organic pollutants (POPs). In arguing that bodies are "first environments," Native activists link reproductive health, environmental health, and cultural survival.[78] POPs are a set of extremely toxic, long-lasting, chlorinated, organic chemicals that can travel long distances from their emission source (often thousands of miles away) and that accumulate in animals, ecosystems, and people. In the 1980s, scientists began to find high levels of toxic chemicals (pesticides, insecticides, fungicides, industrial chemicals, and waste combustion) far from the source of their production. POPs were discovered in the Arctic indigenous populations in the 1980s, when a Nunavik midwife in the Canadian North offered to collect local breast-milk samples. These samples were supposed to be control samples from a "clean" environment in a study of toxic breast-milk contamination. Researchers were astounded to find that Arctic indigenous women had POP concentrations in their breast milk five to ten times greater than in women in southern Canada, and among the highest recorded in the world.[79] Suspected health effects include higher rates of infectious diseases and immune dysfunction, negative effects on neurobehavioral development (adverse behavioral functioning, slowed growth rates and intellectual functioning), and negative effects on newborn height.

The POPs example raises questions of harms and injustices and rights and responsibilities, both for the local Arctic communities and in the global realm. Women struggled with whether to feed their babies contaminated breast milk, and the larger communities debated whether and how to hunt and consume traditional food. Native women, in other words, were cyborgs and hybrids, much like DES animal/female bodies and the polluted babies. Arctic women suffered the worst health effects of production processes that their culture did not create or condone.[80] Despite the unique technological and bodily problems that POPs presented, there have been successful developments aimed at addressing this particular problem. In 1998 and 2001, two binding international treaties were signed.[81] Both ban or limit production, use, release, and trade of particularly toxic POPs, establish scientifically based criteria and specific procedures for establishing controls on additional POPs, and seek to prevent development and commercial introduction of new POPs.[82] The agreements would not have been negotiated without indigenous activism. Sheila Watt-Cloutier, the chair of the Inuit Circumpolar Conference describes the indigenous negotiating position on POPs: "A poisoned Inuk child, a poisoned Arctic, and a poisoned planet are all one and the same."[83] Indigenous activism, particularly by midwives and female tribal leaders, privileges the environmentalist trope and discourse of interconnection and web of life, especially between mother and child, culture and planet. Without sanctioning or celebrating the pollution that turns their bodies into paradigmatic sites of boundary breakdowns, between cultures of global production and local consumption, Arctic activists are engaging in debates about the role of technology on female/indigenous bodies and the politics of what can and should be done in the face of the ubiquitous state of border crossings and cyborgian alterations that define our age.

Lastly, understanding DES through the framework of environmental justice allows us to understand and critique it as one of the three pathways that brought the environmental endocrine hypothesis to light (the other two were discoveries that wildlife reproductive disorders were linked to chemical effluents and pesticides and decline in sperm counts and quality).[84] The environmental endocrine hypothesis is the emerging scientific consensus "that a diverse group of industrial and agricultural chemicals in contact with humans and wildlife have the capacity to mimic or obstruct hormone function . . . by fooling it into accepting new instructions that distort the normal development of the organisms." Synthetic organic chemicals have been linked to two dozen human and animal disorders.[85] By understanding how technologies are linked with bodies and nature, we can understand how to criticize the problems that come with synthetic organic chemicals, rather than accept their ubiquity as an acceptance of a hybrid, postnormal, and postnatural cyborgian state of the world.

Conclusion: Boundaries, Borders, and Blowback

As Aidan Davison suggests in his *Technology and the Contested Meanings of Sustainability*, the contest over "the ideal of sustainability is at once a contest over the future of technology."[86] He describes human practice as the "drawing toward and into ourselves of worldly things: things living and nonliving, artefactual and ecological, human and non-human, earthly and heavenly."[87] Technologically polluted bodies are a dark version of human and corporate practice, a fusion of things living (human bodies) and nonliving (chemicals, plastics, pesticides), human and nonhuman, earthly and a synthetic hell created out of our culture's desire to engineer the natural through the technological. In telling truths that are fictions, in a narrative both documentary and fabulist, *My Year of Meats* and cyborgs, analyzed through an American studies framework, provocatively reframe DES as a story that continues to provide important insights about American values, corporations, and cultures. Although in one sense DES is "in the past," its negative effects on the health and well-being of second and third generations of those exposed to DES remain relevant. DES is also a reminder of the continued dangers of chemical intervention on humans and animals, since it was immediately replaced by other estrogenic, growth-enhancing additives, as well as the widespread use of antibiotics.[88]

Which returns us to the babies with whom we began. Clearly, the notion that babies are pure, natural, and outside culture is false. But perhaps the larger problem is what kind of culture we live in where babies are born with 200 chemicals in their bodies, and that level of pollution becomes acceptable, even as it becomes normalized. If pollution is ever-present, and we are already hybridized and cyborgian through our pollution exposure, what are the possibilities for cultural and environmental change? Knowledge, activism, and narrative make sense of technologically polluted bodies and perhaps help to remediate these health and environmental problems. Narratives by environmentalists and writers such as Ruth Ozeki and Sandra Steingraber have focused on issues of toxicity, gender, and breast milk contamination and are central to illuminating the stories of people, especially women, intimately shaped by corporate pollution.[89] Thus, I want to end by reframing hybrids and boundary crossings through the environmentalist discourse of interconnection and the "web of life." In a testimony about a DES daughter who died from cancer, her siblings recounts that "one thing Betsy taught me is that the environment is a *complex web of life*; everything is *interwoven* . . . I think we need to use our best science with our best intuition to remember that everything is interwoven. If you make a disturbance, it can have consequences far beyond."[90]

Beyond knowledge and activism, we need a cultural and political analysis and a vocabulary that makes sense of DES's roots and impacts, which American studies can provide. This analysis reveals what happens when we look closely at the interconnections, interweavings, and webs: of race, gender, and nature as constructed categories, the porous boundaries and borders between mother and child, between human (woman) and nonhuman nature (animal), between production and consumption, and between the environmental and the technological. Only through such intersectional and interdisciplinary analysis can we better begin to consider where and how justice is at all possible in our complex age.

Notes

Thanks to Sasha Abramsky, Siva Vaidhyanathan, Carolyn Thomas de la Peña, Marita Sturken, anonymous reviewers, and Stacey Lynn. Thanks also to Adria Imada, Richard Kim, and Grace Wang for their comments.

1. Environmental Working Group, "Body Burden: The Pollution in Newborns," July 14, 2005; retrievable at http://www.ewg.org/reports_content/bodyburden2/pdf/bodyburden2_final-r2.pdf (accessed June 14, 2006).

2. Barbara Seaman, *The Greatest Experiment Ever Performed on Women: Exploding the Estrogen Myth* (New York: Hyperion, 2003), 36.
3. Roberta J. Apfel and Susan Fisher, *To Do No Harm: DES and the Dilemmas of Modern Medicine* (New Haven, Conn.: Yale University Press, 1984), 1.
4. Barbara Seaman and Gideon Seaman, *Women and the Crisis in Sex Hormones* (New York: Bantam, 1977).
5. Alan I. Marcus, *Cancer from Beef: DES, Federal Food Regulation, and Consumer Confidence* (Baltimore: Johns Hopkins University Press, 1994), 12–13.
6. DES daughters (females exposed in utero) have an increased risk for infertility, and are at a higher risk for ectopic pregnancy, miscarriage, and preterm labor and delivery. They also have an increased incidence of structural changes in their reproductive organs. DES sons face increased risk for problems with their genitalia. These include cysts, testicular problems, microphallus, testicular varicoceles, hypospadias, and meatal stenosis. A 1993 study of DES mothers confirms a 30 percent increased incidence in breast cancer. For a further explanation, see DES Action, a support group for those who took or were exposed to DES, retrievable at http://www.desaction.org/.
7. Ellul's discussion of technique, technology, and nature are also suggestive. In short, technique and technology stand opposed to nature. Jacques Ellul, *The Technological Society* (New York: Vintage, 1964), 79.
8. Donna J. Haraway, *Simians, Cyborgs, and Women: The Reinvention of Nature* (New York: Routledge, 1991).
9. Ellul, *The Technological Society* (originally titled *La Technique: L'enjen du siècle*, or *The Stake of the Century*).
10. William Cronon, "The Trouble with Wilderness, or, Getting Back to the Wrong Nature," in *Uncommon Ground: Rethinking the Human Place in Nature*, ed. William Cronon (New York: Norton, 1996), 69–90. See Carolyn Merchant's work, specifically "Reinventing Eden: Western Culture as a Recovery Narrative," in the same volume, 132–59. On the erasure of Native Americans from wilderness, see Mark David Spence, *Dispossessing the Wilderness: Indian Removal and the Making of the National Parks* (New York: Oxford University Press, 1999).
11. See *New Perspectives on Environmental Justice: Gender, Sexuality, and Activism*, ed. Rachel Stein (New Jersey: Rutgers University Press, 2004), and *The Environmental Justice Reader: Politics, Poetics, and Pedagogy*, ed. Joni Adamson, Mei Mei Evans, and Rachel Stein (Tucson: University of Arizona Press, 2002), as examples of the emerging literature of environmental justice arising from literary and cultural studies.
12. See Shelley Fisher Fishkin, "Crossroads of Cultures: The Transnational Turn in American Studies—Presidential Address to the American Studies Association, November 12, 2004," *American Quarterly* 57.1 (March 2005): 31, and George Lipsitz, "In the Midnight Hour," *American Studies in a Moment of Danger* (Minneapolis: University of Minnesota, 2001), 3–30.
13. Environmental philosophers and philosophers of technology, such as Andrew Light, have focused much effort on this line of analysis, as have environmental ethicists.
14. He first successfully removed the steroid nucleus to create a substance he called "anol" in 1936, but it was only weakly estrogenic. Dodds then worked with a competitor, Sir Robert Robinson at Oxford. Together, they isolated a white crystalline substance with roughly three times the effect of natural estrogen, which they called Silbestrol. It was a derivative of stilbene. Stephen Fenichell and Lawrence S. Charfoos, *Daughters at Risk: A Personal DES History* (New York: Doubleday, 1981).
15. Cynthia Orenberg, *DES: The Complete Story* (New York: St. Martin's Press, 1981), 11.
16. Fenichell and Charfoos, *Daughters at Risk*, 17.
17. He also called for limits and caution in the area of long-term hormonal treatment. Ibid., 36.
18. Ibid., 19.
19. His concern was demonstrated in a 1933 report to *Nature* titled "Sex Hormones and Cancer-Producing Compounds." Ibid., 16.
20. However, their studies on DES's effectiveness during pregnancy lacked adequate controls, and were not double-blind. Despite inadequate research data, the FDA declared DES to be safe and no longer requiring annual approval in 1952, by administrative fiat. See Diana Dutton's "DES and Drug Safety" chapter in her book *Worse Than the Disease: Pitfalls of Medical Progress* (Cambridge: Cambridge University Press, 1988).

21. According to women's health journalist Barbara Seaman, Dodds was acutely aware of the larger political context of his discovery. The Nazis were seeking secrets to human reproduction, and their hormone research was part of their eugenic program. As Seaman recounts, in Hitler's Germany, "where eugenics and ethnic cleansing were primary goals," sterilization and contraception were "high priorities" toward the achievement of the master race. Seaman, *The Greatest Experiment*, 28.
22. For an extended discussion of the Smiths, see Fenichell and Charfoos, *Daughters at Risk*, and Orenberg, *DES: The Complete Story*.
23. Olive Smith, "Diethystilbestrol in the Prevention and Treatment of Complications of Pregnancy," *American Journal of Obstetrics and Gynecology* 56 (1948): 821–34.
24. Dutton, *Worse Than the Disease*, 54. "More Normal Than Normal" is also the name of Seaman's first chapter in Seaman and Seaman, *Women and the Crisis in Sex Hormones*.
25. However, as mentioned above, the Smiths' studies on DES's effectiveness during pregnancy lacked adequate controls and were not double-blind. Nevertheless, the FDA declared DES to be safe and no longer requiring annual approval in 1952. Apfel and Fisher, *To Do No Harm*, 21.
26. This was research conducted by William Dieckmann from the University of Chicago's study. Later reanalysis of data found that it was actually linked to *increased* miscarriages, premature births, and higher infant mortality. Orenberg, *DES: The Complete Story*, 4.
27. Apfel and Fisher, *To Do No Harm*, 26.
28. Even after DES's link with cancer was shown, it was commonly used as a "morning after pill" on college campuses well into the 1980s, although the FDA never approved it for that use. See Dutton, *Worse Than the Disease*, 80–86.
29. Rachel Carson, *Silent Spring* (New York: Houghton Mifflin, 1962), 1–3.
30. Joe Dumit with Sylvia Sensiper, "Living with the 'Truths' of DES: Toward an Anthropology of Facts," in *Cyborg Babies: From Techno-Sex to Techno-Tots*, ed. Robbie Davis-Floyd and Joseph Dumit (New York: Routledge, 1998), 212–39.
31. DES was banned in chickens under the Delaney Clause, which prohibited the use of any carcinogenic food additive, but was not banned in sheep or cattle because no residues had been found. Its use in cattle grew exponentially, so that, by 1970, nearly 75 percent of U.S. cattle were given DES. Environmental groups revealed that DES residues were found in cattle, but that the United States Department of Agriculture had suppressed the results. Dutton's *Worse Than the Disease* is one of only a few histories that treat the DES story in women and animals together.
32. For Apfel and Fisher in *To Do Know Harm*, DES is "a 20th century medical detective story." DES represents a "paradigm of the peculiarly modern phenomenon in which large-scale destructive consequences of a medical or technological innovation emerge unexpectedly as much as a generation after a benign or inconsequential beginning. A second, even more far-reaching aspect of the DES story is that it encapsulates in a quite remarkable fashion the whole complex history and structure of modern medicine in relation to modern life," 3.
33. For Marcus, the DES story reflects two themes, the late-nineteenth-century identification of science as expertise and the "reconceptualization" of the Progressive dream of experts as objective decision makers transformed into a regulatory and industry "alliance" of interests. *Cancer from Beef*, 1.
34. Marcus begins his story with Wise Burroughs, an Iowa State College ruminant nutritionist who discovered the hormone's cattle growth-promoting properties in 1954, as the founding father of his tale. Fred Lorenz, a professor of poultry husbandry at the University of California at Davis was responsible for DES in chickens. Ibid.
35. For example, "much of the DES story has been portrayed by feminist writers as an example of male doctors doing something thoughtless and harmful to women." Apfel and Fisher, *Do No Harm*, 40.
36. Orenberg, *DES: The Complete Story*, xiii.
37. Ibid, 28. This is also exemplified by one early promoter of DES, Dr. Karnaky of Houston, who gave it to dogs and found "the dang dogs were dying like flies." See Dutton, *Worse Than the Disease*, 48.
38. See Fenichell and Charfoos, *Daughters at Risk*; Orenberg, *DES: The Complete Story*; Deborah Davidson, "Woe the Women: DES, Mothers and Daughters," in *Gender, Identity and Reproduction: Social Perspectives*, ed. Sarah Earle and Gayle Letherby (London: Palgrave Macmillan, 2003).
39. In the 1990s DES activist groups successfully lobbied for federal funding for research on the health risks for DES mothers and children. The DES Research and Education Amendment was signed into law in 1992 by George Bush. It was the first federal legislation for DES research that ordered the

National Cancer Institute to study the long-term health effects of DES exposure. See Margaret Braun, *DES Stories: Faces and Voices of People Exposed to Diethystilbestrol* (Rochester, New York: Visual Studies Workshop, 2001), xvii.

40. Dutton, *Worse Than the Disease*, 75–76.
41. Apfel and Fisher, *To Do No Harm*, 15.
42. Ibid.
43. Another influential, popular advice author-physician claimed that as estrogen decreases, women enter the "world of intersex" (quoted in Seaman, *The Greatest Experiment*, 51). The phenomenon of giving female hormones to lesbians to "improve" them by making them "more female" and, by definition, heterosexual has been recounted by other cultural historians. See David Serlin's *Replaceable You: Engineering the Body in Postwar America* (Chicago: University of Chicago Press, 2004).
44. Despite the warnings about DES's cancer-causing effects, the FDA approved the use of DES in 1941 for four specific uses (menopausal symptoms, gonorrheal vaginitis, senile vaginitis, and suppression of lactation after pregnancy).
45. Robert Wilson's 1966 *Feminine Forever* (New York: M. Evans, 1966), as quoted in Seaman, *The Greatest Experiment*, 55.
46. Ibid., 49.
47. Reproductive technology is most often described through a list that includes assisted fertility technologies, such as artificial insemination, artificial wombs, cloning, embryo testing, embryo transfer, in vitro fertilization (IVF), intracytoplasmic sperm injection, preimplantation genetic diagnosis (PGD), and sperm selection. Others include ultrasound, new contraceptives such as Norplant and Depo-Provera, and surrogacy or contract pregnancy as reproductive technologies. Helen Bequaert Holmes, ed., *Issues in Reproductive Technology: An Anthology* (New York: Garland, 1992), ix.
48. Nancy Lublin, *Pandora's Box: Feminism Confronts Reproductive Technology* (Lanham, Md.: Rowman and Littlefield, 1998); emphasis added.
49. Ellul, *The Technological Society*, 7.
50. Ibid., xxv.
51. Ibid., x.
52. According to Ruth Ozeki's online biography, "*My Year of Meats* was an international success, translated into ten languages and published in fourteen countries." See www.ruthozeki.com. Accessed July 20, 2006.
53. Haraway, *Simians, Cyborgs, and Women*, 150.
54. For Haraway, cyborgs are a "condensed image of both imagination and material reality, the two joined centres structuring any possibility of historical transformation. In the traditions of 'Western' science and politics—the tradition of racist, male-dominant capitalism; the tradition of progress; the tradition of the *appropriation of nature* as resource for the *productions of culture*; the tradition of reproduction of the self from the reflections of the other—the relation between organism and machine has been a border war." Ibid., 150; emphasis added.
55. He analyzes the production of knowledge and "facts" about DES through "writing technologies" (popular pregnancy books and medical texts), or how and where meanings travel and circulate in texts, practices, and speech. Dumit with Sensiper, *Living with the "Truths" of DES*, 216.
56. Ibid.
57. He argues that DES children are cyborgs in five ways. First, their mothers are fusions of human hormones and DES, with known, profound, and "cascading effects" on the body. Second, new anatomical structures are created that are unique to those who were exposed in utero to DES (i.e., T-shaped uteruses unique to DES daughters, as well as other structural alterations of the cervix, described as "hoods," "collars," and "cockscombs"). Thus, DES is like an "additional, pharmaceutical parent." Third, the children's bodies, especially DES daughters, are under constant medical and technological surveillance when it comes to their health and illnesses. Fourth, DES daughters are dependent on the medical community to carefully monitor their own infertility problems and pregnancies. Last, there is a lack of knowledge of DES children about their exposure and a lack of knowledge of most doctors to the ubiquity of DES exposure in women. Ibid.
58. Haraway, *Simians, Cyborgs, and Women*, 154.
59. In particular, "the dualisms of self/other, mind/body, culture/nature, male/female, civilized/primitive, reality/appearance, whole/part, agent/resource, maker/made, active/passive, right/wrong, truth/illusion, total/partial, god/man" are challenged. Ibid., 151.

60. Ibid., 177.
61. Ibid., 176; emphasis added.
62. As *DES Stories*, a photography book by and about DES survivors, notes, "every DES-exposed person has an important story to tell" (3). See Margaret Braun's photographic essays in *DES Stories: Faces and Voices of People Exposed to Diethystilbestrol*; *A Healthy Baby Girl*, directed by Judith Helfand, 1997; and *Philomela's Tapestry*, a play by Alice Cohen.
63. Back cover of Braun's *DES Stories*.
64. Biological invasion discourse has been tied to social ideas and anxieties about place, nature, and culture, particularly in the South. See Joshua Blu Buh, *The Fire Ant Wars* (Chicago: University of Chicago Press, 2004).
65. For a historical example, see the section on "Hybrid Vigor," in *Technical Bulletin* 52,"The Cross-Breeding of Poultry," published by the Agricultural Experiment Station, Kansas State, October 1942, retrievable at http://www.oznet.ksu.edu/historicpublications/Pubs/STB052.PDF (accessed June 14, 2006).
66. The most well-known sterile hybrid is the mule, which is the product of a donkey and a horse (breeding a male donkey to a female horse results in a mule; breeding a male horse to a female donkey produces a hinny).
67. Ruth Ozeki, *My Year of Meats* (New York: Penguin, 1998).
68. Nick Fiddes, *Meat: A Natural Symbol* (London: Routledge, 1991).
69. For example, early pioneers of IVF, such as Jacques Testart (France) and Alan Trounson (Australia), began their careers as animal biologists, who often sexualized their control over female fertility. Janice Raymond, *Women as Wombs: Reproductive Technologies and the Battle over Women's Freedom* (Melbourne: Spinifex Press, 1994), xxvii.
70. Seaman scathingly critiques hormone replacement therapy in *The Greatest Experiment* as an "experiment" done without people's informed consent or knowledge. She argues that the history of estrogen therapies has been "shoot first, apologize later." In other words, there is inadequate proof that the drugs work for what they are supposed to do, little research was done on appropriate dosages and potential short- and long-term risks and benefits, and their links to cancer have been ignored in the pursuit of profits at the hands of the pharmaceutical companies. See also Michelle J. Naughton, Alison Snow Jones, and Sally A. Shumaker, "When Practices, Promises, and Policies Outpace Hard Evidence: The Post-Menopausal Hormone Debate," *Journal of Social Issues* 61.1 (March 2005): 159–79.
71. See United Animal Nations, Anti-Premarin Campaign at http://www.uan.org/antipremarin/rxforcruelty.html and http://www.pmurescue.org/about_pmu.php (accessed June 14, 2006).
72. According to a 2005 Human Rights Watch report, meat-packing is the most dangerous industry in the United States, with few worker protections. It is also increasingly an industry populated by immigrant workers. See "Blood, Sweat, and Fear: Workers' Rights in U.S. Meat and Poultry Plants," retrievable at http://www.hrw.org/reports/2005/usa0105/ (accessed June 14, 2006).
73. Roger Horowitz, "Meatpacking," in *Gender and Technology: A Reader*, ed. Nina Lerman, Ruth Oldenziel, and Arwen Mohun (Baltimore: Johns Hopkins University Press, 2003), 267–94.
74. Feminist vegetarian critic Carole Adams connects animals to nonwhite populations and attitudes tied to colonialism and racism. Carol Adams, *The Sexual Politics of Meat: A Feminist-Vegetarian Critical Theory*, tenth anniversary ed. (New York: Continuum, 2000), 40–41.
75. Roger Horowitz, *Putting Meat on the American Table: Taste, Technology, Transformation* (Baltimore: Johns Hopkins University Press, 2006), 151–52; emphasis added.
76. Scott Ratzan, ed., *Mad Cow Crisis: Health and the Public Good* (New York: New York University Press, 1998); Sheldon Rampton and John Stauber, *Mad Cow USA: Could the Nightmare Happen Here?* (Monroe, Me.: Common Courage Press, 1997).
77. Haraway, *Simians, Cyborgs, and Women*, 177.
78. See *Undivided Rights: Women of Color Organize for Reproductive Justice*, ed. Jael Silliman, Marlene Gerber Fried, Loretta Ross, and Elena R. Gutierrez (Cambridge, Mass.: South End Press, 2004).
79. David Leonard Downie and Terry Fenge, eds., *Northern Lights Against POPs: Combatting Toxic Threats in the Arctic* (Montreal: McGill-Queen's University Press, 2003).
80. Further research showed why and how these high rates of contamination were possible. First, there are the unique properties of POPs and organochlorines in the Arctic. These contaminants are released (low estimates are 70 percent from the United States, 11 percent from Canadian sources, and 5 percent from Mexican sources), then travel by atmospheric and ocean currents to the Arctic, where they are

deposited and enter the food chain. They are lipophilic (fat loving), which means they bioconcentrate in fatty tissues and biomagnify as they move up the food chain. Second, Arctic indigenous populations consume at the top of the food chain (particularly marine mammals), absorbing high levels of contaminants in traditional foods, such as narwhal, walrus, and beluga blubber.

81. The 1998 Arhus Protocol covers North America and Europe and is part of the 1979 U.N. Economic Commission for Europe, Geneva Convention on Long Range Transboundary Air Pollution. The 2001 Stockholm Convention was the first global treaty that seeks to eliminate substances specifically toxic to the environment (by September 2002, 151 countries had signed and 21 countries had ratified it; a minimum of 50 country ratifications are necessary to enter it into force).
82. The Stockholm Convention in the Context of International Environmental Law—the agreement was a milestone in multilateral environmental law. In less than three years, a global treaty was negotiated, which focused significantly on implementation.
83. "The Inuit Journey to a POPs-Free World," in *Northern Lights Against POPs*, ed. Downie and Fenge, 256–72.
84. See Theo Colborn, Dianne Dumanoski, and John Peterson Myers, *Our Stolen Future: Are We Threatening Our Fertility, Intelligence, and Survival? A Scientific Detective Story* (New York: Dutton, 1996), and Deborah Cadbury, *The Feminization of Nature: Our Future at Risk* (London: Hamish Hamilton, 1997). I am also indebted to historian Nancy Langston for her ongoing research on this topic.
85. Sheldon Krimsky, *Hormonal Chaos: The Scientific and Social Origins of the Environmental Endocrine Hypothesis* (Baltimore: Johns Hopkins University Press, 2000).
86. Aidan Davison, *Technology and the Contested Meanings of Sustainability* (Albany: State University of New York Press, 2001), 93.
87. Ibid., 166.
88. Dutton, *Worse Than the Disease*, 80.
89. Sandra Steingraber, *Having Faith: An Ecologist's Journey to Motherhood* (Cambridge, Mass.: Perseus, 2001).
90. This quote is from Susan Wood, talking about her sister, Betsy, in Braun, *DES Stories*, 60; emphasis added.

Markets and Machines: Work in the Technological Sensoryscapes of Finance

Caitlin Zaloom

In the hit 1983 film *Trading Places*, the august banking brothers Randolph and Mortimer Duke bet one dollar on the answer to the question of what matters more, environment or genetics, nature or nurture. The pair wager on the fortunes of Louis Winthorpe III (Dan Akroyd), a pedigreed white commodities trader from whom they strip job, family, and reputation, and Billy Ray Valentine (Eddie Murphy), an African American hustler and con artist, whom they set out to remake into a successful businessman. The Dukes's social experiment unravels as Winthorpe and Valentine learn about the scheme and unite to take revenge on the brothers, conspiring to bankrupt their bosses and enrich themselves. In the film's climactic scene, Valentine takes his place on a trading floor that is teeming with activity. Immersed in the sea of bodies, he begins to sell enormous numbers of orange juice futures contracts, driving down prices, bringing financial ruin to the film's comic villains, and earning Louis and himself enough money for an immediate beachside retirement.

The decade after the film's release found real life financial firms engaged in far more ambitious social experiments—experiments motivated by economic concerns and fueled by technological innovations. During the 1990s, financial exchanges themselves traded places, largely abandoning the wild and boisterous pits where men like Winthorpe and Valentine made their deals face-to-face, instead moving to high-tech digital dealing rooms where speculators work quietly in front of computers, face-to-screen. This transformation—essential to the "Americanization" and globalization of financial markets—required both technological and social engineering at the levels of both individuals and systems. The spread of "American style" or "Chicago style" trading systems depends on the new social dynamics facilitated by technological and architectural changes. More trenchantly, these changes illustrate the ideological power of American neoliberal assumptions that are made material in the design and application of information technologies, architectural designs, and work practices within and among increasingly "virtual" marketplaces. Yet such policy choices are not distinct from corporeal and sensory experiences. In fact, they are intimately connected. What people see and hear—and the technologies that regulate such sensory inputs—have profound effects on the sorts of things that get done, not only in major global financial markets but in office and commercial spaces around the world and at all levels.

Market makers—a diverse cast of financial managers, traders, economists, architects, and software designers—coalesced in the 1990s to craft technological devices, forge workplace conditions, and create new spaces where Chicago-style trading and the competitive individuals who practice it could flourish. Their projects would ultimately affect financial markets throughout the world, marrying electronic technologies and screen-based trading to what critics often refer to as the "Americanization" and "neoliberalization" of the global economy: the process of reconfiguring historically embedded institutional structures to establish novel market forms. However, it is important to note that the Chicago style of trading—by which I mean the practice of individual speculators, each trading for his own account and directly competing with every other dealer in the market—was itself honed through such "practical experiments."[1] The process of imagining a perfect market and refining the technological and physical design of the marketplace to reflect this vision is an integral part of the modern market system that always fails to live up to its ideal state.[2]

Globalization theorists such as David Harvey and Manuel Castells argue that under neoliberalism, capitalist institutions have used information technology to flatten the distinctions among places, reducing the significance of spatial distance and temporal difference.[3] They claim that this is nowhere more true than in the sphere of global finance, since economic networks are bound together with technological systems that permit institutions to coordinate activity in real time on a planetary scale, and that allow modern financial instruments to turn commodities—oranges, grain, Treasury bonds, and the like—into abstractions that are bought and sold thousands of times without ever moving an inch. Even global skeptics, it seems, are willing to admit that the speed and scale of financial markets enabled by innovative communications technologies is something new and profound.[4] Never before have financial actors been able to trade with dealers from Frankfurt to Singapore at any time of the day or night. However, these depictions often overstate the role of technology, leading observers to believe that such interconnections were inevitable, a narrative that obscures the human activities that propel money through technological channels, including the labors, false starts, and complex histories of the social and technical forms of global work.

Global markets do not emerge from the ether in which they seem to operate. City centers such as those in Chicago and London organize the industries that support "flows" of capital.[5] Struggles within specific workplace environments—over the designs of technologies and the best model for hiring workers, for example—give shape to the "the information society" and, at the same time, coordinate the work that compresses time and space associated with fast-moving information and cash. Just as Silicon Valley structured the conditions for software engineering and Hollywood shaped the production of film,[6] Chicago has left its footprint on global finance, and it continues to make an impression, often in surprising ways, in this global and digital age.

Drawing on extensive fieldwork and archival research, this article explores how market actors create the "socio-technical" arrangements that constitute futures markets, focusing on the technologies, organizational strategies, and ideas about competitive individuals that define American-style markets.[7] Financial markets are ideal subjects for analyzing the relationships among material technologies, economy, and the social systems in which they arise. By paying special attention to the kinds of problems that seem to demand technological solutions—in this case, using technologies to free the market by freeing the individuals who work in them from social constraint—we can observe how global trends, such as Americanization of the financial system, depend on the intertwined transformation of social and technological systems.

I explore two moments when the neoliberal value of individuals in competition took on material form through new technological and social arrangements: first, the creation of a new trading floor for the Chicago Board of Trade (CBOT) in 1930; and second, the opening of a Chicago-style dealing firm in London in 2000. I begin in Chicago during the 1920s, where managers at the CBOT were arranging trading technologies and the physical and social space of the dealing floor to foster individual competition, creating an environment where only the quickest responses and sharpest calculations would yield profit. Then I examine a more recent grouping of technological and social innovations as I follow a Chicago trading firm to London in the late 1990s and early 2000s, exploring how its managers attempted to adapt distinctively Chicago-style trading techniques and notions of human capacities to the European marketplace and London's financial environment.[8] This case illustrates how U.S. firms engage in social experiments and build material infrastructures to export their particular styles of capitalism into expanding markets.

Socio-Technical Arrangements

I use the concept of a "socio-technical arrangement" to build a more robust analysis of how technological structures bring to life the neoliberal value of individuals in competition. In markets, the American influences on globalization encourage the development of social and technological systems that enhance competition among individuals and fabricate the social context to support it. By focusing attention not only on a single new technology, but also on the blend of devices, social forms, and human skills that are necessary to make them work, I want to expand both the empirical frame and the analytic categories for cultural studies of technology. Scholars in science and technology studies use the term socio-technical "systems" to indicate the inextricable knot of society and material technology that brings order and stability to both.[9] The concept of "arrangement," however, emphasizes the elements of conscious design (such as trading floor architecture and electronic dealing room planning), and the constraints that existing technologies and social ties exert on the design process. Moreover, it highlights the dynamic interplay between social forms and technological devices that generate *provisional* structures, those which managers and designers are constantly revising in an attempt to construct a marketplace that lives up to the neoliberal ideals generated by American financial culture.[10]

Markets are made up of technological and social materials, not just economic objects.[11] The managers, designers, and traders who inhabit them incorporate emerging material technologies as well as the human skills, expertise, and social systems that put them into action and that shape the transactions that travel through them. Traders worldwide consider the open trading pits of futures markets to be a distinctively American economic form, a tool sharpened in the institutions and history of Chicago's commodities exchanges. As William Cronon has shown, the pit became an anchor for modern economic activity because it integrated market and place into a nexus of face-to-face competition, and nowhere did it integrate and coordinate so much market activity as it did in Chicago, where the CBOT and the Chicago Mercantile Exchange (CME) became key nodes in the global economy long before globalization was a word.[12] In the nineteenth century, the Chicago markets formalized the futures contract to organize and standardize the price of grain throughout the world. And in the 1970s, University of Chicago economist Milton Friedman helped to establish markets on the future price of Eurodollars, currencies, American Treasury Bonds, and the Dow Jones Industrial Average Index in the city, drawing on Chicago traders' well-honed expertise in dealing futures contracts. Chicago has not only created new products for trading; it has also shaped the physical marketplaces for futures trading in cities everywhere. As David Kynaston has shown, London's markets were explicitly built on the Chicago model, with market designers calling on experts from the midwestern metropolis to learn how to create conditions for encouraging open competition among aggressive, risk-taking speculators.

Ethnography of Technology

The instability of technical and social arrangements in contemporary finance raises the question of how to study forms that are constantly in motion. Ethnography based in participant observation and historical research is a key technique for examining processes under constant reform. On the trading floors, dealing rooms, and executive offices that help constitute global financial markets, we can see the ways that neoliberal ideas (such as the central values of individual competition and free markets) are built into the material technologies, physical structures, and human capacities that support global trade. By watching the action both within and behind the marketplace, we can see the moments of construction, success, and failure wherein the values embedded in socio-technical arrangements are forced to light. If American influence is remaking the conditions for global capitalism, ethnography can help us understand how practical, daily activities of market design become part of the broader cultural economy.

Conducting fieldwork in the high-tech landscape poses serious challenges. Financial contracts are immaterial and appear in contexts that seem culturally thin; but the traders, technologies, and forms of organization that collectively constitute the market are prime objects for observation, and by analyzing the daily life of financial exchanges and trading rooms, we gain insight into circulation of forms and ideas, such as those associated with American financial practice.[13] To gain access to the Chicago and London marketplaces, I first took a job as a clerk delivering orders and answering phones at the CBOT. I followed the apprenticeship process that leads from the clerk's desk to the trading pit, gathering advice from willing traders and gaining experience by working inside the market's unending flux. From inside the dealing room, I tracked how the technologies of the trading floor, the physical arrangements of traders in the trading pits, and the social competitions that were intertwined with the search for individual profit helped produce the traders as social beings who, in this environment, formed an extreme version of the everyday economic risk takers that American-style capitalism encourages its subjects to become. Within days of taking the position, however, I learned that the trading floor was no longer the cutting edge of market action. Digital dealing rooms were replacing the pits in market after market, city after city. The future of the CBOT was in question, and savvy financial firms were exploring possibilities for bringing their special style of dealing to electronic markets—including those in Europe that were leading the digital charge.

I followed the Chicago firm that employed me to London, where it was assembling a trading room from the new electronic technologies of European futures markets, trading screens designed in the United States, and skills particular to London traders—both those who had been thrown out of work as their trading pits were replaced by electronic markets and new recruits who were pulled from a pool of university graduates eager to reap the financial benefits of a job in the City, London's financial district. I joined their ranks, dealing on the German-Swiss exchange, Eurex, an advanced electronic market that was challenging Chicago's dominance in futures markets. In the conference room of Perkins Silver, I participated in training sessions and worked alongside the new recruits and more experienced traders.[14] Together we applied the techniques of dealing learned from our Chicago trainers against an abstract market available only through the blinking monitor. In London, the densely human world of bodies and faces on the trading floor—captured so well in *Trading Places*—had already become obsolete.[15]

Designing the Trading Floor

The CBOT's mission is the design and production of markets—creating the instruments, assembling the speculators, and arranging the technologies that connect users to dealing centers. In particular, buildings and rooms are technologies for making markets. Since its founding in 1848, the CBOT has created contracts to buy and sell financial and agricultural goods, but it has also built the material structures with which dealers work—the hangarlike trading room that lies at its heart and the electronic networks that define its digital future. The CBOT has reconstructed its trading floor, the key site of financial exchange, at crucial moments when it could no longer accommodate its rapidly expanding markets. The process of design demonstrates how market managers, engineers, and architects create the material and technological conditions that make competition among individuals possible.[16]

The layout of the trading floor guides the daily paths of the traders, configures whom they can see and hear, defines their access to information, and determines what communications technologies they can use in a flash and which they have to stretch to procure. These technological, social, and spatial arrangements influence vectors of money flows and determine who has what information when. The exchange's members argued over the construction of the "soundscape" and sightlines defined by the trading floor's technological configuration. As Emily Thompson has shown, "soundscapes" were, at the beginning of the twentieth-century, sites of contestation and important influences on qualities of life and work. A soundscape, for Thompson, is a "sonic environment," or more properly an "auditory or aural landscape." The amount of sound is not always as important as the quality of sound and the ways humans can derive signals from that environment.[17] The shouted numbers of financial prices were the key signals for Chicago traders, and the auditory, as well as the visual, environment that conveyed them was well worth fighting over.

The design moment was not only an occasion for compromise among interests, but assembling physical plans was an opportunity for experiment. In the moment of design, the marketplace could be more closely aligned with the ideals of commerce, shaping the dealing floor, especially its central technologies of trading pit, and telephone and telegraph stations, to draw this market closer to principles of individual competition and smooth circulation, making these abstractions durable in material form.

On the trading floor, traders produced futures markets, but producing the Board Room, as it was called, took some feats of engineering. First, it required a vast open space at the building's core. Six huge trusses, each weighing 227 tons, held up the skyscraper over the enormous hall, eliminating the need for support columns that would block the movement and view of the traders. The wide-open arena allowed traders to circulate easily among pits, the telegraph and telephone operators, their offices, and the smoking room where traders met clients. This room, at 165 feet long by 130 feet wide with a 60-foot ceiling, offered a space where each trader could have equal access to the markets and to the information he needed to trade. The design applied the market principles of order and equal access to information.

The trading floor of the older building had been straying from the ideal and had become chaotic as the CBOT grew. In a 1927 memorandum, the architects Holabird and Root observe that the trading floor "looks more dingy, more cluttered up and less orderly than it did some time ago. This is due to the demand for increased facilities." A cotton pit, a trading post for oats, several telephone stations, and a coat room for telegraph operators crowded the floor. More and more Western Union operators packed the desk surrounding their office. Companies overcrowded their telegraph stations with illicit operators. Swarms of clerks also congregated by the telephones, creating "a very unsightly condition" as the messengers elbowed their way to the phone lines.[18] Quotation boards, where the changing prices of commodities were recorded, were raised off the ground to make more room and operated from a balcony. The news bureau was shunted into the smoking room as the markets they reported on ballooned.

The architects, builders, and the CBOT's New Building Committee debated how to turn this haphazard arrangement into an ordered whole. Creating a rationalized Board Room meant both providing good arrangements for the traders, giving each equal access to the market and its sources of information, and providing well-constructed conduits for prices from the Board Room's markets to reach the outside world. Each demanded sharp attention to the arrangement of technologies and people in physical space. Luckily for the room's designers, an experiment in marketplace design was already under way. Before the builders began to wreck the 1885 building, the CBOT reestablished itself in a temporary trading space not far from the new construction site. New arrangements and materials could be tested there. The experiment showed that traders' concerns were centered on their access to other traders' eyes and voices.

In the Board Room, hands, shouts, and visual contact conveyed prices and dealing offers, placing the body at the center of traders' dealing strategies. Floor trading was also taxing physical labor. To take full advantage of the markets, a trader stood for hours a day among throngs of competitors jostling each other for advantage. For the interim trading floor to work, it had to provide a certain level of physical comfort of the traders and allow their eyes and ears to take in the sounds and gestures of their trading partners and the market as a whole. The acoustics of the hall were crucial to the operation of the market and to profits. H. R. Rumsey, chair of the New Building Committee, conveyed this to Holabird and Root in a request to use a flooring material for the permanent trading room that would absorb excess sound and be "easy on the feet."[19] The architects dismissed wood, rubber tile, cork, and linoleum as options. The softer materials would quickly give way under the floor traffic and would, "in a short time present a dilapidated appearance."[20]

With the architects' aesthetic intransigence, the managers turned to a scientist for help. Rumsey hired Professor F. R. Watson, a physicist at the University of Illinois, to analyze the acoustics on the temporary trading floor. There was apparently much room for improvement. The corn pit overflowed with traders' complaints, as the New Building Committee reported in a memo:

> Conditions are terrible, far worse than it was at the old Board; can't make yourself heard across the pit; in the old Board could stand on edge of pit with back turned and pick out and

> distinguish voices, impossible now; majority of traders are experiencing throat trouble since they began trading here; go home at night actually tired from the exhaustion of shouting in order to be heard and from the continual uproar and noise in the trading hall.[21]

Clearly the experimental space had detracted from the operation of the market, dampening access to information that each trader needed to compete. The cacophony even threatened the accuracy of the information coming out of the pits. A Mr. Chronister, who managed the CBOT's Quotations Department and was in charge of recording and distributing prices traded, reported that the new space was "500% worse than the old building. The reporters have difficulty in getting quotations correctly and the traders themselves are unable to hear properly across the pit."[22]

If the quotations were incorrect, false prices would flow out by way of the telephones, telegraphs, and pneumatic tubes that connected locations as near as offices in the building and as far as England and Argentina to the pricing machine of the CBOT trading floor. Demand for information from the CBOT was growing rapidly. For the new building, the architects expanded the electric wiring capacity of the floor and opened larger channels for telegraph cables to run from the Board Room to the offices of the Cleveland Telegraph Company, the Western Union office, and the postal telegraph stations. Although these companies provided the major public access to CBOT quotations, many firms maintained their own dedicated lines to the trading floor.

How to arrange the informational conduits of the trading floor was a hot-button issue. Rumsey and the New Building Committee had to mediate arguments among traders, telegraph companies, and the architects over where the telegraph and telephone stations were to be located on the floor, and how to allot to Western Union and other telegraph companies. Noting that, "at present there is more demand for telegraph facilities than ever before," the Holabird and Root plan provided for sixteen telephone booths in sight of the quotation board.

Holabird and Root were also concerned that market information be available to all participants. They noted that in the chaos that had engulfed the older trading floor, the quotation boards had been obscured. A fair market, in which skill and speed would determine profit, required that each trader have equal access to information. They set out to construct the Board Room so there would be no inherent advantage to place. However, not all participants were willing to give up their privileges so easily. Some member firms tried to manipulate access to telephones to gain advantage in the market by influencing the arrangement of space and technology on the trading floor. They pressured the board of directors to secure extra telephone lines that would support their own business. Responding to these firms, the president pressured Rumsey to accommodate their requests.

Rumsey objected to this departure from the ideals of an apolitical market developed under the direction of experts. He replied to the president's intervention in his plans with ire.

> Your Committee on New Building has been advised of the wishes of your Honorable Board . . . relative to . . . the installation of equipment for fourteen private telephones adjacent to the Wheat pit to the South.
>
> We must respectfully, but nevertheless earnestly, protest against such a plan. Two years of study and thought were dedicated to the main floor arrangement, for handling the business of the entire active membership rather than the few. The best architectural and engineering talent have counseled us in determining the best possible arrangement for our floor facilities, including private telephones.

The committee, too, objected to the placement of the telephones on the grounds of equal access to information. Some firms had begun "flashing" their orders to the pit from the telephone lines, relying on the rapid hand signals that would become an integral and identifying part of financial pits. The New Building Committee saw this as introducing informational disorder. The hand signals made customers' orders visible to attentive traders, who could see them before they reached the open market, providing a privileged source of information unavailable to the rest of the traders.

Rumsey and his committee had worked to create a Board Room where the lines of communication limited the influence of groups and allowed each individual to compete with all the information that the market provided. The sensory landscaping for the traders' eyes and ears constructed an environment for the mutual circulation of information and money.

Chicago-Style Trading in London Markets

The late 1990s found futures markets, both in Chicago and around the world, once again stretching their seams, now as part of a rapidly expanding global "derivatives" markets made up of the complex financial instruments integral to the circulation of capital and the risk-taking strategies of global firms. In the place of trading pits, these newest markets operated through digital networks connecting the world's financial centers. To take advantage of this burgeoning trade, Chicago firms looked abroad to transplant their dealing style honed on the city's trading floors.

Perkins Silver, a mid-sized firm of roughly a hundred traders that was founded by Chicago traders, was maneuvering to take advantage of the electronic upheaval. The two directors, Eric Perkins and Philip Silver, established the company in 1985 to manage the accounts of Chicago independent traders. Perkins, a charming southerner who moved to Chicago for its marketplace and quickly ascended to the top tier of the trading floor, and Silver, a distinguished and successful veteran who parlayed his success in the grain pit into an even more lucrative business in financial futures, recognized how their company could take advantage of the opportunities in the emerging overseas electronic markets. The introduction of electronic trading and new markets created an opening to rework the socio-technical arrangement of the trading pits, shaping futures dealing to the contours of a global marketplace. For Perkins and Silver, this meant two things: first, that electronic trading provided an opening for smaller firms such as theirs to become key players in the largest futures markets, and second, that the anonymity of the trading screen made it possible for them to bring new kinds of players into the marketplace, in particular the women and minorities who had been largely excluded from the face-to-face trading pits.

Trading Populations

The Chicago exchanges, still dominant players in the industry, no longer had a corner on futures trading skill and knowledge. Eric Perkins noted that both São Paulo and London had successfully adopted "the Chicago trading culture" in the past ten years. In London, moreover, the financial futures exchange had been built on the Chicago model and the market was already populated with scores of Chicago-style traders whose work demanded that they manage the highs and lows of risking money in their own accounts. Although the success of world exchanges posed a threat to power and dominance of organizations like the CBOT, for entrepreneurial firms like Perkins Silver, the local trading populations in such financial centers could provide human materials to build trading operations beyond North America. Silver and Perkins positioned themselves to take advantage of this experienced workforce. They intended to hone the skills of the British futures and foreign exchange dealers along the Chicago model and train new ones who would ultimately supplant them according to their plan to profit from social diversity. They began work to set up an electronic dealing room in London that would bring Chicago techniques of trading to the electronic financial frontier.

The new technologies based on the trading screen, the new physical arrangements of traders, and presentation of information they allow seemed to offer the possibility of making more competitive markets. Creating a new trading room designed around trading screens spurred Perkins Silver to reflect on and distill the ideal of Chicago trading and, at the same time, reenvision the Chicago trader. In the experimental process, the Perkins Silver managers reflected on and refashioned the familiar Chicago forms into a novel socio-technical arrangement.

Assembling the Space

They began by building a trading room in the heart of the City of London, Britain's financial center, choosing a site that hovered over the dwindling pits of the London International Financial Futures Exchange (LIFFE). In the trading room, the physical plan is open, but the trading screens and the presentation of the market divide the traders from each other, channeling the traders' energies toward competition with the market. Long desks partition the open floor plan into neat rows, and traders sit four to a desk, each behind a bulwark of computer screens. The terminals display both the markets they work in and predictive charts that project historical price patterns into the present. Each trader spends the workday behind these screens almost wrapped in the information of the market. Behind their screens traders' eyes are fixed mostly on the rise and fall of the prices of the futures contract. Sitting together in the crowded dealing room, each trader acts alone.

Electronic trading individuates the traders in another way, through the abstracted representation of the market available on the screen. On the dealing screen, the market is represented in columns of numbers: prices for the financial product lined up neatly with the number of traders willing to buy or sell at that price. Screen traders are anonymous; they are represented only in the aggregate numbers that their competitors around the world see. By design, the screen limits calculations to the quantitative assessment of price and discreet decisions to buy or sell completely removed from the strategies or mistakes of others. All information about the market is provided within the blinking square. In this abstracted representation, the market as a whole confronts the trader, who must compete against the market without knowledge of the strategies or trading patterns of any individual within it. The anonymity of the screen removes the personal aspect of competition—in a market based only in numbers, grudges, jealousies, or other states generated in personal interactions that might interfere with the pure processing of market information fall away. In the trader's competition against the anonymous market, an individual's economic actions are generated entirely alone.

After assembling the technologies and building their trading room, Perkins and Silver drew up a strategy to staff their electronic trading room with speculators, but with a crucial difference from the older Chicago model. The anonymity of the trading screen allowed the Perkins Silver managers to recruit dealers from populations that had not before been welcomed onto the trading floor.

The CBOT markets remained largely based in tight networks reinforced by friendship and family. In the pit environment, where economic exchange was based in dense and personal interactions, the social homogeneity facilitated trading networks; traders dealt along the social ties that bound together the almost all-white and all-male cohort. This homogeneity was invisible to the designers, who treated arranging competition as if among abstract individuals; however, the disruptions of electronic trading allowed the Perkins Silver managers to rethink this conflation of the white, male competitor with market individualism.

Electronic trading shifts the overlapping nature of market and space, tempering the density of face-to-face markets and its associated homogeneity. New technologies create connections among traders in dispersed trading rooms, no longer centralized in a trading pit, but sitting high in offices in the sky behind computer terminals. This creates an opportunity to transform the social arrangement of the market, bringing in the women and minorities whose identity-based "perspectives" now orient the collectivity of the market.

In the offices of Perkins Silver, managers organized their recruitment strategies around the values of formal education and the social diversity that market advocates touted as the unrealized potential of free exchange—after all, profits based on calculation and acumen, and not personal connections, are the markers of the ideal market. The Perkins Silver managers recruited educated professionals, including women and minorities, to replace the working-class traders who had manned the London pits and who had made up the first wave of online traders. Their working theory posited that traders with these backgrounds and experiences—often the ones that were prevented entry into the CBOT pits—would generate particular and profitable readings of the market. According to the Perkins Silver philosophy, the academic training and social positioning of the new dealers would help them to see new angles for interpreting market activity. The managers' multicultural vision and neoliberal logic organized their hiring practices as they trusted that sound market behavior would produce profits.[23]

New LIFFE in the City

Perkins Silver's experiments were not the first, however, to fuse Chicago trading styles with London's market materials. On the morning of September 20, 1982, the opening bell was set to ring for the first time on the trading floor of the LIFFE. A throng of young men shuffled from foot to foot in anticipation, their frenetic energy and brightly colored jackets overturning the usual aesthetic sobriety of the City.[24] At the sound of the buzzer, traders' hands shot up and they shouted the first deals to buy and sell contracts for future delivery of bonds and currencies. The opening of the LIFFE was a social, technological, and economic experiment, a project to adapt American-style financial markets to European territory. The bankers and City regulators who planned the LIFFE sought a model for development, yet they did not draw on the Royal Stock Exchange, whose old building they now inhabited, or the metals markets they may have passed on their way to the new site. Instead, they looked across the Atlantic to Chicago, the city whose name in the financial world is synonymous with futures trading—and where governments, banks, and speculators from around the world came to do business.

Although London had its own commodities exchanges, for example, for aluminum, copper, and cotton, the Chicago exchanges—the CBOT and the Chicago Mercantile Exchange (the Merc)—were the models for the global trade in financial futures.[25] But there remained important differences between Chicago and London that shaped how London's exchanges and traders took up these models. Historically, trading firms had dominated the London futures markets, and that discouraged individual traders from doing business on their trading floors.[26] In contrast, the Chicago markets were heavily populated by local speculators trading mostly for themselves. The designers of the London market identified these locals and the trading pit as the source of the Chicago markets' strength, and they explicitly set out to develop the socio-technical arrangement of the pit—marrying the physical architecture for ordering competition and individuals who would thrive within it.

The development of LIFFE and the cultivation of new London locals were part of a financial revolution that was taking place in England in the 1970s and 1980s. Margaret Thatcher's policies had initiated a move from "gentlemanly capitalism" to the new order of the "Big Bang."[27] The Big Bang was technically a series of regulatory changes outlined in the Financial Services Bill that were put into practice October 27, 1986, and opened the City to new kinds of firms and traders. This transformation generated conflicts around the new players and new organizations entering the City—particularly the foreign ownership of formerly British banks, such as Warburg and eventually Schroeder's and Barings—and new financial tools and both the companies and individuals that traded them.

The death of an older style of British capitalism was marked by the demise of a figure that represented the values of the old City—the gentleman capitalist who, as political commentator Will Hutton describes,

> does not try too hard; is understated in his approach to life; celebrates sport, games and pleasure; he is fair-minded; he has good manners; is in relaxed control of his time; has independent means; is steady under fire. A gentleman's word is his bond; he does not lie, takes pride in being practical; distrusts foreigners; is public spirited; and above all keeps his distance from those below him.[28]

But while such characteristics may have fit well within finance before the Big Bang era, as Hutton asserts, they did not describe many of the new actors who entered the City, least of all the new traders brought in to staff the foreign exchange rooms of merchant banks. The market in currencies led the way, welcoming in working-class men with low levels of education who dealt yen and dollars over the phone, often screaming into two or three handsets at once. Soon City leaders were considering developing a financial futures exchange staffed by Chicago-style speculators to complement the currency markets thriving in City banks. In 1977, John Edwards wrote a piece in the London-based *Financial Times* called "Speculators Are Made Welcome," lauding Chicago traders and considering the potential of such risk takers for London markets. "It is the 'locals' operating exclusively for themselves, who make the U. S. markets so different from London, where all the business is channeled through member companies of the exchange," he wrote.[29] The article was a challenge to London to build an army of such enterprising individuals to staff the City's markets. Leaving behind their own models of firm-based financial activity and organization, London focused its sights on the American Midwest to find the kind of trader who would populate the open-outcry pits of the newly envisioned London International Financial Futures Exchange.

In July 1981, an article in the *Financial Times* posed the question "Can Financial Futures Traders Out Shout the Old-Timers in Chicago?" The reporter felt that "British bankers and other supporters of the London International Financial Futures Market are confident they can create a respectable complement for the more difficult financial futures markets in the USA." But these new British traders would be operating on Chicago turf. Traders who ended up on the LIFFE floor had "to adjust to an environment firmly based on the Chicago model," especially the pits, which were the key technology for the liquid market in American financial futures.[30] The LIFFE managers, together with consultants who had come directly from the CBOT and the Merc, seeded the floor with a dozen "natives of Chicago" and implemented an educational program in speculation.[31] When the trading floor opened in 1982, a new kind of London trader had been ushered into existence.

The *Mail on Sunday* described this social shift in the walled confines of the City: "The City has produced a new breed of broker. He swaps millions at the flick of an eye in the rainbow-hued Financial Futures Exchange. He's young and brash and sometimes without an O-level to his name."[32] The wild behavior and spending practices of the mostly working-class traders became legendary. From the mid-1980s, the City was no longer driven by English commerce and upper-class ideals. The models for proper financial conduct derived from relations and tensions between City norms and the new organizational forms like the LIFFE and the American- and European-owned banks. With the Big Bang, the City officially liberalized not only its securities markets, but also its social space.[33]

The creation of the LIFFE introduced a new socio-technical arrangement to London built from the architectural materials of the trading pits and the competitive capacities of the Chicago-style traders. But opportunities soon dissolved for the working class traders whose skills were a crucial component of the pit-based system. In 2000, LIFFE abandoned its trading floors under the competitive pressures of electronic exchange and online technologies.[34] The locals dispersed to the trading rooms and unemployment offices of London. By the millennium, the LIFFE floor trader had become the most recent casualty of the ascendance of electronic markets in financial futures.

The Market and Multiculturalism

As LIFFE went digital and the old markets closed, Perkins Silver prepared to supply some of the trade that the London locals had supported in the rough-and-tumble pits. The transition from the trading floor to the electronic dealing room organized around anonymous screen-based trading allowed for new experiments in market design, and the owners and managers of Perkins Silver quickly moved to create novel socio-technical arrangements.

The new traders provided the raw materials for the managers to construct a Chicago-style trading room in London, rationalizing the social content of the trading room to draw profits to the firm. In particular, the Perkins Silver managers sought to take advantage of the trading talent of new university graduates, minorities, and women, groups largely absent from LIFFE trading pits. The managers believed that these new kinds of recruits would bring fresh perspectives to reading the market, allowing Perkins Silver to profit from their elimination of barriers to participation based on race and gender and, at the same time, fulfilling the promise of markets to be open to all individuals regardless of group characteristics. Yet, the Perkins Silver strategy was based on the idea that education, experience, and membership in different racial, ethnic, and gender categories and levels of education shaped the each individual's vision and produced their calculative strategies.

Many of the recruits had never been in a dealing room before and had had little contact with finance. But the Perkins Silver trainers understood that deep knowledge of the financial products was not necessary to trade successfully. According to the Perkins Silver executives, and many other traders I interviewed, a good trader could deal in any product. The particulars of the contract itself were not important; a good trader has mastery over the techniques of speculation that are central to financial risk taking in futures markets. So the Perkins Silver trainers focused on producing speculators, not experts in government debt products. They focused on staffing their trading room with a diverse set of individuals who, together, would give the company a unique "optic" for observing and profiting from the market.

The Perkins Silver managers constructed their dealing room to create a cohort of traders within an American-style multiculturalist paradigm, hiring Asians (British citizens of Indian and Pakistani descent), blacks, and women, all of them educated, to bring in different views of the market. According to this logic, the categories that each trader inhabited would lead him or her to interpret the market differently, providing a range of insights into the market's actions. The new traders equipped with these diverse lenses for reading the market and with Chicago trading techniques would, they hoped, help their fledgling operations prosper.

The Perkins Silver managers used gendered consumer desire and what they called "doggedness" to identify recruits with profit-making potential. Philip Silver was particularly impressed with one young woman. When he asked in their interview why she wanted to become a trader, she responded, "I have very expensive taste in jewelry." One recruiter told me that he demands a certain kind of drive. "Give me a room full of outsiders. Immigrants. People who came to the city with no friends. People who are hungry."

The more experienced traders, who had come from the LIFFE pits, clashed with their new colleagues and their bosses over defining the appropriate economic subjects for a global market. What does the kind of person who operates effectively in electronic financial futures markets look like? These were not obvious questions for Perkins Silver, the CBOT, or the London financial world.

Early in the process of building the financial futures market, Chicago had a dominant role. In the first wave of innovation, the LIFFE copied the Chicago model closely, including working to instill the competitive culture of the Chicago pits. But the electronic environment demanded further adjustments. Perkins Silver set out to correct for the imperfections of the pits by creating a dealing room more closely in line with market ideals of individualism than the Chicago model allowed. The Perkins Silver founders and managers worked to make their trading room conform to the image defined by their market ethic, using the available technological and social materials to mold their trading room using ideas at the intersection of American pluralism and capitalism.

The technological and geographic displacements enabled by electronic markets gave the Perkins Silver managers distance from the norms and practices of the CBOT. The contrast of London traders and electronic markets with Chicago-style practices threw into relief what was unique about Chicago's methods, techniques, and the mushy but significant area that Eric Perkins identified as "Chicago culture"—the specific sets of relationship to the self and to competitors, coupled with trading techniques and orientation to risk rooted in pit trading.

To create their own trading room on electronic technologies and with physical distance from their home institutions, Perkins Silver managers were forced to formulate the key characteristics of Chicago-style speculation, based in developing individual speculators, which they would bring to London markets. In building the London trading room, the Perkins Silver managers refined Chicago-honed techniques for arranging the relationship between technologies and individual competitors. The disruptions of electronic trading inspired the managers to refashion a socio-technical arrangement rather than simply to transfer an existing one intact to a new location, designing their trading floor as a more perfect machine for trading by linking the new technologies to a cohort of speculators of mixed ethnicity, race, and gender.[35]

The shift from face-to-face to face-to-screen markets offered the promise of bringing individual competitors into an immediate relationship to the wider market. The anonymity of the screen separated the traders from each other and from social entanglements—they separate emerging collectivities of friendship, collusion, or simple geography into their individual parts—ordering a presentation of the market based only in numbers. Only one collectivity escapes this division: the market itself—a theoretical grouping more than a collective—constituted of autonomous individuals. Paradoxically, this anonymity also allowed for managers to experiment with the gender and ethnic profile of Chicago's market makers. Anonymity allowed for the Perkins Silver managers to acknowledge and manipulate the unmarked competitive individual.

The Perkins Silver managers were working on a shifting ground between the ideological commitment to treating individuals as ultimately independent of their context, a condition that market designers work to realize, and the daily work of market managers who engage with recruits and trainees who arrive at their door marked by race, ethnicity, gender, class, and the history of the exclusion of these individuals from the financial marketplace.

The anonymous screen seemed a solution to this problem. In the world of online trading, competitors can no longer see the marks of difference in their rivals, seeming to allow for those who had been excluded to enter into competition for profit on equal footing. With this logic, the Perkins Silver managers presented the screen as a medium for democratizing the market, allowing each individual trader to be treated only on the basis of his or her engagement with financial risk taking. At the same time, the managers used a multicultural lens for recruitment that created a trading room that reflected the worldwide marketplace, matching the new dimensions of the market with an array of perspectives that could begin to capture its new complexity, and all for a profit.

Conclusion

Markets are socio-technical arrangements of material devices and the competitive individuals that these arrangements create. From the advent of the trading pits in the nineteenth century to today's cutting-edge digital technologies, spatial and technical designs have been constantly scrutinized to identify departures from liberal and, now, neoliberal ideals of the marketplace. Inevitably, such departures emerge, and market makers get to work adapting the technological and social landscape to bring individuals in line with their visions of how competition should operate. For those who labor in these marketplaces, the technologies and spaces that result shape the sensoryscapes of their work every day. The informational and physical designs of market spaces configure what traders see and hear, which in turn shapes traders' understandings of which economic actions are possible. By focusing on these socio-technical arrangements, the designers, and the process of implementation that bring them to life, critics of the contemporary economy can begin to see the everyday work that transmits the neoliberal values of American capitalist practice, both within and beyond the boundaries of the United States.

Technical devices such as computer screens, which may seem to be freestanding, always operate in relation to human skills and institutional contexts, giving managers and designers the opportunity to organize and reorganize their use to fit ideological goals.[36] With the emergence of electronic technologies, managers of trading firms, with the screen in mind, are retooling their ideas about who will make a good trader—making decisions that potentially redefine who will be included in and excluded from the field. In this way, new technologies provide opportunities for social experimentation at the levels of organization, trading room, and individual. Through a messy set of cultural and political processes, the managers and designers plan and orchestrate the elements of socio-technical arrangements that encourage individual calculation and competition.

Communications technologies are key elements in creating a physical environment where equal access to information is possible, shifting the responsibility for profit making from using social connections to gather information to the quick calculations of each trader. In electronic trading, the technology heightens this process by creating anonymity, stripping traders of their personal characteristics that might mark them as members of collectivities outside the market, and reducing them to only their competitive aptitudes. In addition to increasing the speed and connectivity of markets, technological innovations provide an opportunity to rearrange the social and technical materials that structure speculation, to once again revise the marketplace to reflect the most recent ideals of competition.

In emergent visions of global markets, individuals, regardless of location, social position, or personal characteristics, are liberated to compete for profits. In some places, new projects seek to make use of the ways that ethnicity, gender, and race can shape a trader's perspective and, in turn, his or her mode of competition and calculation. This is not a race-denying strategy that champions the individual, denuded of collective commitments and histories. Rather, it is an understanding of ethnicity, race, and gender that aims to exploit presumptive group-level experiences for corporate profit in a global economy. Marketplace multiculturalism is just one tool that financial managers are using in their recent experiments, though, and it would be naive to believe that it will substantially diversify the trading workforce.

Practical experiments, such as the construction of the 1930 Board Room and the Perkins Silver's electronic trading floor, bring into line the existing and evolving social forms of economic activity with such market ideals. Experiments such as the CBOT's 1930 trading floor and the Perkins Silver dealing room are constantly under revision, and in the near future the arrangements of electronic trading rooms will themselves seem incomplete.

What does the recent exportation of U.S. market models mean for our understanding of the cultural power of economic circulation? The British use of the Chicago model of financial exchange is an important example of the authority of U.S. economic institutions to direct the material and cultural contents of markets. But this American influence goes deeper than we commonly hear. American-influenced markets are more than the "flows" of capital they direct, the transformations in time and space that they affect, and the economic conditions that they inflict on countries whose currencies become objects of negative speculation. Technological infrastructure, organizational forms, and cultural norms concerning competitive individualism are important conduits for the spread of American-style economic action and the dispersion of neoliberal cultural practices. The daily work of designers, managers, and traders collectively make markets out of technologies, human capacities, and social materials bring the American ideals of competitive action to the global stage.

Notes

1. Caitlin Zaloom, *Out of the Pits: Traders and Technology from Chicago to London* (Chicago: University of Chicago Press, 2006).
2. Joseph Alois Schumpeter and Redvers Opie, *The Theory of Economic Development: An Inquiry into Profits, Capital, Credit, Interest, and the Business Cycle* (Cambridge, Mass.: Harvard University Press, 1934). See also Tyler Cowen, *Creative Destruction: How Globalization Is Changing the World's Cultures* (Princeton, N.J.: Princeton University Press, 2002).
3. Manuel Castells, *The Rise of the Network Society* (Oxford: Blackwell, 1996); David Harvey, *The Condition of Postmodernity: An Enquiry into the Origins of Cultural Change* (Oxford: Blackwell, 1989).
4. David Held, Anthony G. McGrew, David Goldblatt, and Jonathan Perraton, *Global Transformations: Politics, Economics, and Culture* (Stanford, Calif.: Stanford University Press, 1999).
5. Saskia Sassen, *The Global City: New York, London, Tokyo* (Princeton, N.J.: Princeton University Press, 2001).
6. Manuel Castells calls such sites "milieux of innovation." Manuel Castells, *The Informational City: Information Technology, Economic Restructuring, and the Urban-Regional Process* (Oxford: Blackwell, 1989). Also see AnnaLee Saxenian, *Regional Advantage: Culture and Competition in Silicon Valley and Route 128* (Cambridge, Mass.: Harvard University Press, 1994).
7. The trading style that the London markets emulated in 1982 was, of course, influenced by the transatlantic travels of mid-nineteenth-century laissez-faire economics, an intellectual movement that lionized the self-regulating market and the autonomous, freely acting self, and that coincided with the founding of the CBOT. The appearance of the Chicago trader in London is, therefore, a return of the British ideals of market activity in flesh and wire. See Daniel T. Rodgers, *Atlantic Crossings: Social Politics in a Progressive Age* (Cambridge, Mass.: Harvard University Press, 1998).
8. Daniel Miller refers to this as "virtualism." Daniel Miller, "Conclusion: A Theory of Virtualism," in *Virtualism: A New Political Economy*, ed. James G. Carrier and Daniel Miller (New York: Berg, 1998), 187–217.
9. Thomas Hughes first coined this term to describe the way that politics and social concerns shaped electricity provision the United States. For Hughes, the term redirected attention away from Thomas Edison's individual genius to the larger forces that both constrained and enabled a certain kind of system. This has since been expanded to include many more elements, but always expanding the understanding that material technologies can be understood only in the context of the hardware, software, technical procedures, human skills, physical environments, and rules and regulations that come together to make them effective.

10. Michel Callon has recently abandoned the idea of the socio-technical "arrangement" in favor of the socio-technical "agencement," a concept that places agency in the social and material form, undermining the notion of a unique human form of agency. In contrast, I use arrangement to emphasize the role of human players, their ideals, and their ability to manipulate existing social and technological configurations, a contrasting move to Callon's.
11. This mutual inflection of technological and social innovation was also at work in the spread of American-style consumer relations. Victoria de Grazia demonstrates how washing machines, refrigerators, and Teflon pans equipped a new kind of woman—Mrs. Consumer—with labor-saving devices that shifted her image from household laborer to the manager of the modern home, which she administered with expertise, good taste, and foresight. American technologies and models of housewifery transformed European households, establishing a crucial piece of the U.S. commercial empire. Victoria de Grazia, *Irresistible Empire: America's Advance through Twentieth-Century Europe* (Cambridge, Mass.: Harvard University Press, 2005).
12. William Cronon, *Nature's Metropolis* (New York: W. W. Norton, 1991); Dan Morgan, *Merchants of Grain* (New York: Penguin Books, 1979).
13. Arguments and narratives of American economic influence and transatlantic commerce often focus on commodity chain, emphasizing the economic and political domination of the traffic in goods, both human and inanimate. For two classic treatments see Sidney Mintz, *Sweetness and Power* (New York: Penguin, 1985); and Eric Wolf, *Europe and the People without History* (Berkeley: University of California Press, 1982).
14. Perkins Silver is a pseudonym, as are the names of the traders and managers I describe.
15. These are not the only materials used to rationalize markets. The standardization of contracts, for instance, is a crucial precondition for the operation of futures markets that I am describing. See Bruce Carruthers and Arthur L. Stinchcombe, "The Social Structure of Liquidity: Flexibility, Markets, and States," *Theory and Society* 28.3 (1999). However, the daily work of rationalization represents a different order of action, one that requires tweaking and adjustment rather than the construction of stable systems.
16. This "heterogeneous design" process resolves conflicts among competing interests—builders, managers, and city planners—at the same time that it creates a building. Thomas F. Gieryn, "What Buildings Do," *Theory and Society* 31.1 (2002): 35–74.
17. Emily Ann Thompson, *The Soundscape of Modernity : Architectural Acoustics and the Culture of Listening in America, 1900–1933* (Cambridge, Mass.: MIT Press, 2002).
18. Building Committee to CBOT Directors, April 5, 1930.
19. Holabird and Root to Rumsey, October 4, 1928.
20. John Holabird to the Secretary of the New Building Committee, Dean Rankin, February 15, 1929.
21. Ibid.
22. Ibid.
23. James Surowiecki, *The Wisdom of Crowds: Why the Many Are Smarter Than the Few and How Collective Wisdom Shapes Business, Economies, Societies, and Nations* (New York: Doubleday: 2004). Surowiecki concludes that the best judgments are made not by individuals or committees of "experts," but instead by loosely connected groups of diverse individuals.
24. As the *Financial Times* reported, "LIFFE opened for business in the old Royal Exchange Building opposite the Bank of England . . . to a flourish of gaily-dressed traders. More soberly the event was accompanied by some reminders from the banking establishment about the need for propriety." "LIFFE Styles," *Financial Times*, November 10, 1982.
25. The connection between Chicago and London futures markets was first initiated in the nineteenth century when the London metals market was modeled on the Chicago futures trade.
26. David Kynaston, *Liffe: A Market and Its Makers* (Cambridge: Granta Editions, 1997).
27. Michael Unseem, "Business and Politics in the United States and the United Kingdom," in *Structures of Capital*, ed. Sharon Zukin and Paul DiMaggio (New York: Cambridge University Press, 1989); Linda McDowell, *Capital Culture: Gender at Work in the City* (Oxford: Blackwell, 1997); Will Hutton, *The State We're In* (London: Vintage, 1996).
28. Hutton, *The State We're In.*
29. Kynaston, *Liffe: A Market and Its Makers*, 10.
30. "Can London Financial Futures Traders Out Shout the Old-Timers in Chicago?" *Financial Times*, July 27, 1981.

31. Kynaston, *Liffe: A Market and Its Makers*, 72.
32. Ibid., 94.
33. Michael Pryke, "An International City Going 'Global': Spatial Change in the City of London," *Environment and Planning D: Society and Space 9* (1991): 197.
34. Under the leadership of Brian Williamson, LIFFE "demutualized" in 1998, switching from a membership organization like the CBOT to a corporate structure citing the demands for swift action required under an electronic regime.
35. James Clifford understands "practices of displacement . . . as *constitutive* of cultural meanings rather than as their simple transfer or extension." James Clifford, *Routes: Travel and Translation in the Late Twentieth Century* (Cambridge, Mass.: Harvard University Press, 1997), 3.
36. Andrew Barry, *Political Machines: Governing a Technological Society* (London: Athlone Press, 2001); Wiebe E. Bijker, Thomas Parke Hughes, T. J. Pinch, eds. *The Social Construction of Technological Systems: New Directions in the Sociology and History of Technology* (Cambridge, Mass.: MIT Press, 1987).

Educating the Eye: Body Mechanics and Streamlining in the United States, 1925–1950

Carma R. Gorman

During the second quarter of the twentieth century, many writers on industrial design noted that, prior to about 1925, attractive appearance was not designers' or consumers' priority, at least not in the product categories that a *Fortune* magazine writer defined in 1934 as the "formerly artless industries": aluminum manufactures, baby carriages, sleds, railroad cars, cash registers, clocks, electrical appliances, food packaging, automobiles, eyeglasses, pens, refrigerators, scales, sewing machines, stoves, and washing machines.[1] Consultant industrial designer Harold Van Doren, writing in 1940, stated that in contrast to the fields of ceramics, glassware, textiles, silverware, jewelry, wallpaper, and furniture, in which products "are sold, and always have been sold, largely on appearance," "in the manufacture of engineered products like typewriters, utility and price were the prime concerns of manufacturer and purchaser alike until a few years ago."[2] Similarly, Van Doren's contemporary Raymond Loewy, writing in 1951, noted that in the nineteenth and early twentieth centuries, American consumers had been satisfied with "engineered as you go" mechanical products that were characterized by a "haphazard, disorderly look" (see, for example, fig. 1a).[3]

Both period commentators and historians have agreed that *after* 1925, however, consumers began to demand beauty even in those products for which there had been, as Van Doren put it, a "lack of an educated demand for attractive appearance in years past."[4] In an era when refrigerators or washing machines in a given price range could be expected to work and to wear about equally well, newly professionalized American industrial designers acknowledged that they were "designing for the eye"—trying to lure consumers with products that were distinguishable from one another stylistically more than technologically.[5] Particularly during the Depression, it was widely acknowledged that "the sales curve would not respond to the old forms of pressure . . . [and] the product had to be made to sell itself" through attractive appearance.[6] The clean-lined

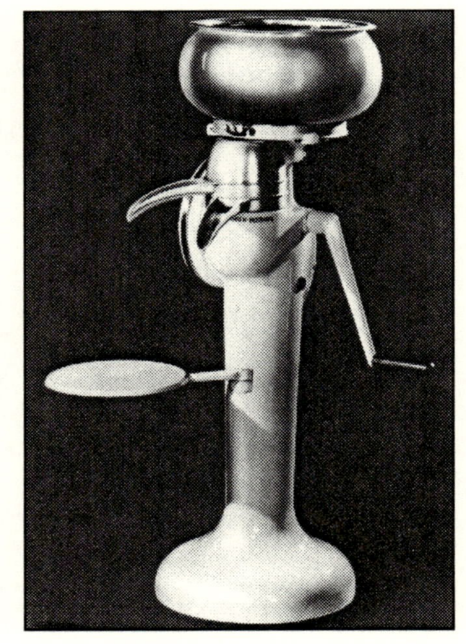

Figure 1.
Raymond Loewy, McCormick-Deering cream separator, before (a) and after (b) Loewy's redesign (1945), from Raymond Loewy, *Industrial Design* (Woodstock, NY: Overlook Press, 1988), 121. Reproduced with permission of Laurence Loewy.

style that consultant industrial designers developed to meet consumers' demands for beauty in the formerly artless industries went by the name "streamlining" (for example, fig. 1b).[7] The origins of the term *streamlining* lie in hydro- and aerodynamics, but most industrial designers—even proponents of scientific streamlining such as Norman Bel Geddes—admitted that in the 1930s, streamlining was primarily an aesthetic device rather than an aerodynamic one, and further claimed that their aesthetic was derived primarily from the form of the human body.[8]

Period observers and historians have both offered many different explanations for how and why this post-1925 shift in tastes and demands occurred, and for the rise of the popularity of streamlining.[9] Many of the well-known explanations are "supply-side" ones, in which particular designers or exhibitions or manufacturers or merchandisers are understood to be the drivers of stylistic change. Consumers, in this paradigm—when they are understood to affect the design of products at all—do so by adopting new styles after seeing them in exhibitions, magazines, movies, and department stores. Designers then are presumed to cater to consumer tastes by making more products that look like the ones that have already sold well. The problem with this model is that it assumes two things: first, that consumers develop taste preferences primarily through informal means (such as skimming a magazine) rather than through formal ones (such as undergoing a required course of study at school), and second, that consumers develop ideas about and tastes for consumer goods only by looking at *other* consumer goods. Both assumptions are unwarranted. Between 1925 and 1950, there were at least two important formal mechanisms for the teaching and acquisition of taste that had implications for the appearance of the artless industries: "related art" and "body mechanics" training. These forms of education, which flourished in elementary, high school, and college classrooms, have clear implications for the history of taste in this period, yet have not been taken into account by scholars of design. Neither related art nor body mechanics training focused specifically on the aesthetics of the formerly artless industries, but both emphasized the personal and social importance of good form, and taught students how to recognize it and analyze it in the decorative arts and in the human body, respectively. Textbooks and works of pedagogy from these disciplines are resources that scholars have not previously related to the history of design and consumption, but an examination of these forms of literature makes it clear that they can provide insight into the shifts in consumer tastes and product styles that occurred after 1925.[10]

This essay thus addresses two different kinds of *technology*: tangible artifacts and abstract systems. First, I attempt to provide a new explanation for how certain technological artifacts—in this case, streamlined machines of the 1930s and 40s, and human bodies that were subjected to the discipline of "body mechanics" during that same period—came to look they way they did. "Technology" in this sense of the word, then, refers specifically to machines, as well as to organisms—human bodies—that period writers construed as mechanical systems. The second kind of technology that this essay addresses, however, is not a tangible artifact, such as a washing machine or a perfectly poised body, but rather a system of practices and intellectual and visual tools that is, to borrow from Herbert Simon's famous definition of design, "aimed at changing existing situations into preferred ones."[11] The subdiscipline of physical education called "body mechanics," together with its attendant standards, charts, photographs, statistics, games, and so on (each tools/artifacts in their own right), constituted a technology of bodily—and ultimately, its proponents hoped, social and economic—improvement. This technological *system*—though seemingly unrelated to the technological *artifacts* that are also the subject of this article—did, I argue, have an important impact on the appearance and reception of streamlined goods in the 1930s and 40s.

By focusing on the ways of seeing and thinking that were promoted by body mechanics instruction, I contend that streamlining, the style that designers created for use in the formerly artless industries in response to consumers' new demand for beauty, owed at least part of its commercial success to an audience of well-educated, middle- and upper-middle-class consumers who had been primed by their schooling to expect bodies and products to conform to similar standards of beauty and efficiency.[12] For as art historian Michael Baxandall has argued, the visual skills that a society values highly tend to be the ones promoted in its system of formal education, and such consciously learned ways of seeing (which are teachable because they rely on a technology of "rules and categories, a terminology and stated standards") often shape "[the] categories with which [an individual] classifies his visual stimuli, the knowledge he will use to supplement what his immediate vision gives him, and the attitude he will adopt to the kind of artificial objects seen."[13] Studying the ways of looking that are promoted by formal education, Baxandall suggests, is one way to better understand the tastes of viewers (or consumers), because "much of what we call taste" lies in "the conformity between discriminations demanded by a painting [or other designed object] and skills of discrimination possessed by the beholder."[14]

Training in visual discernment via body mechanics instruction ultimately encouraged participants not only to understand the body as a mechanism, but also to judge mechanisms (such as formerly artless products) as beautiful to the degree that they looked like bodies. In distinction, then, to the many scholarly arguments that explore the ways in which the machine served as a model for the human body and its behavior in the 1920s through 1940s, I argue that the human body also became a model for the appearance of machines and other formerly artless products due to the rhetoric of, and emphasis on visual discernment in, the body mechanics instruction that was so heavily promoted in U.S. schools and other organizations in the 1920s, 30s, and 40s.[15] Further, I suggest that paying attention to the body mechanics literature explains more directly than many previous theories do why technological products that looked like human bodies (i.e., streamlined formerly artless goods) would have resonated with consumers: they had been taught, through body mechanics instruction, to understand both bodies and formerly artless goods as "mechanical," and as products of similar Taylorist technologies of improvement.

Body Mechanics: Origins and Aims

As dancing and etiquette manuals from past centuries attest, posture has been a preoccupation—and has been justified as a field of study in various ways—for much of Western history. Until the mid- to late 1920s, most writers on human carriage tended to use the word *posture* to describe their field. For example, Jessie Bancroft, founder of the American Posture League and author of an important 1913 book called *The Posture of School Children*, used the word *posture* almost exclusively.[16] However, when posture became a subject of "serious" scientific study in the 1910s and 1920s to physicians such as Joel Goldthwait, Robert Tait McKenzie, Armin Klein, Robert Osgood, Eliza Mosher, and George Fisher, the nomenclature of the field changed.[17] In the 1920s "body mechanics" became the theorists' favored term, because it was understood to be "more inclusive and descriptive" than the term *posture*.[18] Body mechanics referred not only to stance, but also to "the mechanical correlation of the various systems of the body with special reference to the skeletal, muscular, and visceral systems and their neurological associations."[19] Posture, a word derived by way of the French *positure* from the Latin *ponere*, "to place," probably sounded too static, given that body mechanics dealt with the body in motion as well as at rest. It also connoted "posing" and "posturing," which implied artificial, deceitful, and ultimately damaging uses of the body, as opposed to the "natural" and therapeutic ones physicians wished to cultivate. "Mechanics," in contrast, was a useful word not only because it referred to the field of mathematics concerned with motion and equilibrium, but also because it connoted machines and technology, since the word was derived from the Greek root *mechane*, meaning machine or contrivance. Indeed, many body mechanics theorists made explicit comparisons between bodies and machine systems; for example, the authors of a government report noted that "one of the most complicated and yet mechanically efficient products of the age is the automobile. It is incomparably less complicated and less efficient than the human body, yet the driving public are frequently made aware of the fact that slight disturbances of alignment in an automobile's working parts and slight dysfunctions of its electrical organs, may interfere with its function and cause it to develop chronic diseases."[20] The physicians' use of the term "body mechanics," then, suggested a certain willingness to conflate the categories of the biological and the mechanical/technological.

The physicians promoting body mechanics—whose credentials distinguished them from the many other individuals and groups interested in posture during this period (such as Joseph Pilates, F. Matthias Alexander, and chiropractors)—successfully justified the scientific study of body mechanics by arguing that modern, sedentary life was wreaking havoc on the physiques and health of Americans, especially children, to the detriment of personal and social efficiency.[21] Statistics derived from the military draft during the Great War, from examinations made of matriculating students to Ivy League and Seven

Sisters schools, and from the so-called Chelsea Survey funded by the Children's Bureau of the Department of Labor in 1923–24 suggested that approximately 75 to 80 percent of young Americans exhibited poor body mechanics.[22] These findings were troubling, the physicians argued, because proper body mechanics was linked to good digestion and elimination, healthy weight gain in children, increased alertness, emotional well-being, decreased absences from school, efficient combustion by the lungs, and decreased menstrual and gynecological problems, among a host of other benefits. In clinical studies, physicians noted that what were otherwise unexplainable improvements in chronic health conditions seemed to coincide with corrections in posture. Studies showed that groups of students that had received posture training, for example, had lower rates of absences from school than did control groups.[23]

Absence of disease was not the only reason that physicians and other writers promoted good body mechanics, however. As Jessie Bancroft made clear in her 1913 book—and through her inclusion of scientific management experts Frank and Lillian Gilbreth on the board of the American Posture League—a desire for increased bodily and mental efficiency (in the scientific management sense of the term) was another of the reasons for the renewed interest in and respectability of the field of body mechanics in the 1910s, 1920s, and 1930s. Bancroft, who clearly was enamored of Frederick Winslow Taylor's techniques for achieving industrial efficiency and often made reference to efficiency as a primary benefit of body mechanics instruction, contended that "fatigue comes less readily in correct posture, and the energy spent through unconscious muscular action in maintaining a bad position is available in good posture, for other uses. In good posture, also, better circulation, respiration, and digestion keep the stores of energy and sense of well-being at a higher level, and the efficiency and even the spirits of the individual are thereby placed on a loftier plane."[24] Similarly, Marguerite Sanderson of the Boston School of Physical Education claimed in 1922 that for men, "the strong, erect figure is desired not only for military fitness, but is [also] being demanded more and more in industry."[25] Even character and moral qualities were understood to be linked to good form; Henry Eastman Bennett, author of *School Posture and Seating* (1928), went so far as to argue that "the highest human traits and the finest moral perfections are inseparably associated in our thinking with erectness of carriage and posture. Our very language testifies that our concepts of moral qualities are derived from physical bearing: witness such terms as 'uprightness,' 'poise,' 'well-balanced,' 'level-headed,' 'backbone,' 'chesty,' and a host of others. We inevitably judge character from postural evidences. We ascribe poise, dignity, confidence, courage, self-reliance, self-respect, leadership, aggressiveness, and dependability to those whose posture expresses such traits."[26]

As Anson Rabinbach has shown in the case of Germany, and as Carolyn Thomas de la Peña has discussed in the case of the United States, even before the publication of Taylor's book *The Principles of Scientific Management* in 1911, late-nineteenth- and early-twentieth-century physical educators and physicians were busily preoccupied with developing technologies for minimizing fatigue and maximizing physical and mental energy.[27] But Taylorism proved a particularly useful and persuasive technology for conceptualizing and maximizing bodily efficiency. Both theorists and educators argued, for example, that the purpose of body mechanics training was "to develop the highest physical efficiency for the largest number of people" and "to prepare the body for the greatest usefulness in the world"—phrasings that seem directly inspired by Taylor's ideas, but that also echo the sentiments of eugenicists of the period.[28] However, although some of the people who promoted body mechanics in the 1920s and 1930s were probably also sympathetic to eugenics, most seemed dedicated to the proposition that the majority of postural defects was caused by environmental rather than hereditary factors.[29] The rhetoric of body mechanics suggested that everyone—except possibly "idiots and mental defectives," as Bancroft bluntly termed them—could achieve good (or at least improved) body alignment and thus better health and greater personal efficiency, if only they were willing to work hard enough at it.[30] Willpower, physical training, and visual discernment, rather than good genes, were touted as the sole endowments that most people needed to achieve a more beautiful, efficient body. As Bennett claimed, "postural habits are controllable" and are "as definitely subject to educational direction as are habits of language, of thought, of manners, of conduct, or other objectives in teaching."[31] It was through this faith in the efficacy of education and in the malleability of human bodies and minds that body mechanics promoters most differed from eugenicists. The point of body mechanics, as of Taylorism and other similar technologies, was that it could be systematized and replicated, taught and learned.

The Reach of Body Mechanics Instruction

Although much of the theory of body mechanics was established by physicians in the 1900s and 1910s, the push to have body mechanics taught in schools gained real momentum in the 1920s and the early 1930s. The federal government, for example, published pamphlets through the Government Printing Office in the late 1920s on body mechanics theory and pedagogy, and under President Herbert Hoover sponsored the 1932 White House Conference on Child Health and Protection.[32] At this conference, four (male) physicians and one (female) physical education professor constituted the Subcommittee on Orthopedics and Body Mechanics (hereafter "the Subcommittee"), which was part of the conference's Committee on Medical Care for Children.[33] Not surprisingly—given the importance they attributed to good posture—the Subcommittee members recommended in their final report (hereafter "the Report") "that steps be taken to make it not only possible but compulsory for all the children of the United States to receive instruction in good body mechanics."[34] They particularly stressed that "body mechanics should be made the basic principle of all physical education," and that all games and sports should be used to develop good posture.[35]

Although it is difficult to assess precisely the reach of body mechanics education across the nation, a number of kinds of evidence point to its ubiquity by the 1930s. First is the number of states that mandated physical education. In 1900, only four states had laws mandating physical education in public schools, but by 1930, thirty-nine states had passed such laws, and twenty-two states had state directors of physical education.[36] Granted, training in physical education could, in the 1930s, comprise anything from military drill to gymnastics to calisthenics to dance to competitive sports to body mechanics, but even before the Subcommittee published its recommendations in 1932, California, Delaware, and Utah boasted physical education programs in which "training in posture and body mechanics is given as an integral part of the health and physical education program in all the preparatory schools of the state."[37] Similarly, at the time the Subcommittee wrote the Report, in thirteen other states, "there is some recognition of the importance of body mechanics, but no coordinated state program," and in "many localities where health education is provided in the public schools, the subject of body mechanics is considered and given a place of varying importance on the program" (the Subcommittee members singled out Los Angeles and Boston public schools for their particularly impressive programs).[38] Physical education teachers were also being schooled in body mechanics at this time; in response to a Subcommittee questionnaire sent to 223 schools of physical education, 82 percent of those schools that replied (the response rate was 76 percent) answered "yes" to the question "Does your school give any specific formal instruction in the theory and practical application of Body Mechanics or posture . . . in its relation to the health of the individual and distinguished from 'Calisthenics' or corrective work for poor muscular development?"[39] Presumably the number of states and local programs offering such instruction both to schoolchildren and to physical educators increased after the publication of the Subcommittee's report, and flourished in tandem with the growth of physical education programs generally. Second, although body mechanics/posture training is usually mentioned in passing, if at all, in most histories of physical education, the prominence of American Posture League (APL) incorporators and board members in shaping physical education theory and pedagogy in the public schools, in the military, and in civilian organizations such as the Boy Scouts suggests that body mechanics training had a wide reach in American society.[40]

The Pedagogical Literature on Body Mechanics

Although the 1932 White House Conference Report spelled out very clearly the nature of the posture problem, it was neither a motivational tool nor a practical guidebook for the teaching of good body mechanics. Before the 1932 conference, there were only a few texts available that focused on the pedagogy rather than the theory of body mechanics. One of these was Bancroft's 1913 book *The Posture of School Children, With Its Home Hygiene and New Efficiency Methods for School Training*. The content of Bancroft's book was split about equally between theory and pedagogy, but—as its subtitle made clear—her notion of pedagogy primarily involved applying the principles of Taylorism to classroom management and record-keeping, and secondarily explaining physical exercises that teachers could prescribe to their students. Her method was similar to that of "social efficiency" educators such as John Franklin Bobbitt and W. W. Charters, who modeled their curricula and their language on Taylor's theories of industrial production (they referred to school superintendents as "educational engineers" and—as Bancroft did—to school buildings as "plants").[41] Bancroft dedicated only a couple of pages to other methods of teaching, such as drawing, studying pictures, and so on.[42]

However, a somewhat kinder and gentler approach to body mechanics education was promoted by Leah Thomas in the following decade. Thomas, a professor of physical education and the sole woman (and nonphysician) on the Subcommittee, was probably appointed to the group because the second edition of her book *Body Mechanics and Health* (1929) was essentially the first published text that served both as an overview of the theory of body mechanics *and* as a thorough resource for its teaching.[43] Thomas's book not only contained suggestions for exercises, but also included games, rhymes, songs, and pictures that she believed were motivational. Thomas's book set the tone for subsequent works of body mechanics pedagogy, and later writers emulated and expanded upon it.

One of these authors was Ivalclare Sprow Howland, an associate professor at Battle Creek College in Michigan, author of the 1936 book *The Teaching of Body Mechanics in Elementary and Secondary Schools*.[44] Reacting to the relative dearth of pedagogical literature on posture, Howland wrote her book specifically in response to the 1932 White House Conference report (4). The scope and length of her book, its engaging approach, its eminently respectable and scientific underpinnings (namely, the Report), and its balancing of the social-efficiency agenda with Deweyism, made it a particularly useful resource for teachers, who generally reviewed it favorably.[45] Thomas and Howland, then, more than any previous authors, promoted "fun" approaches to learning good body mechanics, including plays, games, pictures, and discussions, as well as or in lieu of corrective exercises.

Bancroft's, Thomas's, and Howland's books were some of the first widely published, comprehensive books on body mechanics written expressly for teachers, and upon their publication probably had a fairly wide impact on the way body mechanics (and physical education generally) was taught in classrooms across the nation. Because physical education was a school subject in which textbooks were usually not used, these works of pedagogy are some of the best remaining indicators of what posture instruction in schools of the 1920s and 1930s was like. Not only were these books widely distributed (most major university libraries still own copies today), but their impact is also evident in the subsequent pedagogical literature on body mechanics.[46] The authors' methods thus merit careful examination.

Educating the Eye: A Technology for Biomechanical Improvement

Although one might expect a physical education instructor to stress foremost the importance of physical training to the development of good posture, as indeed Bancroft and a number of other writers did, Thomas and Howland instead proceeded on the assumption that physical exercises alone were not a conducive means of developing good body mechanics in students. Although Howland noted that "to express himself by 'doing and experiencing' is truly educational and inspirational" for the student of body mechanics, and although she did prescribe many games and physical activities for her students, at heart her method was visual. "Body Mechanics," she argued, "can best be taught by educating the 'eye'" (37). Howland and her colleagues thus employed a many-pronged strategy for developing their students' "artistic and esthetic sense" through "recognizing and practicing good body alignment and posture" (112). Their method was thus a logical extension of the physicians' assumption that the quality of the body's "mechanical correlations" was visible in its contours.

Primary among the educators' techniques was teaching students to recognize the difference between good and poor body mechanics, which they did through reference to government-published standards that were available in six varieties: thin, intermediate, and stocky for both boys and girls (fig. 2).[47] In all of the standards, the "grade A" body was compactly disposed around a central vertical axis that ran from approximately the ear to the arch of the feet, displaying only restrained convex curves at the shoulder blades, buttocks, and chest/stomach. In contrast, the bodies with poor posture were characterized by greater deviance from the vertical axis via "exaggerated" curves in the spine, as well as inward slumping of the chest and an outthrust, lowered chin. Part of the body mechanics educators' strategy, then, lay in teaching students to be connoisseurs of curves, and to distinguish between "normal" curvature and "exaggerated" curvature. It is easy to see how this kind of aesthetic education could play a role in shaping consumers' expectations for beauty in the formerly artless industries, since in effect, the skills that Howland and her colleagues asked students to apply to the human body were similar to those one would use to "appreciate" and criticize a designed object or a work of art, especially a streamlined one.

Much of the body mechanics educators' method depended on having students analyze their own profiles and compare them to the posture standards. "'To see ourselves as others see us,'" Howland wrote, "is profoundly important in taking personal inventory of one's bodily attitudes" (23). Physicians and early body mechanics educators like Bancroft had devised a number of ways of gauging the quality of body mechanics (and by implication, of bodily health and efficiency) from the contours of the body seen in profile—the view in which "others see us," but which we cannot ourselves see directly. Bancroft promoted the use of what she called the "vertical line test," which involved holding a window pole next to the body to help gauge "the whole figure at a glance."[48] But ideally, according to Howland, teachers would have had equipment available with which to take silhouettograph photos, which yielded permanent, accurate profile images that students could use to compare their own posture to that of the standards.[49] For those schools that could not afford the equipment required to make these indexical records, however, body mechanics assessments had to take other forms. "Posture stamps" developed by American Posture League member Lillian Drew and available to educators from the Robey French Company, Boston, could be used to cheaply rubber stamp index cards with diagrams and checklists useful in student posture inspections (fig. 3).[50] The teacher judged the posture of partially disrobed students for common flaws (weak arches, hyperflexed knees, round upper back, low chest, protruding abdomen, etc.), and made marks on the cards indicating the severity of each imperfection. Follow-up inspections, recorded on the same card in different colors of ink, charted students' progress visually and verbally.

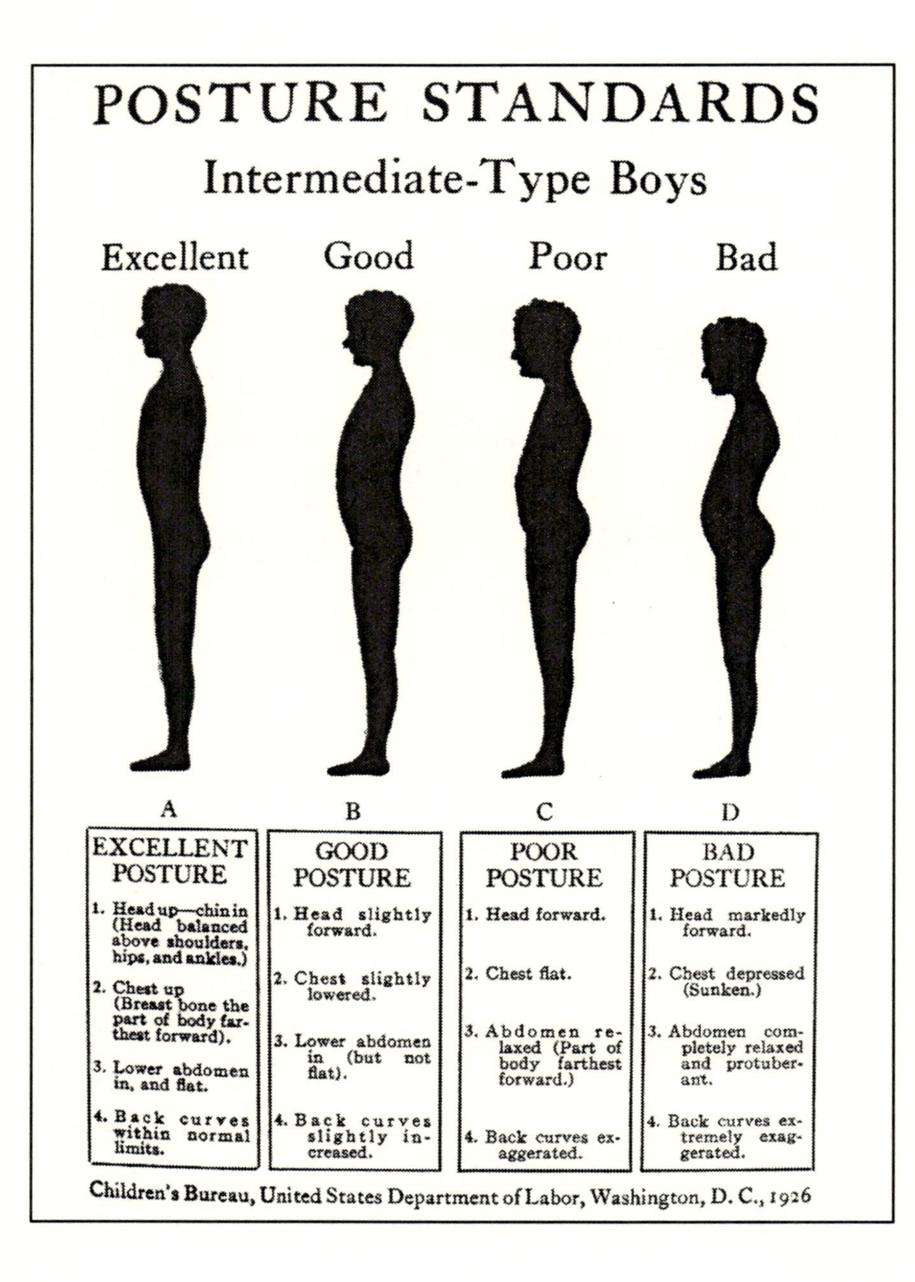
POSTURE STANDARDS
Intermediate-Type Boys
Excellent
Good
Poor
Bad
A
B
C
D
EXCELLENT POSTURE
1. Head up—chin in (Head balanced above shoulders, hips, and ankles.)
2. Chest up (Breast bone the part of body farthest forward).
3. Lower abdomen in, and flat.
4. Back curves within normal limits.
GOOD POSTURE
1. Head slightly forward.
2. Chest slightly lowered.
3. Lower abdomen in (but not flat).
4. Back curves slightly increased.
POOR POSTURE
1. Head forward.
2. Chest flat.
3. Abdomen relaxed (Part of body farthest forward.)
4. Back curves exaggerated.
BAD POSTURE
1. Head markedly forward.
2. Chest depressed (Sunken.)
3. Abdomen completely relaxed and protuberant.
4. Back curves extremely exaggerated.
Children's Bureau, United States Department of Labor, Washington, D. C., 1926

Howland urged students to take advantage of other chances for self-assessment, too; she encouraged self-study in the mirror at home, and also sideways glances in store windows, which she claimed were particularly useful because they showed the disposition of the body in motion (the significance of her recommendation that students view an image of their bodies as reflected in a pane of glass that also displayed consumer goods will become apparent shortly). She also recommended having the instructor "mirror" or mimic students' posture as they marched in single file past her, no doubt to the great amusement of all. She noted that, not surprisingly, "a second march by the mirror may reveal a change in the 'every-day tallness' of the students" (112). The public nature of this last kind of assessment was apparently part of what was believed to make it successful. The "mirror" exercise, which was performed publicly and which had the potential to be embarrassing if one's posture were poor, was only one example of Howland's use of the airing of imperfections as an inducement to faster learning (a technique she may have adapted from Bancroft, who believed that peer pressure was an excellent motivating tool).[51]

Figure 2.
"Intermediate-Type Boys," from *Body Mechanics: Education and Practice. Report of the Subcommittee on Orthopedics and Body Mechanics of the White House Conference on Child Health and Protection* (New York: The Century Company, 1932), 15. Originally published with hand-lettering in Armin Klein, M.D., "Posture Clinics: Organization and Exercises," U. S. Department of Labor/Children's Bureau Publication No. 164 (Washington: Government Printing Office, 1926), 12.

If helping students to see themselves as others did was one of the body mechanics educators' most important strategies for educating the eye, having students learn to critique the body mechanics of other people was nearly as important. One of the ways in which Howland gauged students' progress was through their development of critical faculties: "Do they report recognizing poor body mechanics in others?" (14). To that end, she devised games and competitions in which students, rather than teachers, were the judges of bearing, and in which they were required to use their skills of visual discernment on their peers. She suggested that "one group for a month at a time may serve as cleanliness inspectors for their classrooms. . . . One method is to have the children sit in their seats while the inspectors pass down the aisles; another method is to have the children pass in file by the inspectors. Charts and records may be kept for individuals or by groups" (57). Howland further suggested that the groups take on different names, such as the "Spicks" and the "Spans," and that the competition in inspection be arranged between the said groups, or between boys and girls, or between different classes (57). The winners, she suggested, should be treated by the other half of the class to simple awards such as bookmarks, tags, or arm bands. She also suggested a similar peer-monitoring

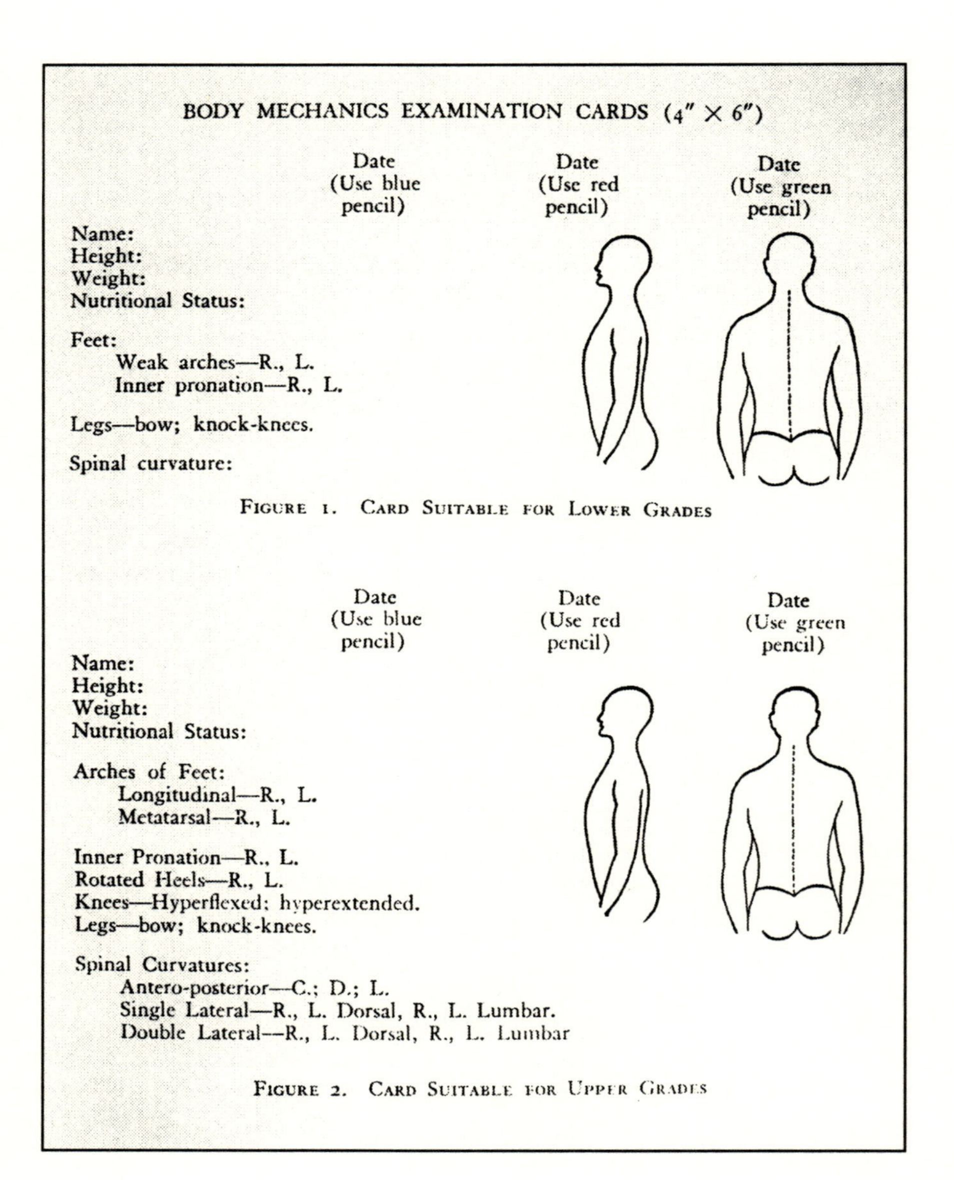
BODY MECHANICS EXAMINATION CARDS (4″ × 6″)

Date (Use blue pencil)
Date (Use red pencil)
Date (Use green pencil)

Name:
Height:
Weight:
Nutritional Status:

Feet:
Weak arches—R., L.
Inner pronation—R., L.

Legs—bow; knock-knees.

Spinal curvature:

FIGURE 1. CARD SUITABLE FOR LOWER GRADES

Date (Use blue pencil)
Date (Use red pencil)
Date (Use green pencil)

Name:
Height:
Weight:
Nutritional Status:

Arches of Feet:
Longitudinal—R., L.
Metatarsal—R., L.

Inner Pronation—R., L.
Rotated Heels—R., L.
Knees—Hyperflexed; hyperextended.
Legs—bow; knock-knees.

Spinal Curvatures:
Antero-posterior—C.; D.; L.
Single Lateral—R., L. Dorsal, R., L. Lumbar.
Double Lateral—R., L. Dorsal, R., L. Lumbar

FIGURE 2. CARD SUITABLE FOR UPPER GRADES

activity for a schoolwide Good Body Mechanics and Posture Week: "Posture cops—selected to tag all good postures each day" (59).

These kinds of games—especially the ones that emphasized the keeping of "charts and records"—again seem clearly derived from Taylorist forms of education like Bancroft's. It is also hard not to be reminded of Michel Foucault's discussion of surveillance and discipline when reading about these "games"; they were clearly intended to develop a culture of surveillance in order to enforce properly disciplined carriage.[52] Indeed, Bancroft noted (with, one feels, a certain degree of glee at the success of her motivational techniques) that "the boys in one class waylaid a classmate after school and pommeled him because his poor posture kept the class from one hundred per cent."[53] Both official and unsanctioned monitoring activities like these communicated to students not only that they were being constantly watched and judged based on their form, but also that they should in turn feel free—even obligated—to judge the health, efficiency, and character of others based on their form.

Figure 3.
"Body Mechanics Examination Cards" (developed by Lillian Drew and produced by the Robey French Company, Boston), from Ivalclare Sprow Howland, *The Teaching of Body Mechanics in Elementary and Secondary Schools* (New York: A. S. Barnes, 1936), 17.

Fortunately for their students, though, Thomas and Howland did suggest many less-public techniques for cultivating visual discernment. Once students had learned how their own bodies looked and were able to track their improvement (through the use of Taylorist technologies such as silhouettographs, cards, mirrors, charts, competitions, etc.), they were encouraged to compare their own bodies not only to the posture standards but also to other exemplary bodies. These other bodies could include those of athletes, actresses, and physical education teachers, all of whom were presumed not only to have excellent posture, but also to be people whom students were anxious to emulate.[54] Or these exemplary bodies could come from the fine arts: "Let the children choose a statue or picture in the school or city park as an example of good posture," Howland advised (113), claiming that "the selection of famous paintings and sculpture that serve as examples of good body mechanics is very stimulating and inspiring" (50). Bancroft argued that "with pictures of Washington and Lincoln before them, of kings and queens, of pioneers and heroes, who figure in history and literature, it should not be hard to inspire any child with a desire for the best carriage he can cultivate."[55] Thomas provided a list of paintings—"Beatrice d'Este, Elizabeth, Countess of Derby, Queen Mary, The Blue Boy"—that she considered to be "examples of excellent posture," and also illustrated a John Faed painting of George Washington and an Augustus Saint-Gaudens sculpture of Abraham Lincoln in her book. Howland provided a long list of "suitable pictures" that she believed demonstrated good body mechanics, although she did not illustrate any of the items. It included the following works of art and architecture, which she stated could be acquired from art extension societies or art publishing houses (50):

George Washington by Stuart
Song of the Lark by Breton
Sistine Madonna by Raphael
Signing of the Declaration of Independence by Trumbull
Pilgrims Going to Church by Boughton
Sir Galahad by Watts
Spirit of 1776 by Willard
Washington Crossing the Delaware by Leutze
Windmills of Holland by Hencke
Pueblo Indian by Amick
Blue Boy by Gainsborough
Bridge of Avignon by Garrison
Milan Cathedral, Italy
Rheims Cathedral, France
Roman Forum
Roman Coliseum
Appeal to the Great Spirit by Dallin

As Howland made clear through this list of recommended images, yet another kind of visual comparison that she encouraged her students to make was between bodies and buildings. Though the point of comparing one human form to another (whether actual or painted/sculpted) is fairly clear, the usefulness of comparing the body to the architectural monuments toward the end of Howland's list is more difficult to discern. What was there about Milan cathedral, for example, that could inspire good posture? Were students really expected to make a connection between vertical spires of stone and upright spines of bone? Although in the West there is a long history of comparing the human body to a column and the body of Christ to the plans of churches, neither of these comparisons has typically been drawn in the interest of fostering good posture, nor is either comparison a good parallel for what Howland did. But apparently students were expected to make the connection without difficulty, for Howland made such architectural comparisons frequently. She suggested the phrase "Economic Architecture" as a body mechanics poster slogan (39); she captioned her book's frontispiece—which may have depicted her daughter Gloria—as "Dynamic Architecture" (fig. 4); she suggested that students "tag trees and buildings demonstrating poise and beauty in perfect alignment" (62); she recommended that senior high students discuss the sentence "each one of us is the architect of his own body" (160); and she assigned first- through fourth-graders to "discuss the posture of buildings. . . . Illustrate with pictures. . . . Contrast strong, well-balanced structures with those about to fall over" (113). These comparisons all point to her belief in a widely shared understanding of the usefulness of buildings—whether classical temples or skyscrapers—as models for human bearing on both formal and metaphorical levels. The implication is that both buildings and bodies are "designed" objects and can be evaluated similarly, in that they are technological artifacts that in turn are used as tools for other purposes.

However, real bodies, painted and sculpted bodies, and buildings were not the only designed objects to which Thomas and Howland and their colleagues urged students to compare their own bodies. Although many of the games that Thomas recommended compared the human form to animals (camels, rabbits, giraffes, horses, etc.) and to plants (poplars, apples, hollyhocks), some of her exercises compared the human body to machines and other "formerly artless" consumer products. Her "Story Play" called "Change! Change! Change!" directed students to "[sit] at desks with one hand on chest; pretend chest is an elevator; raise and lower chest; sometimes stop halfway to allow passengers to step out; other times it is express to top floor" (123). A game called "Lights on, Lights off" compared children to lightbulbs:

> Two members of the class with habitual good posture are chosen to be the men who want to buy electric light bulbs. The rest of the class are the bulbs. They are in the shop; that is, they are lined up along the walls of the room. If they are very straight against the wall, they are good bulbs. The men choose or "buy" alternately, and as each is bought it finds its socket, which is a chair, or a place on the floor, if the game is played in the gymnasium. One "man" says, "Lights on." All stand. The teacher is "tester." She names any one who is in poor posture. That one is a poor bulb, and must go back to the shop for repairs. The other "man" says, "Lights off," and all sit. The teacher again names any who are in poor posture. These go for repairs. She then names those that are "mended," and they return to their places. At the end of five minutes, or any time agreed upon, the side having the greatest number of lights in place is declared the winner.[56]

Howland made many similar body/machine comparisons in her book. For example, echoing the Report, she claimed that "the body is like the machine; its working parts must be accurately adjusted to one another and, if any one part is out of position, the machine does not work perfectly" (114). Further, in dedicating her book "to my little daughter Gloria[,] and to all children[,] in the hope that their bodies may become dynamic mechanisms of beauty

DYNAMIC ARCHITECTURE

and symmetry, capable of supreme service for lives of happiness and worth," Howland used a machine metaphor that seemed not only to emphasize the use a well-tuned body could give its owner, but also to emphasize the ways in which that individual body could serve industry, "the race," and so on (iii). And echoing some of Thomas's games, an exercise Howland recommended for fifth- and sixth-graders was to "discuss the beauty and efficiency of inanimate machines such as automobiles, machines in factories, the corn popping machine, the doughnut fryer, etc." (115).

The point of Thomas's and Howland's many body/machine comparisons seems to have been twofold: that "good" bodies and machines should be efficient at operation/production, and that they should be characterized—like the posture standards—by a taut disposition of forms around a central axis and an absence of "exaggerated" curves or bulges. These authors' intriguing comparison of bodies to formerly artless products such as lightbulbs, doughnut fryers, and corn poppers casts an interesting light on Howland's aforementioned recommendation to students to view their bodies in the reflections of store windows—where their images presumably would have been juxtaposed, ghostlike, with exactly these kinds of products. Also, comparisons of human bodies to productive technologies such as elevators, lightbulbs, and factory machines—artifacts designed to manipulate conditions and/or do work—makes it clear that body mechanics educators conceived of the human body not only as designed, but also as mechanical, even technological. Although many historians have argued that the point of machine/body comparisons in this period was that the machine served as a model for the modern body, a notion that the body mechanics literature supports, these comparisons could cut both ways: that is, even when the comparisons were meant to serve as lessons for how to manage and form the body, they also suggested that perhaps bodies could serve as models for machines and products.[57]

Figure 4.
"Dynamic Architecture," from Ivalclare Sprow Howland, *The Teaching of Body Mechanics in Elementary and Secondary Schools* (New York: A. S. Barnes, 1936), frontispiece.

Designers on Streamlining and the Body

Indeed, many consultant industrial designers of this period made exactly the same point about streamlined products: that their form was derived from the appearance of the ideal human body. Renowned designer Raymond Loewy's before-and-after pictures of McCormick-Deering cream separators (figs. 1a, 1b) show particularly effectively the ways in which streamlining "humanized" the form of products by making their skins or contours tauter, more compact, and more axial. Loewy not only implicitly compared the form of the female body to different forms of architecture and design in his well-known 1930 "evolution charts" of design (figs. 5, 6), but also made such comparisons explicit using photographs. In his autobiography *Never Leave Well Enough Alone* (1951), he compared an automobile chassis to a human skeleton (a female one, in this case; her bracelets, rings, and high heels are visible on the X-ray), as well as comparing good and poor automobile design to what he considered to be beautiful and ugly female forms (figs. 7, 8). Loewy described the poor auto design as "bulbous, fat, and without character: what we call the 'jelly mold school of design,'" a phrasing that closely echoes the body mechanics educators' assessments of the physical and moral qualities of bodies that displayed poor posture.[58] On the other hand, Loewy's chosen model for good design, the sculpted female figure (fig. 7, lower left), could easily have been a poster girl in Thomas's or Howland's body mechanics classes. Her tautly contoured, not-too-curvy body was very similar to the bodies illustrated in the posture standards, and to other sculptures shown as examples in body mechanics books. Loewy concluded that good automobile body design "obeys the same aesthetic canons of slenderness and economy of means as the human figure," and claimed that "even when you look at these illustrations upside down or sidewise, the results are equally pleasing or repellent," suggesting that the ideal human body could be used as a model for objects of all orientations and sizes.[59]

Figure 5.
Raymond Loewy, evolution chart of design: trains (1930), from Raymond Loewy, *Industrial Design* (Woodstock, NY: Overlook Press, 1988), 74. Reproduced with permission of Laurence Loewy.

Loewy's use of the human body as a source for his streamline aesthetic was shared by many other consultant industrial designers. Adrienne Berney has noted that "industrial designers as well as advertising companies recognized the parallel aesthetics between automobiles and women's bodies," and quotes Paul Frankl as saying that continuity of line "was characteristic of the modern style as we find in the stream-line body of a car or in the long unbroken lines in fashions."[60] Similarly, Walter Dorwin Teague claimed that "the human body [was] the source of balance and symmetry in design," and that "the healthy, vigorous human body in action is the most productive field of study we can find."[61] J. Gordon Lippincott claimed that "much of our appreciation of the abstract is based on our inherent sensitivity to the nude form. The flowing rhythmic lines of nature have been man's natural environment, and therefore it is not surprising that he applies this feeling to his expressions in abstract creation."[62] Industrial designer Egmont Arens, who assuredly knew the technical

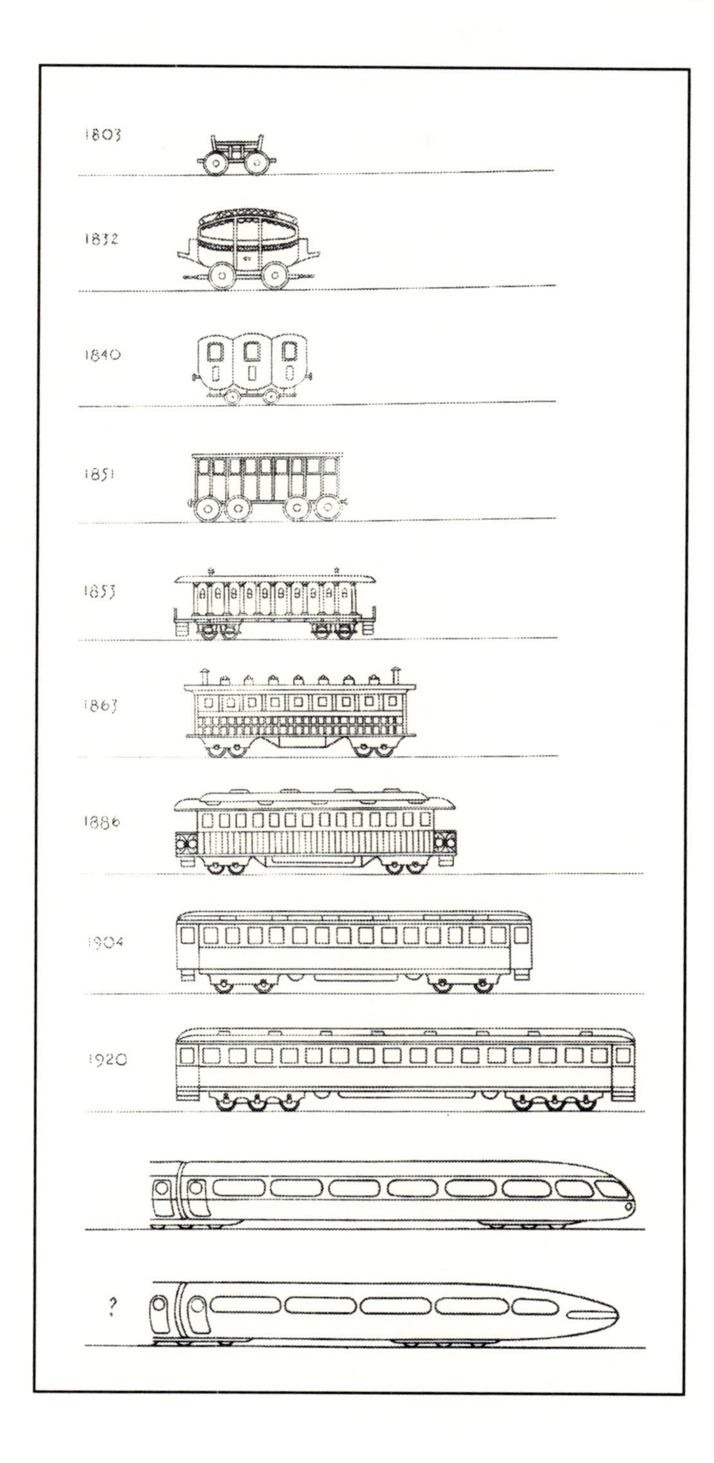
1803
1832
1840
1851
1853
1863
1886
1904
1920
?

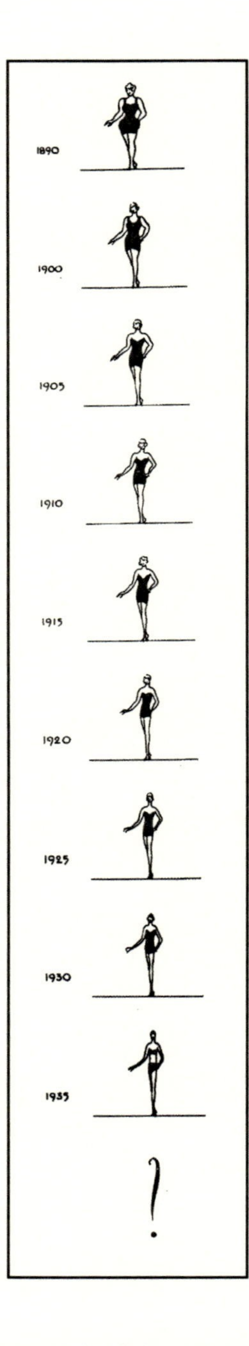

Figure 6.
Raymond Loewy, evolution chart of design: women's bathing suits (1930), from Raymond Loewy, *Industrial Design* (Woodstock, NY: Overlook Press, 1988), 76. Reproduced with permission of Laurence Loewy.

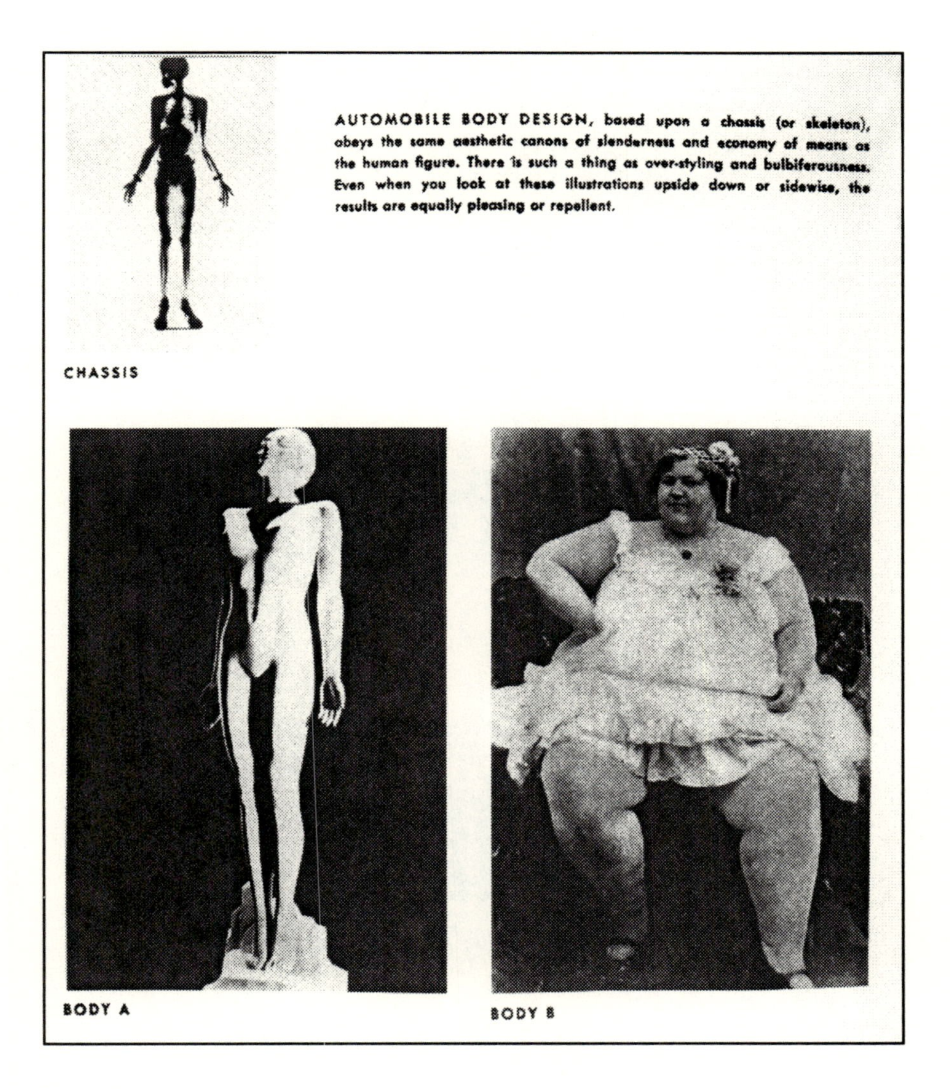

Figure 7.
Human bodies, from Raymond Loewy, *Never Leave Well Enough Alone* (New York: Simon & Schuster, 1951), 312f. Reproduced with permission of Laurence Loewy.

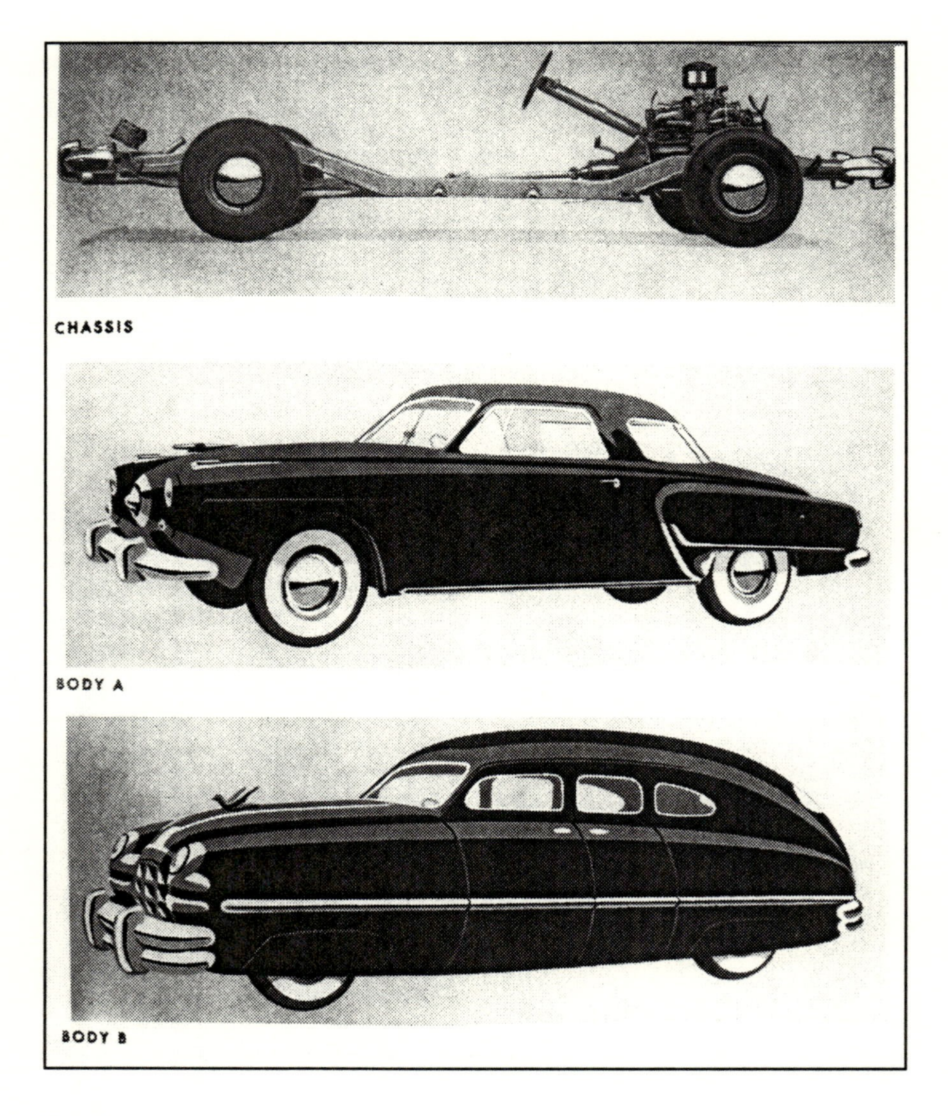

Figure 8.
Automobile bodies, from Raymond Loewy, *Never Leave Well Enough Alone* (New York: Simon & Schuster, 1951), 312f. Reproduced with permission of Laurence Loewy.

meaning of the term streamlining, nonetheless often used it as a synonym for "sleek" or "beautiful." In a 1934 telegram to President Franklin D. Roosevelt, for example, he described a slide lecture he had given in which he showed images of "streamlined Trees, Flowers, Whales, Diving Girls, Refrigerators, Houses, Gadgets, [and] Women[']s Fashions."[63] What he meant by streamlined trees and flowers is debatable, but by streamlined women's fashions and diving girls he no doubt referred to the restrained, taut curves that are seen in Loewy's sculpture (fig. 7). Claude Bragdon, whose Theosophist architectural theories were known to many industrial designers (most notably to Teague), stated that "a study of the human figure with a view to analyzing the sources of its beauty cannot fail to be profitable. Pursued intelligently, such a study will stimulate the mind to a perception of those simple yet subtle laws according to which nature everywhere works, and it will educate the eye in the finest known school of proportion, training it to distinguish minute differences, in the same way that the hearing of good music cultivates the ear."[64] Bragdon's assertion, in particular, echoes very closely the rhetoric of body mechanics writers, even down to the phrase "educating the eye," which he also envisioned doing through study of the human body. It seems, then, that designers were well aware that their audience expected bodies and machines to conform to similar standards of beauty and efficiency.

Conclusion

The shift in consumer tastes and demands in the late 1920s that so many period commentators noted but could not fully explain—and the rise of streamlining seemingly in response to it—was likely related to the aesthetic training that many young people received between the 1920s and the 1940s in body mechanics coursework. Although the visual training that the discipline of body mechanics provided young women and men is of course not the only factor in the rise of consumers' "object consciousness," certainly the effect of body mechanics training should not be ignored as a potentially significant factor in consumers' increased demand for "good form" in the formerly artless industries, and for the particular configuration of good form—streamlining—that became popular during that era.[65] Most students who attended elementary school, high school, and/or college between the 1920s and 1940s would have been required to develop, at least to some extent, an "educated eye": a vocabulary for describing form, an "artistic and esthetic sense" (as Howland put it), and a mind-set that encouraged them to see parallels between bodies and designed goods, especially between bodies and machines (37, 115). This significant segment of the population constituted a group of people who shared specialized visual and cognitive habits and vocabulary. These consumers' skills at assessing the form of bodies were available to be used even for the analysis of those categories of goods whose value had formerly been judged almost exclusively by their function: the formerly artless industries.

Not coincidentally, streamlining, the style that designers applied to the formerly artless industries after 1925, was by all accounts most popular with a white upper-middle-class audience (the same demographic group that would have received the greatest number of years of posture training in high school and/or college).[66] It was admirably suited to the visual proclivities of consumers of the 1930s and 1940s who had undergone body mechanics training. The taut "skins" and restrained curves of post-1925 streamlined formerly artless products lent themselves to precisely the same kind of analysis that body mechanics educators urged when they asked students to compare their own posture to lightbulbs, buildings, cars, doughnut fryers, and corn poppers, and indeed, designers claimed that their products were modeled on the body. Given that young Americans were taught to interpret an erect, compact, subtly curved body as being possessed of positive physical and even moral qualities, and that they were repeatedly asked to compare bodies to certain kinds of products, it is entirely likely that they transferred standards of and ideas about bodily good form to the assessment of objects, and came to expect machines and other artless products not only to reveal their efficiency and "character" through their contours, but also to conform to human standards of beauty.

The body mechanics literature—like other forms of educational literature that promote the teaching of visual discernment—is thus useful to American studies scholars in a number of ways. First, it provides insights into early-twentieth-century conceptions of the relationship between bodies and machines that examination of either the bodies or the machines alone would not. Second, it provides a means of theorizing stylistic change as consumer- rather than designer-driven. And third, it helps us understand why early-twentieth-century consumers might have been drawn, very specifically, to purchase mechanical products that looked like human bodies: they had been taught, through body mechanics instruction, to understand both bodies and formerly artless goods as "designed," as mechanical in nature, and as products of Taylorist-inspired technological systems.

Notes

Many people have helped me with the research, writing, and images for various versions of this essay over the years, and I am grateful to them all. However, I would particularly like to thank Cortney Boyd, Carolyn Thomas de la Peña, Paul Groth, Phil Howze, Kathleen James-Chakraborty, Ken Friedman, Jiawei Gong, Laurence Loewy, Margaretta Lovell, Stacey Lynn, Sharon Marcus, Eric Peterson, Marita Sturken, Siva Vaidhyanathan, the members of the Berkeley Americanists' Group, the School of Art and Design at Southern Illinois University Carbondale, and an anonymous reviewer for *American Quarterly* for their kind assistance.

1. "Both Fish and Fowl," *Fortune*, 9 February 1934, 98.
2. Harold Van Doren, *Industrial Design: A Practical Guide* (New York: McGraw-Hill, 1940), 4–5.
3. Raymond Loewy, *Never Leave Well Enough Alone* (New York: Simon & Schuster, 1951), 11.
4. Van Doren, *Industrial Design*, 43.
5. "Designing for the Eye" is the title of an article by Franklin E. Brill, *Product Engineering* 4.1 (January 1933): 16–17.
6. "Both Fish and Fowl," 40.
7. By "streamlining," I do not mean the teardrop-shaped, aerodynamics-inspired transportation designs by Norman Bel Geddes and Buckminster Fuller, but rather, the more restrained and commercially successful kind of streamlining that was applied to "formerly artless" products of the period.
8. Norman Bel Geddes, "Streamlining," *Atlantic Monthly*, November 1934, 553, 556–58. The derivation of streamlining from the form of the human body is discussed later in this essay.
9. For period observers' explanations, see, for example, Christine Frederick, *Selling Mrs. Consumer* (New York: The Business Bourse, 1929), 355–57; Van Doren, *Industrial Design*, 43; and Earnest Elmo Calkins, in the introduction to *Consumer Engineering: A New Technique for Prosperity*, by Roy Sheldon and Egmont Arens (New York: Harper, 1932), 4–6. For historians' explanations, see, for example, Jane N. Law, "Designing the Dream," in *Streamlining America: A Henry Ford Museum Exhibit*, ed. Fannia Weingartner (Dearborn, Mich.: Henry Ford Museum and Greenfield Village, 1986); Dianne H. Pilgrim, "Design for the Machine," in *The Machine Age in America, 1918–1941*, ed. Richard Guy Wilson, Dianne H. Pilgrim, and Dickran Tashjian (New York: Brooklyn Museum in association with Abrams, 1986); Karen Davies, *At Home in Manhattan: Modern Decorative Arts, 1925 to the Depression* (New Haven, Conn.: Yale University Art Gallery, 1983); Donald J. Bush, *The Streamlined Decade* (New York: George Braziller, 1975); Christina Cogdell, "The Futurama Recontextualized: Norman Bel Geddes's Eugenic 'World of Tomorrow,'" *American Quarterly* 52.2 (June 2000): 193–245; Cogdell, *Eugenic Design: Streamlining America in the 1930s* (Philadelphia: University of Pennsylvania Press, 2004); Ellen Lupton and J. Abbott Miller, *The Bathroom, the Kitchen, and the Aesthetics of Waste: A Process of Elimination* (New York: Kiosk, 1992); Adrienne Berney, "Streamlining Breasts: The Exaltation of Form and Disguise of Function in 1930s' Ideals," *Journal of Design History* 14.4 (2001); Jeffrey Meikle, *Twentieth-Century Limited: Industrial Design in America, 1925–1939* (Philadelphia: Temple University Press, 1979); and Neil Harris, "The Drama of Consumer Desire," in *Cultural Excursions: Marketing Appetites and Cultural Tastes in Modern America* (Chicago: University of Chicago Press, 1990).
10. For a discussion of related art pedagogy and its relation to consumer tastes and design, see Carma R. Gorman, "'An Educated Demand': The Implications of *Art in Every Day Life* for American Industrial Design, 1925–1950," *Design Issues* 16.3 (Autumn 2000): 45–66.
11. Herbert Simon, *The Sciences of the Artificial*, 3rd ed. (Cambridge, Mass.: MIT Press, 1998), 112.
12. For varying scholarly assessments of the commercial success of streamlining, see Regina Lee Blaszczyk, *Imagining Consumers: Design and Innovation from Wedgwood to Corning* (Baltimore: Johns Hopkins University Press, 2000); Shelly Nickles, "Preserving Women: Refrigerator Design as Social Process in the 1930s," *Technology and Culture* 43.4 (October 2002): 693–727; and Herbert J. Gans, "Design and the Consumer: A View of the Sociology and Culture of 'Good Design,'" in *Design Since 1945*, by Kathryn B. Hiesinger, George H. Marcus, and Max Bill (Philadelphia: Philadelphia Museum of Art, 1983).
13. Michael Baxandall, *Painting and Experience in Fifteenth-Century Italy: A Primer in the Social History of Pictorial Style* (New York: Oxford University Press, 1972), 37, 40.
14. Baxandall, *Painting and Experience*, 34.
15. Useful discussions of 1920s–1930s body/machine comparisons can be found in Joel Dinerstein, *Swinging the Machine: Modernity, Technology, and African American Culture Between the World Wars*

(Amherst: University of Massachusetts Press, 2003); Mark Seltzer, *Bodies and Machines* (New York: Routledge, 1992); Terry Smith, *Making the Modern: Industry, Art, and Design in America* (Chicago: University of Chicago Press, 1993); Cecelia Tichi, *Shifting Gears: Technology, Literature, Culture in Modernist America* (Chapel Hill: University of North Carolina Press, 1987); Martha Banta, *Taylored Lives: Narrative Productions in the Age of Taylor, Veblen, and Ford* (Chicago: University of Chicago Press, 1993); and Carolyn Thomas de la Peña, *The Body Electric: How Strange Machines Built the Modern American* (New York: New York University Press, 2003), esp. 16–17, 22–25, 42, and 98.

16. Jessie H. Bancroft, *The Posture of School Children, With Its Home Hygiene and New Efficiency Methods for School Training* (New York: Macmillan, 1913). See also Henry Ling Taylor, M.D., "The American Posture League: Its History, Work and Future," *Modern Medicine* (December 1920): 777–79; and George J. Fisher, M.D., "The American Posture League," *Journal of Health and Physical Education* 6 (October 1935): 16–17.
17. Drs. Lloyd Brown and Roger Lee initiated the use of the term "body mechanics" in their assessment of the Harvard freshman class in 1916. George T. Stafford, *Preventive and Corrective Physical Education* (New York: A. S. Barnes and Company, 1928), 59.
18. *Body Mechanics: Education and Practice. Report of the Subcommittee on Orthopedics and Body Mechanics of the White House Conference on Child Health and Protection* (New York: Century Company, 1932), 5.
19. *Body Mechanics: Education and Practice*, 3. Armin Klein and Leah C. Thomas's report on the Chelsea Survey, *Posture and Physical Fitness*, Children's Bureau Publication No. 205 (Washington, D.C.: Government Printing Office, 1931), is reprinted as Appendix I of *Body Mechanics: Education and Practice.*
20. *Body Mechanics: Education and Practice*, 21–22.
21. The aspects of modern civilization that the body mechanics theorists understood to inhibit children's natural development included cramped city dwellings, polluted air, adulterated food, child labor, ill-fitting clothing and shoes, and enforced inactivity in schools—the last made even worse by chairs that "violate[d] the fundamentals of posture hygiene" by being the wrong size and shape for most children. Henry Eastman Bennett, *School Posture and Seating: A Manual for Teachers, Physical Directors and School Officials* (Boston: Ginn and Company, 1928), iii.
22. *Body Mechanics: Education and Practice*, 19–20. Members of the survey team studied the posture of 1,708 children of five to eighteen years of age in the city of Chelsea, Massachusetts, in 1923–24.
23. *Body Mechanics: Education and Practice*, 23–26.
24. Bancroft, *Posture of School Children*, 108.
25. Marguerite Sanderson of the Boston School of Physical Education, in the introduction to *Body Mechanics and Health*, by Leah C. Thomas and Joel E. Goldthwait (Boston: Houghton Mifflin, 1922), 7.
26. Bennett, *School Posture and Seating*, 53–54.
27. Anson Rabinbach, *The Human Motor: Energy, Fatigue, and the Origins of Modernity* (Basic Books, 1990); Thomas de la Peña, *Body Electric.*
28. Thomas and Goldthwait, *Body Mechanics and Health*, 26; Goldthwait, in the introduction to Leah C. Thomas, *Body Mechanics and Health*, rev. ed. (Boston: Houghton Mifflin, 1929), 5.
29. An example of a body mechanics writer who would surely have been familiar with eugenic theory (though who seems not to have adopted it, at least not wholesale) is Ivalclare Sprow Howland; she was employed at (the second) Battle Creek College, which had been founded by J. H. Kellogg, a promoter of eugenics who hosted the First National Conference on Race Betterment (1914) and the Third National Conference on Race Betterment (1928) in Battle Creek. For a discussion of these conferences, see Robert Rydell, *World of Fairs: The Century-of-Progress Expositions* (Chicago: University of Chicago Press, 1993), 40–41. W. H. Sheldon and E. A. Hooton, who conducted posture studies at Ivy League universities in the middle of the century, were more obviously interested in eugenics. See Ron Rosenbaum, "The Great Ivy League Nude Posture Photo Scandal," *New York Times Magazine*, January 15, 1995, 26ff. In contrast, examples of body mechanics writers who clearly stated the importance of environmental rather than hereditary factors on body mechanics are Josephine Rathbone, *Corrective Physical Education* (Philadelphia: W. B. Saunders, 1934), 88; and Mabel Lee and Miriam Wagner, *The Fundamentals of Body Mechanics and Conditioning: An Illustrated Teaching Manual* (Philadelphia: W. B. Saunders, 1949), 156.
30. Bancroft, *Posture of School Children*, 5.

31. Bennett, *School Seating and Posture*, 4.
32. Two notable government publications of this period were Armin Klein, M.D., "Posture Clinics: Organization and Exercises," U. S. Department of Labor/Children's Bureau Publication No. 164 (Washington, D.C.: Government Printing Office, 1926), and Armin Klein and Leah C. Thomas, "Posture Exercises: A Handbook for Schools and for Teachers of Physical Education," U. S. Department of Labor/Children's Bureau Publication No. 165 (Washington, D.C.: Government Printing Office, 1926).
33. *Body Mechanics: Education and Practice*, ii, xi. The members of the Subcommittee on Orthopedics and Body Mechanics were Robert B. Osgood, M.D., Professor of Orthopedic Surgery at Harvard University (chair); Lloyd T. Brown, M.D., Instructor in Orthopedic Surgery at Harvard; John B. Carnett, M.D., Vice Dean and Professor of Surgery at the Graduate School of Medicine, University of Pennsylvania; Armin Klein, M.D., Assistant Professor of Orthopedic Surgery at Tufts College Medical School; and Leah C. Thomas, Assistant Professor in the Department of Hygiene and Physical Education, Smith College.
34. *Body Mechanics: Education and Practice*, 41.
35. Ibid., 42.
36. Mabel Lee, *History of Physical Education and Sports in the U.S.A.* (New York: Wiley, 1983), 166.
37. *Body Mechanics: Education and Practice*, 28.
38. Ibid.
39. Ibid., 33–34.
40. A particularly striking example of the lack of attention to the history of body mechanics in the physical education literature occurs in Lee's *History of Physical Education and Sports*. Lee, who had herself previously coauthored *Fundamentals of Body Mechanics and Conditioning*, made almost no mention of posture or body mechanics in her history. For a list of incorporators and directors of the American Posture League, see George J. Fisher, "The American Posture League," *Journal of Health and Physical Education* 6 (October 1935): 16–17. Prominent members and leaders/directors of the APL included Robert Tait McKenzie, M.D.; George J. Fisher, M.D.; Thomas A. Storey, Ph.D., M.D.; and Frank and Lillian Gilbreth. Biographies of McKenzie, Fisher, and Storey can be found in Lee, *History of Physical Education and Sports*; Paula D. Welch, *History of American Physical Education and Sport*, 2nd ed. (Springfield, Ill.: Charles C. Thomas, 1996); and C. W. Hackensmith, *History of Physical Education* (New York: Harper & Row, 1966).
41. Herbert M. Kliebard, *Forging the American Curriculum: Essays in Curriculum History and Theory* (New York: Routledge, 1992), chap. 7.
42. Bancroft, *Posture of School Children*, 247–50.
43. The first edition of Thomas's book, coauthored with Joel Goldthwait (Boston: Houghton Mifflin, 1922), included little of the pedagogical material that characterized the second edition.
44. Ivalclare Sprow Howland, *The Teaching of Body Mechanics in Elementary and Secondary Schools* (New York: A. S. Barnes, 1936). When Howland's authorship is made clear by the context, subsequent references to pages in her book will be made in parentheses in the body of the text. On the second Battle Creek College, see James C. Whorton, *Crusaders for Fitness: The History of American Health Reformers* (Princeton, N.J.: Princeton University Press, 1982).
45. On social efficiency education, see Herbert M. Kliebard, *The Struggle for the American Curriculum*, 98, 119, 121. For reviews of Howland's book, see, for example, Mary Fread, review of *The Teaching of Body Mechanics* by Ivalclare Sprow Howland, *Teachers College Journal* 7.6 (June 1936): 136. Even a detractor such as Jay B. Nash, in his review of Howland's book in *The Elementary School Journal* 37 (June 1937): 792–94, praised her inclusion of the posture play "The Slump Family."
46. See, for example, Margaret H. Strong, "Hutchinson Students Become Posture Conscious," *Minnesota Journal of Education* 20.9 (May 1940): 360–61; Lee, *Fundamentals of Body Mechanics and Conditioning*; Ellen Davis Kelly, *Teaching Posture and Body Mechanics* (New York: A. S. Barnes, 1949).
47. A weight- and sex-neutral standard was available from Harvard University; Howland illustrated it, rather than the six government standards, in her book.
48. Bancroft, *Posture of School Children*, 9.
49. "To furnish the child with a silhouettograph picture of his habitual poor postural position and another of an improved or corrected position," Howland claimed, "serves as a worthy incentive for practice in body mechanical improvement" (23).

50. The use of these cards is discussed in Howland, 16–17.
51. See Bancroft, *Posture of School Children*, 187–88.
52. Michel Foucault, *Discipline and Punish: The Birth of the Prison*, trans. Alan Sheridan (New York: Vintage Books, 1995).
53. Bancroft, *Posture of School Children*, 208.
54. Howland argued that photographs of athletes in action could "teach many lessons that words cannot accomplish" (50). And Janet Lane, author of a popular (rather than pedagogical) text on posture, said of actresses that "time after time, they prove our points about the beauty and reliability of a well-adjusted physical machine." Janet Lane, *Your Carriage, Madam! A Guide to Good Posture*, 2nd ed. (New York: John Wiley and Sons, 1947), 135.
55. Bancroft, *Posture of School Children*, 250.
56. Thomas, *Body Mechanics and Health*, 120–21.
57. See note 15 for references to this literature.
58. Loewy, *Never Leave Well Enough Alone*, 213.
59. Ibid., 312ff.
60. Berney, "Streamlining Breasts," 336. Berney also suggests that "breast ideals helped to shape design styles for machinery" (340n7).
61. Walter Dorwin Teague, *Design This Day* (1940; repr., New York: Harcourt, Brace, 1949), 174.
62. J. Gordon Lippincott, *Design for Business* (Chicago: Paul Theobald, 1947), 94.
63. Arens, draft of a telegram to President Franklin D. Roosevelt, November 14, 1934, cited in Meikle, *Twentieth-Century Limited*, 164. According to Meikle, secretary M. H. McIntyre acknowledged receipt in a letter to Arens of November 15, 1934.
64. Claude Bragdon, *The Beautiful Necessity: Seven Essays on Theosophy and Architecture* (Rochester, N.Y.: Manas Press, 1910), 50.
65. The term "object consciousness" is from Harris, "Drama of Consumer Desire," 175.
66. See note 12.

Farewell to the El: Nostalgic Urban Visuality on the Third Avenue Elevated Train

Sunny Stalter

> "Toll a bell for the El/For the last clientele/With a stubborn and sporting *esprit*/Not a cinder around, yet my vision is blurred/Sentimental? Yes, that I may be./Nothing ever again can be quite what the Third Avenue El was to me."—This metropolitan dirge, composed by Michael Brown, is the current hit of Julius Monk's suave revue in the new nightclub cellar, The Downstairs. Mr. Brown's wail poignantly expresses the general lamentation for the passing of Third Avenue as we knew it.
>
> Shedding a tear for Third Avenue, or any city antique, is a popular sport. Nobody in town is kept as busy as the nostalgia-minded. Their eyes are never dry from weeping for another lost cause. New York, chameleon-like, changes with extravagant regularity; the city's old familiar song is the ra-ta-ta of the riveter. But Third Avenue is a special field day for the nostalgic.[1]

May 12, 1955, was truly a "special field day for the nostalgic"; it was the last day to ride on the Third Avenue elevated train. The El had ceased to be an important form of public transportation in New York City by the mid-twentieth century, carrying a mere 70,000 fares per day compared to the subway's 4.5 million.[2] Although there were still elevated trains throughout the city's boroughs, the trains that ran down Ninth, Sixth, and Second Avenues in Manhattan had been torn down by 1942.[3] The Third Avenue El was the last of its kind, and riders were eager to memorialize this dying technology. An account of the last ride on the Third Avenue El describes the train being overrun by "souvenir collectors" who stripped the car of "everything that wasn't nailed down": "The black metal destination markers proved the most popular items and were quickly removed from their window slots. Several persons carried six or eight signs tucked under their arms and for a while there was a brisk trading of 'duplicates.'"[4] This hunger continued for less concrete souvenirs that could communicate the experience of the train ride. The New-York Historical Society opened "Exhibit of the City's Elevateds" the day the Third Avenue El closed, replacing the literal ride with a figurative one.[5] A decade later, the New York Public Library tracked "The Rise and Fall of the Elevated Railroad"; along with the usual photographs, sketches, and contracts, this retrospective included a "1 1/2 hour program of old [E]l movies (and some new ones too)."[6] Elevated trains (and their tracks, stations, and passengers) were common in cultural representation of the New York cityscape throughout the late nineteenth and early twentieth centuries.[7] However, the Third Avenue El's imminent destruction made it a particularly resonant subject for postwar artists and writers. Popular and avant-garde filmmakers shot on its cars and beneath its tracks in the shadowy world of the Bowery.[8] Photographers preserved the flowery ironwork on the girders and the potbellied stoves and stained glass in the stations.[9] In the 1950s, the El traveled nearly empty while subways were packed and cars jammed the streets, but this anachronistic transportation technology gained a new hold on the city's cultural imagination. The Third Avenue El's end was the beginning of its afterlife as an object of aesthetic contemplation.[10]

New Yorkers mourned the El because it was a link to the city's past that was about to disappear—one of many in a period of rapid urban development. New York has always understood itself as a city that changes "with extravagant regularity," as Emory Lewis suggests above.[11] The 1950s were no exception, a decade "highlighted by a construction boom that once again transformed the cityscape. High-rise apartments and slum clearance projects were going up virtually overnight."[12] However, I suggest that the El was not just a metonymic stand-in for "the wonderful old world that's passing."[13] Instead, it structured New Yorkers' knowledge of urban space in a way that was no longer possible on other modes of transportation. Personal and collective views on American life often arise from machine-mediated views of American space, even once the technologies that made those views possible are deemed outdated or razed altogether. We tend to think about technological innovations as they suggest new social arrangements and philosophical understandings.[14] However, in the 1950s, the El was an obsolete technology in a modernizing city, and it emphasized movement through the city as mediated movement through the urban past.

Recent critical attention to the elevated train has emphasized its panoramic quality in nineteenth- and early-twentieth-century literature and the distanced relationship to the city sights (particularly the lives of the poor) that resulted.[15] For these critics, technology is a shield and a frame: it protects New Yorkers who can afford the fare and aestheticizes encounters that would otherwise be experienced more authentically at street level. However, my argument considers how the El's visuality changes over time and in relation to other, later transportation technologies.[16] In the last days of the El, it was no longer a novelty or a mover of the masses. Instead, it had a more symbolic purpose as an alternative to the decontextualizing force of the subway or the isolation of the car.[17] The El showed postwar artists a city dissociated from progress, haunted by machines of the past and visions of modernity from seventy-five years ago.

The progress associated with newer transportation technologies could not erase the satisfactions of riding the El: it offered the urban subject embodied connection and a sense of lived history. Nostalgic representations of the Third Avenue El in the 1950s argued that these pleasures should continue to be a part of New York's visual landscape. Janelle Wilson describes nostalgia as "the opportunity to observe and juxtapose past and present identity"; the nostalgia for the city as seen from the El posited continuity and discontinuity between past and present forms of urban space and experience.[18] A later verse of the "metropolitan dirge" described above compares the danger and excitement of driving on an icy street under the El tracks with the boring experience of driving on Third Avenue after the tracks have been taken down: "Now it's dull and it's drab to go by in a cab/with a feeling approaching ennui"—the anachronistic thrill is gone.[19] This new ease of movement led to an historically specific ennui. The temporal and spatial order of postwar New York idealized an efficient flow that reduced the dangerous (and interesting) proximity of relics of the city's past.[20] The connection to city space made possible by the El stood in sharp relief to the frictionless movement through the city embraced by urban developers, architects, and the like: Robert Bennett describes their form of development as a "postindustrial dematerialization of urban space."[21] We can understand the "nostalgia-mindedness" of artists of the 1950s as an attempt to *re*materialize urban space, creating a concrete, embodied relationship to the city. Like nostalgic cyborgs, artists of the 1950s inhabited anachronistic machines in order to see what modes of perception they offered and how they could be maintained in a world without the material technology itself.[22]

The El's anachronistic visuality collapsed binaries between distance and intimacy, public and private space, past and present. Visuality, what Hal Foster calls "how we see, how we are able, allowed or made to see, and how we see this seeing or the unseen therein," is central to understanding technology's importance in American culture.[23] Transportation technologies situate viewers in literal and symbolic relation to the passing landscape: here we can think of the car's importance as an individualistic technology, for example, arising from its literal separation of the driver both from the surroundings and from the crowd.[24] The defining aspect of the El, particularly in comparison with later forms of public transportation, was its openness to city space. The "shad belly" design included a low-slung center so the cars didn't tip over, and gave extra height to ceilings (they were 9'6"); its double row of windows in the center provided extra light and extended the view.[25] Windows could be opened, and on summer days the back door of the last car was opened to take in the breeze. This connection to the urban environment was impossible on the subway, a space of physical and psychological dissociation.[26] El passengers experienced a clear relationship to the city, while subway passengers could not see where they were going.[27] A 1938 editorial from the *New York Herald Tribune* doubted that the El would ever be torn down precisely because of the aesthetic pleasure and spatial mastery riders felt in their navigation of the city: "Powerful nationalistic smells of cooking, as well as odors from the leather, spice, chemical, coffee and other industries reach the traveler who is seeing New York for a nickel, and if he is a regular, he can tell with his eyes shut about where he is on his journey."[28] Here, regular riding of the El is not associated with the daily grind of commuting but with the sensual experience of smelling fresh ground coffee. The city may be a site of shifting neighborhoods and alliances, but it is legible (even smellable) to those habituated to movement through it. The embodied connection to city space should not suggest that El riders experienced a more *real* version of the city than subway riders did—it was a wholly different one. Subway car views alternated between dark tunnels and stations distinguishable only by name; lone riders retreated into internal fantasy through reading, daydreaming, or dozing.[29] El passengers, on the other hand, experienced fantasies that were directed outward into external city space. Technologies of transportation reinforce urban subjectivity by literalizing views of the city: moving underground, New York became a space of invisibility and isolation; moving aboveground and in close proximity to city life opened up otherwise invisible spaces to fantasies of communion. While the cross-country train offered the possibility of connecting different parts of America, the elevated train suggested this imaginary connection could take place on an individual level as well.[30]

Artists of the 1950s latched on to this externally directed gaze as a way of preserving not only the sights of the El but also the way of seeing from it. American surrealist Joseph Cornell examined this way of seeing more extensively than any of the other artists I discuss here, using assemblage, film, and writing to examine the technology's possibilities for projecting a reverie of connectedness onto urban space. Cornell's prominence in the art world rests on the reputation of his boxes, which were filled with found and purchased objects arranged as they would be in a cabinet of curiosities.[31] His diaries from the 1920s through the 1950s are filled with notes of the people and scenes he saw while riding the El, as well as the magical sense of identification that they evoked. One diary entry describes a "woman adjusting [her] window" seen from a Queens elevated train. The "'flash' view of [the] moving car leav[es] this image imbued with the magic that used to come so strongly in commuting days and following."[32] Cornell often attempted to visually evoke the viewer's past in his artwork just as he experienced his own past while watching the woman described in his diary. One series of shadowboxes, known as the Window Façade boxes, is made up of vertically oriented grids of windows: the box suggests the façade of an apartment building seen from a passing elevated train, and viewers expect that similar isolated scenes will be glimpsed inside (fig. 1).

Throughout this article I return to the works of Joseph Cornell because their organization of the El's visual space illustrates an understanding of public transportation not only as a historically specific technology to be mourned and fetishized but as a way of seeing that can be examined and problematized. Throughout Cornell's corpus, old-fashioned technologies make the persistence of the past a concrete and embodied—but not easily perceived—presence. This is especially true of his film *Gnir Rednow*, produced in collaboration with avant-garde film pioneer Stan Brakhage. Brakhage shot the initial footage on the Third Avenue El and edited it for his own film, *The Wonder Ring* (1955); however, Brakhage's final product did not evoke the sense of encounter between different spaces and times that Cornell had envisioned.[33] Reversing Brakhage's intentions as well as his title, Cornell edited *The Wonder Ring* into *Gnir Rednow* (precise date unknown), a film that re-created the El's visuality as Cornell understood it.[34] While this seems to be an example of what Susan Hegeman calls "salvage ethnography," I argue that Cornell is not merely a nostalgic preservationist.[35] For Cornell, the Third Avenue El is not just a structure; it is a structure of feeling. To know New York through this anachronistic technology is to realize the historical locatedness of urban visuality, and yet to want it to persist outside of its specific history.

Passing Acquaintances

The El oriented passengers like Cornell to the cityscape in two ways: it presented broad vistas that connected neighborhoods as well as intimate close-ups that segmented vision into brief glimpses. The first section of William Dean Howells' novel *A Hazard of New Fortunes* (1890) shows readers how the close-up views affected nineteenth-century passengers' visual relationship to the city. Basil March and his wife, Isabel, ride the elevated trains back and forth across Manhattan looking for an apartment. They glimpse theatrical scenes of the urban poor through windows of the apartment buildings near the tracks:

> It was better than the theatre, of which it reminded him, to see those people through their windows: a family party of work-folk at a late tea, some of the men in their shirt-sleeves; a woman sewing by a lamp; a mother laying her child in its cradle; a man with his head fallen on his hands upon a table; a girl and her lover leaning over the window-sill together. What suggestion! What drama! What infinite interest![36]

Here the El both connects and separates. The respectable bourgeois protagonists share the train with other classes and ethnicities, but the immigrants and poor people glimpsed in passing are intriguingly distant. The scenes are fascinating because of the technological frame provided by the El: the images flow seamlessly one into another, and the passengers experience a panoptic sense that the apartment dwellers are unaware of being observed. By contrast, when the Marches are confronted with a scene of poverty at street level, the experience is far less aesthetically pleasing.[37] The El presents a theatrical display of interiority to an invisible audience that can instantaneously perceive and possess the meaning in the sight. Unlike the scenes described in Carrie Tirado Bramen's "The Urban Picturesque," the El does not display the particularity and variety of working-class life.[38] Each of the tableaux is separated from the others by semicolons, creating a sense of self-enclosure and evoking an entire story. The unbroken stream of images from the El window displayed the passing city as a spectacle, connecting it to earlier forms of visuality instantiated by the passenger train.[39]

Figure 1.
Untitled (Window Façade) by Joseph Cornell, circa 1953. Art © The Joseph and Robert Cornell Memorial Foundation/Licensed by VAGA, New York, NY. Reproduction of the image outside the article is prohibited.

The El's usefulness as a technology that defended passengers from the immediacy of their encounters was not an absolute one. Over time, the habitual exposure to detailed interior spaces moved even the nineteenth-century bourgeois spectator beyond distanced appreciation into what Michael Taussig calls a "tactile appropriation" of city life, a lived connection that transcends typicality.[40] When a sight is repeatedly encountered, its picturesqueness spills over into the intimate, the embodied, and even the confrontational. To return to Howells, "What suggestion!" and "What drama!" both seem praises well within the realm of the picturesque aesthetic, but "What infinite interest!" suggests meaning that cannot be contained by the surveillant gaze of the El rider. When seen once, a scene glimpsed from the El seems to stand in for all immigrants, all working women, or all urban poor; when seen regularly, the glimpse's ability to stand in for a larger whole breaks down. This repeated and immediate view made the El a particularly urban form of transportation, one that offered the fascination of intimacy along with the comfort of anonymity.[41] More contemporary transportation technologies cannot offer both at once:

the subway is intimate, but you are just as visible to the other passengers as they are to you; the car allows for personal anonymity, but the gaze from a car cannot penetrate interior spaces.

As public transportation moved past the windows of private homes, a peculiar kind of sociability was created between these border-crossing spaces—a form of "passing acquaintance" that was intimate without creating social obligations. On the same page as the theatrical scenes in *A Hazard of New Fortunes*, Isabel describes "the fleeting intimacy you formed with people in second and third floor interiors, while all the usual street life went on underneath[, which] had a domestic intensity mixed with the perfect repose that was the last effect of good society with all its security and exclusiveness."[42] Here, separated from the street below, both the El car and the apartment seem to interact. The very depth of the interior space refuses the flattening of a sketch; it is the inability to reduce the vision seen to a picture that creates a sense of the city as intimately social. Of course, this intimacy is fleeting, still feels unidirectional, and takes place in neighborhoods and with people who are not actually part of "good society." But it has an important purpose, particularly for Isabel as an upper-class wife isolated from the Boston milieu to which she is accustomed. The furtively glimpsed wholeness of another city dweller restores the individual passenger's lost sense of social wholeness. In addition, the "security" and "exclusiveness" bring the viewer into contact with lower-class life as a comforting social ritual without the social obligations of a more upper-class "good society." Instead of a separation through technology, the spatial proximity made possible by the El effects an asymptotic easing together of American classes, perhaps suggesting a gradualist model of assimilation. The visual structure of the El welcomes the spectator into an intimate crowd, one defined not by nicety but merely by visual proximity. This proximity was especially striking on avenues where the distance from the tracks to the adjacent buildings was less than thirty feet. (See fig. 2, from *Third Ave El* (1955), for a sense of just how close the trains were to apartment windows.) The El's technological possibilities, rather than creating containment, begin to include letting otherness in. The imaginative connection between physical and psychic interiors creates a kind of dream space of urban intersubjectivity.

In the 1950s, the Third Avenue El's intimate proximity to city life was the theme most often examined in literature, art, and film.[43] A later verse of "Third Avenue El" describes how the structure "looked in the windows of avenue wives/Until it became a part of their lives" and insists that, in spite of its destruction, "a family of six hundred thousand survives."[44] The visual relationship between the El train and the apartment window begins voyeuristically, but it

Figure 2.
A still from *Third Avenue El* shows the proximity of the elevated train tracks to the surrounding buildings. Courtesy of Creative Commons.

becomes a familiar, even a familial sharing of city space.[45] This intersubjective connection reinforced urban community in a period when the car became a more popular form of transportation, one that reinforced the separation of passengers and drivers from city space and from each other.[46] Although critics tend to discuss technology's effects in terms of connection to or alienation from the community, the El was a technology that bridged the two: it created a deep sense of imaginary connection without making people feel responsible for each other.[47]

Cornell's film, *Gnir Rednow*, uses surrealist techniques of appropriation and reversal to defamiliarize these intimate encounters with people seen from the El—he uses most of Brakhage's footage, but shows it upside-down and backward. This reversal articulates the difference between Brakhage's and Cornell's ways of seeing the city from the El, and also makes clear the two possible ways of relating to the city implicit in the technology. *The Wonder Ring* narrows the panorama of the city to a subjective view of a person seeing, while *Gnir Rednow* shows the difficulty of *seeing something*, of visually making sense of city space.[48] Brakhage focuses on the field of vision rather than any particular object in it: his film begins with squares of light falling through the

El platform to the pavement below and an ascent to the platform up stairs that repeat the same abstract rectangular patterns. Within most shots the camera pans from left to right, mimicking the movement of the train as well as the scanning gaze of the flaneur, always in search of the best sight. *Gnir Rednow*, by contrast, forces viewers to interact with city sights by including new shots of people gazing directly into the camera, framed by windows or the train doors. Perhaps most striking is Cornell's addition of an extended shot of a man in a fedora seen through the door of one train car and into the next. The man stares straight into the camera and moves toward it; seen upside down and backward, his movements are both engrossing and disorienting. Brakhage's film does not show any other faces directly: any portraits are partial, oblique reflections in the window, again emphasizing the safely impenetrable eye of the camera—we can see them, but they can't see us. Cornell presents faces that are more centered. However, they are upside down and have to be consciously humanized through a mental flipping of the image.

The reversed faces, like the dramas Howells describes or the faces Ezra Pound saw in a station of the metro, move the commuter to pause, to make a lateral glance, to be momentarily transfixed by something that interrupts the eye's commuting journey. Despite its avant-garde technique, *Gnir Rednow* joins other representations of the El in affirming technology's power to connect the viewer with the urban community. This connection enmeshes passengers in city space by reflecting back a mixture of their desires and the unknowable difference of other lives. The glass of the train and apartment windows separates the El passenger from the scenes glimpsed. However, glass does not separate absolutely, as Bernard Herman points out; it both reflects and refracts, showing the desires of the collector and gently distorting the objects it encloses.[49] Technologically-mediated urban visuality leads to a kind of opacity: the urban subject seen from the El is always visible and socially knowable, but never totally legible, totally interpretable.

Reinserting intersubjective encounter into Brakhage's El ride, Cornell also emphasizes another aspect of the El's visuality that was seen in many nostalgic representations. From Howells on, artists and writers pointed to the interaction between public and private space as the most notable experience of riding the El. The description of an El ride from the 1930s shows passengers and picturesque denizens of city space reaching out even further toward one another than Howells would have dreamt:

> From the car windows one can almost touch the buildings—buildings lined with fire escapes and strung with the family washing. Frowzy women lean from the window sills exchanging

> gossip and bawling at their offspring playing in the gutter three floors below. During the heat of summer, thousands of the residents pull mattresses out onto the fire escapes and sleep the nights through, oblivious of the traffic flowing past just a few feet from their heads.[50]

The scenes of lower-class life glimpsed from the El train are often picturesque to the point of cliché—the same "frowzy women" lean out of their windows in Ashcan school paintings and popular dramas such as Elmer Rice's *Street Scene* (1929). But when passengers "can almost touch the buildings" and the train passes "just a few feet" from sleeping residents' heads, there is an undeniable immediacy to the experience for both groups. The ensemble of El train and apartment building shaped American vision through the power of proximity. In the late nineteenth century, the heyday of the El, apartments had only recently become acceptable alternatives for the members of New York's middle class who could not afford the financial burdens of a private home. Middle-class residents observed a clear distinction between public and private domestic spaces, but apartment buildings still had semipublic areas that were used by the less genteel tenants as extensions of their homes.[51] The woman leaning out of her window and yelling at her child "playing in the gutter" is not observing the niceties of apartment life, since she is engaging in a (quite literal) dialogue between inside and outside. The structure and proximity of the elevated train enacted a similar kind of dialogue, a dialogue that undid the separation between the public, semiprivate domestic, and private spheres.[52]

The adjacency of train tracks and apartment buildings broke down the defined uses of city space. However, by the 1950s, the development projects of urban czar Robert Moses (including parks, bridges, and highways) had moved New York City toward a model of spatial separation and unimpeded movement—the frictionless, dematerialized city previously mentioned. Interaction between residential spaces and spaces of transportation was strictly taboo: Jones Beach, a recreational park on Long Island that Moses helped establish, is separate from the city, and buses had to obtain special permits to enter its grounds.[53] Jane Jacobs described planning of the postwar era as it separated cities into zones with different purposes, creating "business districts," "culture districts," and the like.[54] When nostalgic artists mixed public and private spaces, they were going against the grain of urban planning and popular notions of progress.

Looking and Moving Backward

In addition to creating a visual dialogue between self and other, public and private, the Third Avenue El created dialogue between past and present. The

El's circular movement and old-fashioned appearance made a journey on it seem counterproductive. It was literally an older space: entering through an old-fashioned station with a pot-bellied stove and stained-glass windows, riding around the city in "antiquated wooden coaches," riders spatially inhabited the past.[55] This historicized place was a space of inefficiency made possible by anachronistic technology. In spite of the guidebooks' insistence that "the speed of the El is substantially the same as that of the subway," the El provided a slowed-down relation to city space for artists of the 1950s.[56] The perceived slowness connected passengers to the city's past and gave them time to examine their surroundings. The El's temporal and spatial relationship to the city contrasts dramatically with that of streamlined trains.[57] Streamlined transportation's appearance underscored its ease of movement; its visual beauty, modernity, and emphasis on forward motion "stimulated public faith in a future fueled by technological innovation."[58] In comparison, the El's old-fashioned and inefficient space emphasized the lateral pull of memory and urban history on the forward motion of progress.[59] The proponents of razing the Third Avenue El spoke about the city in terms of unceasing forward movement. The circulation of the El through the city, rather than a parallel forward movement, became an impediment. According to a *New York Times* editorial,

> elimination of the Manhattan El . . . would mean restoration to full use of one of this island's main north-south thoroughfares, which for seventy-six years has been encumbered and blighted by its forest of pillars and its roof of ties and tracks and stations and trains. City Construction Coordinator Robert Moses has urged in the interests of traffic flow that the El be torn down.[60]

The El is an odd, liminal element of the urban landscape—its pillars are naturalized as trees, but the "roof" domesticates it. The diction in this editorial even enacts a thwarted forward motion: the "ties and tracks and stations and trains" throw up one barrier after another, and the roof does not actually encumber movement *on* the north-south street, but it adds to the claustrophobia of the space. The "interests of traffic flow" are humanized and given authority by the name of Robert Moses—a man who embodied efficiency. Even when people were nostalgic about the El, they often couched their arguments in terms of efficiency. The week after the Third Avenue El's last run, the *New Yorker* described the need for destruction of the structure in more poetic terms:

> But cupolas and stained glass adorning green villas have nothing to do with the nineteen-fifties; they are so old and out of place that to us they must be either hideous or quaint. Until the villas have been torn down and half forgotten and then slowly rebuilt in memory, we

> can't be sure how much they may have meant to us. . . . Twenty more years and a brand-new "L," the "L" of recollection, will go darting among the rooftops, at a speed the old "L" never reached, through a city fairer than any of us has ever seen.[61]

The imagined effects of this ghostly technology are curiously similar to the progress-based discourse of unimpeded forward movement that led to the El's destruction. Romanticizing the El as it will be remembered, this description removes any trace of the anachronism that artists of the 1950s embraced: its old-fashioned cars become "brand-new"; its inefficiency is transformed to speed; even its contact with the city is imagined as contact with a utopian city of the future. While the iron grip of Robert Moses would not begin to relax until the late 1950s, nostalgic artists used the El to articulate negative or mixed feelings about what had until then been New York's steamroller style of modernization.[62] Compared to these visions, nostalgia for the El as it actually existed was literally and figuratively movement in the wrong direction.

Perhaps it is no coincidence then that Cornell's film about the Third Avenue El is projected upside down and backward.[63] Cornell's reuse of Brakhage's footage in *Gnir Rednow* foregrounds one of the most old-fashioned and inefficient aspects of movement on the Third Avenue El—its standardization. Unlike foot and car traffic, movement on elevated and passenger trains is predetermined and can follow only a limited set of paths. From the 1920s on, both Els and railroads suffered in relation to more flexible and decentralized forms of movement.[64] Cornell did not feel that movement on old-fashioned transportation technologies was inherently limiting.[65] He often made boxes that included Baedeker guidebook pages, and an entire box of his library is filled with guidebooks for movement by train, car, bike, bus, and cruise ship.[66] By using shots that came from someone else's journey, Cornell emphasized restriction and repetition, key aspects of elevated train commuting that other artists of the period ignored. His interest in preestablished paths and sights also suggests a self-consciously modernist take on urban visuality as something borrowed, cited, but not owned. Instead of a director's credit, *Gnir Rednow* ends with the much-discussed title card, "The end is the beginning."[67] Most critics read this card, obviously and rightly, as a nod to T. S. Eliot; it also underscores the counterproductive or backward-looking movement associated with the El in the 1950s.[68] His citation echoes the model of urban nostalgia espoused by the *New Yorker*, but instead of destroying something in order to idealize it, Cornell argues that one has to *reuse* it, to reimagine past structures in new contexts.

Several critics have noted Cornell's increased interest in urban preservation during the postwar period without pointing to the wider context of artistic nostalgia in which it occurred. Deborah Solomon says "buildings, vistas, and entire blocks that Cornell had explored since the 1920s were disappearing under a wave of glass-and-steel construction. Streetscapes seemed to be changing beyond recognition every time he looked."[69] Lynda Roscoe Hartigan sees a "nostalgic drive to preserve" in Cornell's films, such as *Gnir Rednow* and *Centuries of June* (1955).[70] But it is the desire to make the past present that distinguishes *Gnir Rednow* from the other short films made about the Third Avenue elevated train between 1953 and 1958. Hal Freeman's film *Echo of an Era* (1957) explicitly places this transportation technology in the past when the film's narrator calls it "indifferent to the rhythm of a changing city." This contrast was particularly evident through the choice of music: nearly all of these films were silent (perhaps suggesting the relative peacefulness of an El ride compared to one on the subway or in a car). Sometimes an old-fashioned score emphasized the El's pastness: Carson Davidson's film, *Third Avenue El* (1955) shows modern images of the El set to harpsichord music. D. A. Pennebaker's film *Daybreak Express* (1953) was set to the Duke Ellington song of the same name, suggesting the El's connection to the passenger train in more explicit terms than most. David Amram, the composer for *Echo of an Era* said, "During some of the film that showed elegant old horses and carriages, I wrote some of the wildest jazz, and somehow it worked in relationship to the picture."[71] It "worked," I argue, because the intention of these filmmakers was to emphasize the El's connection to other old-fashioned forms of transportation and its *disconnect* from contemporary life. Truly "salvage ethnographers," these filmmakers portrayed the El as purely anachronistic, a holdover from the past before "modern" art and architecture had taken over the city.[72] Instead of using technology to articulate forms of perception that should be maintained, this contrast made the images being salvaged seem even stranger and more distant.

Cornell, by contrast, articulated both the discontinuities and the continuities between the past he used and the present he reimagined. Under his defamiliarizing gaze, even the present becomes a sign to be read rather than something that can be taken for granted. In *Gnir Rednow*, the process of looking becomes one of searching out the few signs that can be recognized at all. Because they are upside down (and because the El had been torn down by the time Cornell was recutting the film), the advertisements that scroll past the windows can no longer be read as part of the commuter's daily life. When a poster spotted on the El platform is lingered on long enough to be read as *The Blackboard Jungle*, the viewer experiences a certain amount of relief in seeing a sign that establishes the film in a particular place and time. Cornell's anti-lyrical film shows the difficulty of visually possessing an ever-changing city: without that poster, the viewer's connection to the outside world would be lost. He renders urban sight more difficult in order to foreground it as a historical process. Reversing the film takes even the encounters with people out of the realm of personal vision: the face is abstracted, flattened, and emptied. The reversal makes the spectator work at recognizing what can be seen, acknowledging the cutting off of possibilities for poetic sight in the stripped landscape of Robert Moses's New York and encouraging the kinds of views of the city that were being shut down by urban development. Here, the film explicitly articulates the anxiety of midcentury artists' nostalgia: if we keep destroying these structures from the past, how will we even be able to see the present in the dematerialized and dehistoricized space of constant urban renewal? Working in a time and place that championed the ostensibly unmediated subjective views of abstract expressionism, Cornell uses the technology of the El as a way of reconnecting the viewer to the external world and suggesting the difficulty of seeing and recovering history in the American cityscape. Art, he suggests, is one way of retaining traces of the technological past relieved from the burden of efficiency.

Echo of an Era

New Yorkers still navigate the material environments of outdated transportation technologies, as well as the arguments about urban spaces that surround them. Since the mid-1980s, property owners, residents, and rail enthusiasts have debated the fate of the High Line, an elevated freight railway that ran between factories on Manhattan's Far West Side.[73] The history of the High Line is a history of urban American transportation technology in miniature. Freight was initially delivered on a street-level railroad on Tenth Avenue. This made crossing so dangerous that the street was popularly known as "Death Avenue" and men on horses, called the "West Side Cowboys," rode in front of the trains on horseback waving flags and warning passers-by. The High Line replaced this train line in the 1930s through funding from the state and city of New York. Interstate shipping by truck rose in popularity in the 1950s, and the High Line became a victim of the standardized and centralized movement of rail technology, just as the passenger and elevated trains were.

The High Line's shifting meanings, like those of the elevated train, suggest that technologies of transportation are ideal conveyances for the changing material and emotional needs of the city. These parallels are all the more striking since designers consciously tried to avoid them: the "Friends of the High Line" Web site explains that the "structure was designed to go through the center of blocks, rather than over the avenue, to avoid creating the negative conditions associated with elevated subways."[74] Nevertheless, as it fell into disrepair and weeds overran the tracks, the High Line's role as a space for imagining the city echoed that of the Third Avenue El in the 1950s. It mediated opposing visual states through a combination of proximity to and distance from its urban surroundings (see fig. 3). A wilderness penetrating a heavily industrial area, an island of solitude near the busy West Side Highway, the High Line transmuted a space of urban transportation into a space for transportive meditation—albeit for the few who chose to brave the abandoned structure, rather than for the masses who rode (or could ride) the El.

In the 1980s and 1990s, the fate of the structure seemed destined to be the same as that of the Third Avenue El: this time, it impeded economic progress rather than literal movement. However, a group called "Friends of the High Line" formed in 1999. With the help of some celebrities and positive media coverage, the group persuaded the city to preserve the structure and set aside funding for development of a mixed-use open space on the former freight tracks.[75] A design team led by Field Operations and Diller, Scofidio + Renfro was chosen from more than 720 applicants to design the new park space. A new debate has arisen in the wake of these designs, between those who embrace the public space that will be developed and those who fear the new development will eliminate the way of experiencing the city that is possible on the structure in its current state. In some respects, this is a debate about class: the value of businesses and apartments near the High Line will increase exponentially after the development is completed. One critic suggests that "many of the plants and artifacts now flourishing there won't likely find a home on a polished walkway, made for yipping terriers and baby strollers."[76] Many of the outsiders who roam the High Line in its wild state feel they will also lose a home.

"In the end, this treetop world, as we know it, will disappear," says the same critic wary of dogs and strollers, using the language of the "metropolitan dirges" sung fifty years before. City dwellers continue to feel nostalgia for modes of perception that will be lost to urban development. The fear of longtime New York residents—that the city has utterly transformed from a bastion of grimy authenticity to a slick simulacrum—has led them to embrace another anachronistic perch for orienting artistic vision to urban space. The Third Avenue El and the High Line both suggest that anachronistic technologies offer visual alternatives to the top-down organization of American urban space since World War II. Here, the concern is not the imaginary connection between city dwellers, which, after all, would be enhanced by an additional promenade. Instead, it is a connection to wilderness, nature in a state not planned by the likes of Fred-

Figure 3.
The High Line as an elevated urban wilderness. Photo by Joel Sternfeld (c) 2000.

erick Law Olmsted. The utopian desire surrounding the High Line is an opposition to planning itself, observed even in the development's planned preservation of "wild spaces" of weeds and tall grasses. The fact that the High Line *spontaneously* evolved into a nature walk and graffiti- and garbage-strewn outdoor sculpture garden suggests that other things can still spontaneously develop in the city: subcultures, neighborhoods, ideas. The High Line is one of New York City's few popularly embraced wastelands, places that Stephen Carr and Kevin Lynch call "freer than parks . . . places on the margins."[77] Its very marginality—like that of the El—makes it a site of possible perceptual freedom. "The Plain of Heaven," a 2005 exhibit by the Creative Time arts organization, used the redevelopment of the High Line as a starting point for art that considered "how we imagine, and long for, inaccessible spaces [and] the way in which we re-mystify the world we already know."[78] Nostalgic urban visuality articulates this longing for inaccessible urban space and unknowable urban life—both as a mystification and as a form of protest and inquiry.

I do not want to suggest that old technologies produce inherently resistant, liberal, or positive ways of seeing the American city. Instead, they point out an important, and often forgotten, aspect of theorizing visuality—discontinuities in technology's use and meaning. Critics have understood visuality as it arises through public habituation to technological innovation, but Paula McDowell's argument about oral and print culture reminds us that "binary models of media shift have never done justice to the complexity of actual lived experience."[79] New technologies and new visions of city space always compete with old; we can consider more local and less deterministic versions of urban visuality when we think about how technologies hang around urban space, insisting on the persistence of past relationships to the city. The product of multiple technological obsolescences, the High Line forces us to consider America's relationship to technology as one that moves beyond innovation and obsolescence. The aesthetic needs fulfilled by technology proceed at a different pace than its planners, producers, or even consumers intend. The temporality of the Third Avenue El as a symbol, imaginative view, and way of knowing city space persisted long after the last passenger disembarked. Urban development can never bring about a total erasure of the past, even in a city that changes with the "extravagant regularity" of New York. Considering technology's aesthetic residue gives us a more complex map for navigating urban space, tracing the trains of thought that shaped the American city as they cross, trail off, and eventually stop.

Notes

1. Emory Lewis, "Elevated Third Avenue: Famed Street, Widened and Lined with Trees, Will Be Fashionable Boulevard of Luxury Apartments, Offices," *Cue*, March 31, 1956, 12.
2. "The Antiquated El," *New York Times*, April 8, 1955, 20.
3. "The Rise and Fall of the Elevated Railroad, 1867–1967," New York Public Library pamphlet, Exhibit June–September 1967. There were still elevated trains in the boroughs of Brooklyn and the Bronx and, most important, an elevated train from Queens that Joseph Cornell took into Manhattan almost every day.
4. James S. Barstow, "800 Aboard Last Train Strip Car of Souvenirs," *New York Herald Tribune*, May 13, 1955.
5. "Exhibit of City's Elevateds," *New York Times*, April 25, 1955.
6. Unfortunately, the library no longer has records of what movies were shown in this program. However, both the small number of films about the elevated trains and the familiarity of the current staff with the Brakhage and Cornell films and suggest that one or more of the films I discuss in this article were included on the program.
7. See Rebecca Zurier, Robert W. Snyder, and Virginia M. Mecklenburg, *Metropolitan Lives: The Ashcan Artists and Their New York* (Washington, D.C.: National Museum of American Art, 1995).
8. Popular films featuring the El were usually fairly bleak visions of the city, including King Vidor's *The Crowd* (MGM, 1928) and Billy Wilder's *The Lost Weekend* (Paramount, 1945). Avant-garde films, which I discuss later in this article, are more optimistic.

9. Berenice Abbot and Arnold Eagle were the best known photographers of the El. See Michael Brooks, *Subway City* (New Brunswick, N.J.: Rutgers University Press, 1997), 51. Abbot's book, *Changing New York* (New York: E. P. Dutton, 1939) shows the wider results of New York's rapid urban development in a slightly earlier time period.
10. I thank my second anonymous reader for the formulation that my essay's subject was "the aesthetic afterlife of the El."
11. See Max Page, *The Creative Destruction of Manhattan, 1900–1940* (Chicago: University of Chicago Press, 1999), 1–15.
12. George J. Lankevitch, *American Metropolis: A History of New York City* (New York: New York University Press, 1998), 191.
13. Paul V. Beckley, "Reginald Marsh Honored, Pines for City of Yesteryear," *New York Herald Tribune*, March 10, 1954, 7.
14. Leo Marx's *The Machine in the Garden: Technology and the Pastoral Ideal in America* (New York: Oxford University Press, 1964) is a foundational text in this tradition.
15. See Brooks, *Subway City*, 36–38; Cecelia Tichi, *Shifting Gears* (Chapel Hill: University of North Carolina Press, 1987), 247–48; Sabine Haenni, "Visual and Theatrical Culture, Tenement Fiction, and the Immigrant Subject in Abraham Cahan's *Yekl*," *American Literature* 71 (1999): 493–527.
16. Claude Fischer explains how "the role of the telephone unfolded over time" from novelty to business communication medium to popular technology. Claude S. Fischer, *America Calling: A Social History of the Telephone to 1940* (Berkeley: University of California Press, 1992), 24.
17. The car did not have as pronounced an effect on perception in New York City in this period as it did in the suburbs. See Mark Foster, *Nation on Wheels: The Automobile Culture since 1945* (Belmont, Calif.: Wadsworth, 2003).
18. Janelle L. Wilson, *Nostalgia: Sanctuary of Meaning* (Lewisburg, Penn.: Bucknell University Press, 2005), 35. Extending Fred Davis's theory of nostalgia as a pursuit of continuity in concepts of self and community, Stuart Tannock suggests that nostalgia argues for continuities while also positing discontinuities "between a prelapsarian past and a postlapsarian present." Stuart Tannock, "Nostalgia Critique," *Cultural Studies* 9.3 (1995): 457.
19. "Third Avenue El," *Ben Bagley's The Littlest Revue* (New York: Painted Smiles Records, [197-]), New York Public Library for the Performing Arts, Rogers and Hammerstein Archives of Recorded Sound.
20. Svetlana Boym considers nostalgia to be inherently opposed to progress: "At first glance, nostalgia is a longing for a place, but actually it is a yearning for a different time—the time of our childhood, the slower rhythm of our dreams. In a broader sense, nostalgia is rebellion against the modern idea of time, the time of history and progress." *The Future of Nostalgia* (New York: Basic Books, 2001), xv.
21. See Robert Bennett, *Deconstructing Post WWII New York City: The Literature, Art, Jazz, and Architecture of an Emerging Global Capital* (New York: Routledge, 2003), 61.
22. This is not merely a fanciful comparison: many of the films shot from the El in the 1950s had extended sequences from the "point of view" of the train, shot from the front car's window with the tracks stretching out through the cityscape.
23. Preface, *Vision and Visuality*, ed. Hal Foster (Seattle: Dia Art Foundation, 1988), ix. For other influential theorizations of visuality, see Martin Jay's "Scopic Regimes of Modernity" in the same text; Jonathan Crary's *Techniques of the Observer: On Vision and Modernity in the Nineteenth Century* (Cambridge, Mass.: MIT press, 1992); and Anne Friedberg, *Window Shopping: Cinema and the Postmodern* (Berkeley: University of California Press, 1993). Crary and Friedberg in particular articulate the means by which technologies shape new ways of seeing.
24. James A Ward, *Railroads and the Character of America, 1820–1887* (Knoxville: University of Tennessee Press, 1986), 17; Mark S. Foster, *From Streetcar to Superhighway: American City Planners and Urban Transportation, 1900–1940* (Philadelphia: Temple University Press, 1981), 91.
25. See Robert Reed, *The New York Elevated* (New York: A. S. Barnes, 1978).
26. "As we spend more of our lives in interior environments, we are deprived of many natural clues to the passage of day and season. Office and factory buildings, long corridors, and subways are timeless environments, like caves or the deep sea. Light, climate, and visible form are invariant." Kevin Lynch, *What Time Is This Place?* (Cambridge, Mass.: MIT Press, 1972), 69.
27. For descriptions of the subway's effect on urban subjectivity in London, see Andrew Thacker, "Imagist Travels in Modernist Space," *Textual Practice* 7 (Summer 1993): 224–46; for its effect in New York City, see Brooks, *Subway City*, 122–205.

28. "It Won't Be Very Soon," *New York Herald Tribune*, November 19, 1938.
29. Like the passenger train before it, the subway's regular view and social proscriptions against interpersonal contact led passengers to read for the duration of the journey. For a description of reading on the train as it developed in nineteenth-century Europe, see Wolfgang Schivelbusch, *The Railway Journey: The Industrialization of Time and Space in the 19th Century* (Berkeley: University of California Press, 1977), 64–69; for an amusing account of American reading practices in the subway (albeit thirty years before the period dealt with in this article), see "Those Who Read in the Subway," *New York Times*, June 30, 1929, SM9.
30. The initial discourse in favor of American railroad construction emphasized its connective possibility as an ability to unite the nation. Ward, *Railroads and the Character of America*, 17.
31. Cornell is even periodized in the past: although he worked from the 1930s to the late 1960s, he has been called Romantic, Symbolist, and Victorian. See Dore Ashton, *A Joseph Cornell Album* (New York: Da Capo Press, 1974), 10; Mary Ann Caws, *Joseph Cornell's Theater of the Mind: Selected Diaries, Letters, and Files* (London: Thames & Hudson, 1993), 29; Lynda Roscoe Hartigan, "Joseph Cornell's Dance with Duality," in *Shadowplay/Eterniday* (London: Thames & Hudson, 2003), 15. While he worked and exhibited with surrealists who regularly borrowed images and forms from the past (Picasso's primitivism, say, or Max Ernst's Victorian etchings in *Femme 100 Têtes*), Cornell's appropriations have until quite recently been seen as pure, innocent, magical. Michael Moon's article "Cornell's Oralia" does a good job of refuting what he calls the "enchanting innocence line of criticism," arguing that it is a mode of suppressing the complicated ways in which the boxes work out different forms of (often queer) contemporary desire surrounding fandom in particular. See Michael Moon, "Cornell's Oralia," in *A Small Boy and Others: Imitation and Initiation in American Culture from Henry James to Andy Warhol* (Durham, N.C.: Duke University Press, 1998), 153. In a similar vein, more recent discussions of Cornell's work have examined his relationship to contemporary art and popular culture. See Jodi Hauptmann, *Joseph Cornell: Stargazing in the Cinema* (New Haven, Conn.: Yale University Press, 1999).
32. Cornell, Diaries, Smithsonian Institute, AAA, Reel 1059, Frame 657, January 15, 1953. The "commuting days" Cornell refers to were days when he worked as a fabric salesman in the 1920s and 1930s, riding the Third Avenue El up and down the Bowery with his samples.
33. Brakhage "shot a film that remained almost entirely inside the cars of the train . . . Cornell, dissatisfied with the interiorization of what he saw as a highly voyeuristic experience, found another use for the footage." Marjorie Keller, *The Untutored Eye: Childhood in the Films of Cocteau, Cornell, and Brakhage* (Rutherford, N.J.: Fairleigh Dickinson University Press, 1986), 250.
34. P. Adams Sitney, "The Cinematic Gaze of Joseph Cornell," in *Joseph Cornell*, ed. Kynaston McShine (New York: Museum of Modern Art, 1980), 80. Notably, Sitney is one of the few critics to explicitly refute perceptions of Cornell as nostalgic (69): "Cornell was not a nostalgist, a recluse, or a naïf, even though he knew how to play those roles expertly. He was a dialectician of experience."
35. Hegeman describes salvage ethnography as "recording ways of life that are seen to be dying out in the face of encroaching assimilation and modernization." Susan Hegeman, "The Dry Salvages," in *Patterns for America: Modernism and the Concept of Culture* (Princeton, N.J.: Princeton University Press, 1999), 34.
36. William Dean Howells, *A Hazard of New Fortunes* (New York: The Modern Library, 2002), 76.
37. Ibid., 65–67.
38. Carrie Tirado Bramen, "The Urban Picturesque and the Spectacle of Americanization," *American Quarterly* 52.3 (September 2000): 446.
39. For the most influential discussion of the passenger train's panoramic visuality, see Schivelbusch, *The Railway Journey*, 52–69; for a description of American panoramic train travel, see Ward, *Railroads and the Character of America*, 131, where he says, "Americans were becoming a blur as they sped past one another." For discussions of panoramic visuality as it affected modern subjectivity, see Friedberg, *Window Shopping*, 20–29; Mary Anne Doane, "When the Direction of the Force Acting on the Body Is Changed: The Moving Image," *Wide Angle* 7 1–2 (1985): 42–44; Lynne Kirby, *Parallel Tracks: The Railroad and Silent Cinema* (Durham, N.C.: Duke University Press, 1997), 7.
40. Michael Taussig, *The Nervous System* (New York: Routledge, 1992), 144.
41. See Georg Simmel, "The Metropolis and Mental Life," in *Georg Simmel: On Individuality and Social Forms—Selected Writings*, ed. Donald N. Levine (Chicago: University of Chicago Press, 1971), 324–39, for a description of this interpersonal relationship as a particularly urban one.

42. Howells, *A Hazard of New Fortunes*, 76.
43. This attention to the horizontal gaze may be particularly pronounced in the later period because the Third Avenue El chiefly ran through residential areas and was quite close to the apartment windows in those areas.
44. "Third Avenue El."
45. There is an important caveat to this feeling of connection—in all of the films, songs, and novels that I've encountered, it is felt *only* by people riding the trains. In the few instances when the El is represented from the perspective of a Bowery apartment, it is generally an oppressive urban force (see Kirby, *Parallel Tracks*, 165). The relationship of apartment dwellers to the technology of the El is a subject for another study—one that pays as much attention to economic realities as it does to aesthetic-political construction of fantasy spaces. However, this dramatic difference reminds us that technologies do not have a singular meaning, even within a fairly circumscribed population like Manhattan residents.
46. Glenn Holt points out that "commuting, which had been among the most public of daily experiences in the nineteenth century, became one of the most private in the age of the automobile." "The Changing Perception of Urban Pathology: An Essay on the Development of Mass Transit in the United States," ed. Kenneth Jackson and Stanley K. Schultz, *Cities in American History* (New York: Alfred Knopf, 1972), 338.
47. Fischer, *America Calling*, 266, suggests that the telephone "solidified and deepened [preexisting] social relations."
48. This interest is epitomized by the last shot's rack focus, which begins on the sight seen through a train window and becomes more and more blurry until the focus is on the dusty and scratched glass of the window itself.
49. Bernard L. Herman, "The *Bricoleur* Revisited," in *American Material Culture: The Shape of the Field*, ed. Ann Smart Martin and J. Ritchie Garrison (Knoxville: University of Tennessee Press, 1997), 45.
50. J. W. McCloy, "East Side, West Side, All Around the Town, Part II: The Salad Days, and the Present," *Aera*, May 1931, 285.
51. Elizabeth Collins Cromley, *Alone Together: A History of New York's Early Apartments* (Ithaca, N.Y.: Cornell University Press, 1990), 156.
52. Carolyn Marvin discusses the telephone as having a similar effect in breaking down the boundaries between public and private space. *When Old Technologies Were New: Thinking About Electric Communication in the Late Nineteenth Century* (New York: Oxford University Press, 1988), 68.
53. Robert Caro, *The Power Broker* (New York: Alfred A. Knopf, 1974), 318–19.
54. Jacobs, an architect and advocate who often directly opposed Moses's plans, valorized the sidewalk as a site of urban social contact and interaction between diverse spaces and people. Jane Jacobs, *The Death and Life of Great American Cities* (New York: Vintage Books, 1961), 165–70.
55. *The WPA Guide to New York City* (New York: Pantheon Books, 1939), 404.
56. *Ibid.*
57. On streamlining as a design strategy, see Kathleen Church Plummer, "The Streamlined Moderne," *Art in America* 62 (Jan.–Feb. 1974); Donald J. Bush, *The Streamlined Decade* (New York: George Braziller, 1975). On streamlined trains and their psychological effect on the American public, see Jeffrey L. Meikle, *Twentieth Century Limited: Industrial Design in America, 1925–1939* (Philadelphia: Temple University Press, 1979), 162, 179; Joel Dinerstein, *Swinging the Machine: Modernity, Technology, and African American Culture between the World Wars* (Amherst: University of Massachusetts Press, 2003), 137–81.
58. Meikle, *Twentieth Century Limited*, 162.
59. The Third Avenue El was an important "residual" technology, to use Raymond Williams's term. *The Raymond Williams Reader*, ed. John Higgins (Oxford: Blackwell Publishing, 2001), 171.
60. "End of the 'El'?" *New York Times*, June 4, 1954.
61. "Notes and Comment," *The New Yorker* 31, May 21, 1955, 27.
62. See Caro, *The Power Broker*, particularly chapters 41 and 42, for details of Robert Moses losing ground after struggles in the 1950s over public housing and the Tavern on the Green.
63. It also suggests a close reading of the title's temporality: *Gnir* could be read as "near" and *Rednow* reveals "now."
64. The railroad industry suffered particularly in relation to road-based freight because of the latter's flexibility and decentralization. See Albro Martin, *Railroads Triumphant: The Growth, Rejection, and*

Rebirth of a Vital American Force (New York: Oxford University Press, 1992), 358–59. Foster says that planners were consciously rejecting the subway and the El as solutions to problems of urban congestion because of their lack of flexibility compared to the car (64). Curiously, the subway's centralization and standardization did not impede its popularity.

65. Cornell's Hotel boxes are the first works in which he explores the imaginary escape from self at the heart of the nineteenth century's regulated forms of travel. Called "Hotel" boxes because each includes a scrap of paper with a hotel name collaged on one of the walls, these boxes connect Cornell's interest in travel with his exploration of interiority.
66. Box 77, Smithsonian American Art Museum.
67. *The Wonder Ring*, in comparison, begins with its title and Brakhage's name scratched into the leader. This gesture can be read both as a mark of possession as well as an introduction to the tactile and perceptually *personal* quality of the film.
68. Joseph Cornell loved T. S. Eliot. His library at the Smithsonian includes a handmade book with all of the *Four Quartets* written in painstaking calligraphy.
69. Deborah Solomon, *Utopia Parkway: The Life and Work of Joseph Cornell* (New York: Noonday Press, 1997), 223.
70. Lynda Roscoe Hartigan, "Joseph Cornell: A Biography," in *Joseph Cornell*, ed. Kynaston McShine (New York: Museum of Modern Art, 1980), 109.
71. David Amram, *Vibrations: The Adventures and Musical Times of David Amram* (New York: Macmillan, 1968), 243.
72. Pennebaker's film fits somewhere between these two categories: while his use of swing music suggests an interest in the El as anachronism, he also uses the train as a visual analogue for cinema itself, focusing on the flickering images seen through a moving El train and the kaleidoscopic views of the city made possible by its elevation.
73. For a history of the High Line, see J. A. Lobbia, "One Track Mind: Chelsea Group Works to Save an Abandoned Rail Line," *Village Voice*, December 27–January 2, 2001, http://www.thehighline.org/press/articles/122701_voice/ (accessed January 7, 2006).
74. See http://www.thehighline.org/about/highlinehistory.html (accessed January 9, 2006).
75. Thomas De Monchaux, "How Everyone Jumped Aboard a Railroad to Nowhere," *New York Times*, May 8, 2005, http://www.thehighline.org/press/articles/050805_nytimes/ (accessed January 6, 2006).
76. Amy Braunschweiger, "Paradise Lost: When the High Line above Manhattan's West Side Turns into a Park, a Secret World Will Disappear," *Village Voice*, November 16–22, 2005, 38.
77. Stephen Carr and Kevin Lynch, "Open Space: Freedom and Control," in *Urban Open Spaces*, (New York: Cooper-Hewitt Museum, 1979), 9.
78. Creative Time, "The Plain of Heaven," exhibit brochure, October 14–November 2005, New York.
79 Paula McDowell, "Defoe and the Contagion of the Oral: Modeling Media Shift in *A Journal of the Plague Year*," *PMLA* 121.1 (January 2006): 104.

Flexible Technologies of Subjectivity and Mobility across the Americas

Felicity Schaeffer-Grabiel

> With Mexican men . . . they don't like that the woman improves herself.
>
> —Jessica, Cowboy del Amor[1]

Jessica, a thirty-three-year-old dermatologist from Chiapas, Mexico, explains her reasons for searching for a U.S. husband in *Cowboy del Amor*, a recent documentary that follows, over a period of sixteen years, the tactics used by an Anglo-American cowboy to match U.S. men with Mexican women. Jessica's middle-class background is evident as she shares how her father was killed defending his large ranch from the Zapatista guerilla fighting that began during the 1994 uprising of the landless poor in Chiapas. Her pragmatic approach to finding a U.S. husband reflects her philosophy of survival more generally: "Rather than dream, I set goals. I am a person who works toward feasible goals not dreams." Like Jessica, many women from Mexico and Colombia depict Latin men—as well as the economic, social, and political situation in Mexico more broadly—as curbing their desire for social and economic security and mobility. Many of these women turn not only to U.S. men but also to communication technologies to access transnational mobility and opportunities they find lacking in their local milieu. These women articulate their desire for U.S. men alongside the language of professionalism and the marketplace, as they recount the process of finding romance and/or marriage through ideals of hard work, self sacrifice, and individual struggle that afford them the opportunity to express their desire for self-improvement, to *superarse*, to better themselves and the lives of their families. This form of moral mobility justifies personal gain in the face of collective struggles for resource distribution, such as articulated by the Zapatistas.

Through interviews with Mexican and Colombian women, I found connections among unlikely technological couplings, such as how participants used both the Internet and cosmetic surgery to find what they defined as their "true" self by flexibly altering the inner and outer meanings of their subjectivity. Women who rely on Internet dating described unraveling their inner selves with foreign-minded others as a more authentic expression of their identity that contrasted with their local meanings and opportunities. Some women from Cali, Colombia, capitalized on cosmetic surgery to improve their chances of finding a foreign partner and emphasized the importance of the visual rendering of their bodies (as reflective of their inner essence) in accessing dreams of a foreign lifestyle and pliable citizenship, regardless of language, class, or race. The inequalities of mobility, invisible in corporate marketing narratives, are quite apparent in the transnational marriage industry. While many men's global class status is assumed to be at least middle-class, some Latin American women participating in this transnational search for spouses feel that they have to play up the ways they are attractive within the global economy of desire on Web sites, in their e-mail descriptions, interviews, and practices.

In this essay, I connect women's discourses of self-improvement with the larger landscape of corporate advertising of technologies that promise modernity and mobility. For the middle classes of Latin America (or those aspiring to be middle class), modern subjectivity is increasingly imagined through representations of global class mobility offered by values of the free marketplace. Corporate marketing of business, leisure, and travel depict the ideal moral citizen as one who shuttles around the world, taking risks while capitalizing on global opportunities. As women explore Internet communication with others far away, they practice their desire for flexible subjectivity. As promised by Internet advertising, they find they can become someone new in the crevice of global opportunities conveniently at one's fingertips. For more than a decade, Internet marketing campaigns have celebrated the unfettered mobility of identity and travel in cyberspace, invoking the idea that one can leave behind the drudgery of life and enter cyber-worlds where bodies, race, gender, and class no longer matter. Some Internet and technology scholars replicate this corporate marketing ethos by arguing that the transgressive qualities of electronic mediation reside in the ability of individuals to flexibly recombine their identity and social power in cyberspace, de-linked from the constraints of the body.

For Latin American women, unlike their U.S. counterparts, the use of technology is structured by the marketplace of desire in which they understand the need to augment the ways their bodies gain currency in the global economy, such as through hyperfemininity, domesticity, malleability, sexuality, and family values. In this context, many of these women use technologies, such as the Internet and, for some, cosmetic surgery, to reconfigure their bodies in ways that translate across transnational imaginaries and places, and in ways that incorporate neoliberal values of democracy and unfettered mobility in both exciting and dangerous ways. Thus, rather than turn to these technologies for their promise of disembodied citizenship, these women deploy technology based on a desire for a corporeal affiliation to the promises of democracy and transnational mobility. There are mutual scripts shaping the virtual play of identity in highly eroticized fantasies across borders.[2] For example, U.S. men are valuable to these women because of the perception that they are hard workers who enjoy a high quality of life and because they are seen as potentially equitable partners in the domestic sphere. Latin American women, on the other hand, gain currency to U.S. men as traditional, family-oriented, passionate, sexual, and feminine, whether or not the meanings attached to these stereotypes correspond. I draw upon ethnographic research from my project on the cyber-marriage industry between Latin America and the United States to flesh out the possibilities and dangers of theorizing flexible subjectivity and mobility via technology.

While many U.S. Internet scholars celebrate the ability of individuals to mold themselves into various modes of subjectivity dislodged from the stable signs of the body, my research with Mexican and Colombian women demonstrates that the desire for pliable subjectivity is not merely an arena for play, but one that women embrace in the hopes of mobilizing safe migration across national borders. Women do find ways to use technology and representations of their bodies as de-colonial tools that help them improve their lives and those of their families. Yet what is oftentimes unexamined in theories of the body in cyberspace is the salience of the boundaries of cyber-subjectivity, such as how women from the global South negotiate the expectations of a U.S. palate in ways that eroticize their difference from U.S. women. I argue that the concept of transcending the body or recombining one's identity in cyberspace is a privileged position that elides the labor of the body and asserts neoliberal values of choice and democratic notions of upward mobility. Behind the desire for mobility by some Latin American women searching for U.S. husbands are disparities of movement, since heightened border control, political unrest, and economic recession leave few options for either migrating or remaining in place. Women from Latin America utilize technologies of modern subjectivity to creatively transform the master's tools yet also perpetuate a dangerous terrain for resolidifying their objectified status and use value as malleable objects for Western desire.

In my larger project on cyber-marriage across Mexico, Colombia, and the United States, I analyze interviews and ethnographic data with over a hundred men and women at "Vacation Romance Tours,"[3] through Internet chat rooms, via e-mail conversations, in restaurants, and in their homes to understand the process by which participants turned to technology and the foreign "other" to mobilize dreams across borders. While the majority of women at the Guadalajara Tour are well educated, from the professional class, and have access to the Internet, the majority of female participants at the Cali Tour are more racially mixed, from the working class, and have less frequent access to the Internet. For this reason, and because beauty is a form of social mobility in Colombia, many women turned to cosmetic surgery to attract men on the tours.

Through this research, I discovered that embedded within the use and understanding of technology were new notions of the body organized around pliable subjectivity, new race and class formations of mobility across space, and ideals of neoliberalism, development, and the march of social progress across borders. Through their use of the Internet to find romance and marriage, many women described this process as a form of self-help, of discovering their "true" self that transcended local meanings of their identity. Similarly, women from Cali turned to cosmetic surgery to emphasize their desire to become more "authentic" in the eyes of foreign courters to bring into harmony the outer signs of the body with the inner self.

Modern Technologies and the Body

The potential for identity play and self-transformation on the Internet have been eagerly theorized by Internet scholars, especially by those who find refuge in the anonymity of cyberspace. A prominent early scholar of the Internet, Sherry Turkle, raises the possibility of escaping social hierarchies on the Internet, to transform the self, and to play with fluid and multiple identities. She states, "The anonymity of MUDs [multi-user domains] . . . gives people the chance to express multiple and often unexplored aspects of the self, to play with their identity and to try out new ones. MUDs make possible the creation of an identity so fluid and multiple that it strains the limits of the notion."[4] Other feminist scholars have turned to the Internet as a space for theorizing alternative articulations of political affiliation that exceed the confinement of the body and identity politics. For example, Shannon McRae says that all the "things that separate people, all the supposedly immutable facts of gender and geography, don't matter quite so much when we're all in the machine together."[5] Theoretically inspired by Judith Butler's work, many feminist and queer scholars celebrate the queering of cyber interactions that contribute to the dislocating of sex and/or gender from a natural location in the body and detaching the visual cues of the body (race, sex, gender) from the inner realm of the self. Allucquére Stone asserts, "In cyberspace the transgendered body is the natural body. The nets are spaces of transformation, identity factories in which bodies are meaning machines, and transgender—identity as performance, as play, as wrench in the smooth gears of the social apparatus of vision—is the ground state."[6] The radical potential placed in cyborg identity play is inflated through the notion that, as Mimi Nguyen states, "nothing in a cyborg body is essential."[7]

Internet interactions do, in fact, raise some fascinating possibilities for denaturalizing the assumed coherence between the inner and outer self. However, the emphasis on play and fluid identities has led to a theorizing away from the materiality of the body and the actual borders that limit one's flexibility, mobility, and expression of subjectivity on- and off-line. For example, Teresa, a confident forty-two-year-old living in Guadalajara, described taking a break from her career as a journalist to dedicate herself to finding a romantic foreign partner online. She defended her use of the Internet as the place she could be playful and witty in order to judge how men responded to her intelligence, rather than simply her looks. Teresa said, "I can read between the lines in Internet conversations and quickly judge whether someone is open-minded and whether they respect a woman's confidence and intelligence."[8] Interestingly, through various e-mail relationships, Teresa found that many U.S. men want a more family-oriented woman and less of an intellectual mate than do European men, whom she found to be more cultured, liberal, and open-minded. For this reason, she chose to use various online dating agencies and to target European men rather than attend the U.S.-based "vacation tour" introduction parties.

For many women from Latin America, success with Internet technology is mediated by their use value in the global economy as erotic subjects for first world consumption and as laborers in the service economy and domestic sphere. I focus on women's voices and reasons for turning to new technologies (and foreign men) as actions that mark an empowering shift in their lives. Yet, I also acknowledge that their stories are complicated by men's expectations and the broader neoliberal advertising arsenal that celebrates technology's capacity for unfettered mobility and pliable subjectivity. This does not mean that women themselves do not play with their identity, but that we need to be cautious when addressing questions of mobility and flexibility in ways that erase the continued salience of borders, the marketplace of desire, neocolonial legacies of sexuality and race, and state immigration control.

Embedded within theories of cyberspace are shifting understandings of the role of the body. Traditionally, the body was associated with the natural and immutable, with predetermined qualities assigned by "god" and the "fatedness" of human life. Through modern science, the body has been treated as the biological bedrock of theories on self and society—the "only constant in a rapidly changing world."[9] More recently, Internet scholars, again influenced by postmodern thinkers such as Judith Butler, view the body as a social construction.[10] This concept of the body as social construction is also affirmed and reproduced in corporate advertising campaigns, which present the "natural" body as obsolete and the consumer body as an infinitely improvable and malleable entity. Popular discourses that accompany genetic manipulation, cosmetic surgery, and Internet cyber-sociality rearticulate the notion that the body is increasingly of less importance and is instead a malleable and democratic surface that can be changed at will in the search to become someone new.

Internet Romance and Marriage

The marketing of women's bodies on company Web sites—many of which also advertise tours—reinforces the shaping of women's bodies within tropes of family, nature, and malleability, while men are often the unmarked consumer whose bodies matter less than their association with mobility, affluence, and benevolent power on the global stage. International Internet romance and marriage encourages discourses of self-help and development, as participants are persuaded to join matchmaking companies with the promise that taking the plunge abroad will dramatically improve one's identity and life. For example, Web sites depict women as stuck in place and locked in nature, yet also as flexible or adaptable to new challenges when given the opportunity (presumably by U.S. men and capital). This is visually apparent in Web site images of young, modern women set against ancient landscapes such as pyramids, marking their present identity through their *mestizaje*/mixed blood that is rooted in a past breeding of Spanish and indigenous genes.[11] María Josefina Saldaña-Portillo argues that we must be critical of the labor that *mestizaje* does in Chicana/Latina and Latin American studies, reproducing notions of development through its formation as a biological and cultural form of national identity.[12] She scrutinizes the foundation of *mestizaje* in biology for its depiction of indigenous people as the dying breed of an evolutionary process in which *mestizas/os* figure as the present and future. Throughout scores of company Web sites I analyzed, the history of colonial contact and the idea of the frontier are cleansed of violence and instead imagined as a morphing of cultures in which women "choose" which cultural and genetic traits they will carry into the future.

This notion of choice is central to the technological ability of the Internet to reassemble history, lineage, and one's sense of self. *Mestizaje* was crafted during Mexico's independence from Spain as a way to build a common lineage for the making of a national citizenry, a way to unify Mexico's starkly different populations of indigenous peoples and Spaniards through the national symbol of the *mestiza*. Today, Internet marriage Web sites use the idea of *mestizaje* to highlight women's flexible affinity for adaptability, or their ability to assimilate into foreign culture. It is women's long history of contact with other cultures and ideas that cyber-narrations define their ability to blend into a wide variety of cultural spaces, thus orienting their history and genetic lineage toward a natural affinity for traditional domestic life untainted by feminism and global capitalism in the United States.

Internet dating sites promise viewers that new opportunities can be found in the global marketplace of love and that participants will encounter economic and social mobility lacking at the level of the nation. For men and women from the professional class, cyber-travel promises self-transformation and a change in status. Web companies advertise to a broad spectrum of male clients by stating that women care more about a man's values than their physical traits (such as age, race, body fitness, and attractiveness). Women's bodies on the other hand, mark their need for development and simultaneously situate their moral capital in their body, the locus for spiritual rejuvenation, family values, unpolluted nature, and femininity thought to be lost within U.S. capitalist society. Women who are barred from mobility across national borders must thus appeal to their role in the global labor force as caretakers, docile laborers, and malleable subjects, traits that garner more respect and value from U.S. men, but that also resign their mobility to the trends and inequalities of the global marketplace.

These women's use of their bodies as a means of accessing mobility raises questions about the inability of global theories to account for the ways women bypass uneven geographies of power. While I acknowledge that women's understanding of themselves extends national borders, the role of the border continues to shape women's choices for upward mobility. It is through the technology of cyberspace that the new languages of global flows and borderlands grate against each other.

Despite the popular construction of the Internet as a space without a grounding in place, individuals are wired to particular places by what Manuel Castells calls nodes and hubs, places with differential access to flows. In the case of Internet dating, technological networks and cyberspace bring places and people into spatial affinities of desire.[13] Neocolonial geographies of desire between the United States and Latin America put into question the purported equality that the language of connectivity—defined as being joined across time and space—assumes. Latin America and the United States share a history as two colonial interactive places that continue to discipline exchanges as being between tradition and modernity, between nature and culture.[14]

The social force of globalization and technology has been theorized via the language of flows, simultaneous vectors of global circulations across time and space. Both Arjun Appadurai and Manuel Castells argue that flows constitute the processes of exchange between multiple levels of interaction, from information, people, technology, images, capital, and communication.[15] Rather than describe current global processes in terms of uneven and contentious movements across national borders, Castells defines the network society as "the material organization of time-sharing social practices that work through flows."[16] Through an analysis of networks and the social force of the imaginary, both Appadurai and Castells theorize the need to analyze horizontal or elite connections across borders. Yet the language of networks and flows levels hierarchical renderings of modernity versus tradition. I argue that cyber-contact zones continue to be mediated by discrepant understandings of tradition and modernity. Castells claims one of the distinctive features of networking to be flexibility: "Not only processes are reversible, but organizations and institutions can be modified, and even fundamentally altered, by rearranging their components."[17] The corporate model of rearranging parts resembles the promise of technologies such as the Internet and cosmetic surgery that promote the ability to flexibly alter one's body, identity, and social position. The revolution of technology in drastically altering corporate business has trickled into the intimate realm of subjectivity, where broader transformations rest on flexibly altering oneself.

Women's search for foreign husbands must be analyzed alongside neoliberalism in Latin America. Neoliberalism describes an economic, social, and, I would add, personal strategy in which governments and people turn to the foreign marketplace for economic recovery. This neoliberal climate has infiltrated individual desire and strategies for improving one's life through the consumption not only of foreign goods, but also of foreign bodies.[18] Technology, as a tool and ideology associated with modernity, is used by women to transform themselves into modern subjects, to share these identities with others online, and, in the case of cosmetic surgery, to improve their return on the symbolic weight of their bodies as they gain currency in the global arena. Neoliberalism also defines the process whereby government-imposed restrictions on movements between countries are lifted in order to create an open marketplace and "free" trade, ushering in a borderless world economy. The turn to foreign marketplaces, bodies, and imaginaries delineates the influence of neoliberal policies that have shifted economic, cultural, and personal strategies away from a reliance on the state and instead toward the foreign marketplace.[19] While it is important to foreground Latin American women's perspectives and agency, we cannot forget the fact that the percentage of Internet use in Latin America and outside the West is statistically small.[20] Thus, women's use of the Internet as a strategy for mobility must be tempered by larger questions about access.

Alongside questions of access, also missing from Internet and technology studies more broadly is the labor Latinas' bodies perform as the erotic content of electronic currents. Men's use of the Internet to find Latina wives makes visible the emotional and erotic labor Latinas contribute across global circuits. When I interviewed men as to why they traveled to Mexico to find a wife, many described the Latinas/os they knew in the United States (their maids and contract workers) to be faithful, passionate, hard working, and committed to family and the institution of marriage (unlike U.S. "feminists"). The relationship between the Americas and technology continues to harken back to a colonial imaginary in which otherness and modernity are defined against the bodies of (erotic) pleasure for the first world cyber-imaginary, evident in the gendered landscape of Web sites catering to Latinas' erotic, sexual, and domestic appeal through pornography, sex tourism, and cyber-marriage. Similarly elided is how influential surplus capital accrued from outsourced labor to the South is in fueling the postindustrial technological economy. This is evident in the flight of technological production to offshore sites employing young "docile" Mexicanas/Latinas who line up on the factory floor to piece together consumer products such as televisions and computers.[21]

Through Latin American women's turn to technology, they aspire to alter their relationship within the global economy from producers to consumers, from those who provide service for others to visual consumers and actors who participate in the spectacular life of mobility, consumption, and leisure. Increasingly, participation in modernity, globalization, and citizenship demands one gain visibility through participating in a cosmopolitan lifestyle, including limitless mobility and the consumption of goods. This departs from an understanding of class and citizenship through the production of labor, as contemporary capitalism conflates the buying of goods with leisure. Capitalism perpetuates U.S.-style democracy around the globe through the spread of marketplace values of equal access and consumer choice. Thus, the technological boom has created a two-tiered social imaginary between those who become visible through the consumption, use, and creation of technology and those who perform the invisible labor.

Corporate Advertising and the Glamour of Modernity

In the world of marketing, technology is synonymous with global mobility as well as transcultural identities and opportunities, influencing women's desire to assert themselves as modern and future-oriented. In 2000 America On-Line (AOL) launched a series of Internet advertisements and campaigns across Latin America exploiting the global appeal of individual performers such as Salma Hayek and Ricky Martin. In exemplifying the crossover success of open markets through a cultural nationalism that is less rooted in territory and instead extends across borders, the success and glamour of previously local artists typifies the progressive routes of global circulation. AOL commercials and ads depicted those who signed on to the Internet alongside those who have access to the U.S. marketplace. In fact, the launching of the AOL marketing campaign in Guadalajara, Mexico, coincided with the shift of the region's identity as solely steeped in tradition (known for mariachi music and rodeos) to its burgeoning image as "The Silicon Valley of Mexico." Dialing into technology was depicted in the campaign as synonymous with joining "America On-Line" and an elite class who benefit from "free" trade and their more lucrative positioning in the U.S. global marketplace. In the United States and Latin America, the middle and elite classes are envisioned as those who transcend geographic places, or those who are connected to the communicative and interactive pathways of the United States. Geographies across Mexico and Latin America bear witness to the effect of foreign capital as entire neighborhoods have been profoundly upgraded by remittances from migrant communities and connections with the United States.

It is important to juxtapose the corporate marketing of mobility alongside women's own narratives and desire to travel across borders as this foregrounds the differential value of men's and women's bodies within the global marketplace. The women I interviewed described feeling out of step with local affiliations and imagined themselves as more akin to a global middle class that was much more glamorous, educated, and modern. These women hope to move safely across geographies as middle- to upper-class citizens rather than illegal immigrants. The promise of unfettered mobility via technology thus actually translates into segregated mobility where the global North offers mobility for the global South in exchange for cultural alterity. From the U.S. side, corporations such as Continental Airlines offer technological modernity in their advertising campaigns targeted to the Latina/o diaspora who demand culture and tradition alongside modernity. An ad that appeared in multiple issues of *Hispanic Magazine* brings to light questions regarding race, class, sexuality, and identity in relation to development and technology. In bold letters that stand out in front of Continental's signature logo of the blue globe crossed by white equator lines are the words: "You have Roots, We have Routes." In this simple ad campaign, the "you" refers to the Latina/o readership and the "we" refers to the airline company. This relationship between the consumer and the producer also functions as a metonym for a broader relationship of Latinas/os to the technological prowess of the mobile white male elite. Continental Airlines embodies corporate nationalism, whose global reach models the heroic image, limitless mobility, and benevolent reach of universal U.S. citizenship. Through this ad, one can visualize the hierarchy of spatial dimensions of race. On the one hand, "roots" conjures an image of Latina/o migrant bodies gripping the earth in a slow undulating path that spans a great distance across geographic borders. On the other hand, "routes" demonstrates the speed, efficiency, and decisive characteristic of the mobile class imagined from a gaze way up above.[22] The body and race relations continue to be embedded in the technological and disembodied gaze from a mobile roving eye high above while those close to the earth continue to justify the need for technological uplift. The maternal image of roots as located in the earth versus the thrusting masculinity taking off into space are echoed in the broader advertising arsenal that creates an affinity between global manhood and the language of penetration into virgin marketplaces.[23] These spatial hierarchies of mobility tend to slip out of discourses of technology focused on connectivity and networks that assume horizontal geographies of access and mutual empowerment, erasing the vertical relationship between the those who enjoy unfettered mobility via the technological gaze and those whose bodies naturalize particular kinds of placements.

The juxtaposition of roots and routes, however, is more complex than this Cartesian split between the mind and body, or the binary relationship between Latinas/os and white, corporate America. *Hispanic Magazine* pitches itself to upwardly mobile Latinas/os who see themselves as connected to a common pan-Latina/o culture—hybrid subjects sutured to flows of traditional culture as well as a galactic vision linked to the corporate business class. Pan-Latinidad is reproduced in the magazine's content, ads, and sections on traditional culture, business, technology, and some version of the "Most Successful Latinos" section. Continental's advertising campaign capitalizes on this middle-class market by appealing to this target group's sensibilities for cultural rootedness and the reliance on technology to maintain one's "roots." Technology offers Latinas/os the flexibility and play to construct their identity as part of the U.S. nation-state, as authentically Latina/o, and as part of a global class differentiated from that of "illegal" migrants. Thus, the relationship between consumer

You have roots.
We have routes.

We cover Latin America.

We don't want you to lose your roots, so we take you to them. And each time you travel, earn valuable miles on OnePass®, our award-winning frequent flyer program. For reservations, call your travel agent or Continental at 1-800-231-0856, visit AOL Keyword: Continental, or ticket online at continental.com.

Continental Airlines

Work Hard.
Fly Right.

and buyer is posed as reciprocal. Latinas/os enter the future time and space of transnational consumption while the corporation brands itself through the metaphor of roots, a company with a long legacy (and future) based in a mix of conservatism and innovation across the globe.

The contrasting image of curvy rootedness as the alterity of technological precision helps to justify techno-progress as necessary for the disciplining of nature/bodies characterized as spontaneous, loyal, malleable, and out of control. This depiction of nature relates to what Donna Haraway calls the "New World Order" in which corporate interests in techno-science turn nature into a hybrid breeding of the human and the artificial (or culturally enhanced). She defines nature as the foil for culture, "the zone of constraints, of the given, and of matter as resource; nature is the necessary raw material for human action, the field for the imposition of choice, and the corollary of mind."[24] The depiction of Mexican (and Asian) bodies as naturally docile, malleable, and dexterous is part of this machinery to link bodies and the techno-power of progress as a mutually beneficial system in which the North's discipline and the South's malleability converge in perfect harmonized nature. This logic has been central to the implementation of *maquila* shop-floor logic, trade agreements, and development programs across Latin America. The reconfiguration of technology in the form of nature, as flexible, was also central to the respatializing of capital in the 1970s when a U.S. recession and new developments in technology enabled corporations to move to offshore sites as the new frontier for higher return on capital investments, flexible shop-floor strategies, and an abundant flow of cheap labor.[25]

Transcending Social Norms via Internet Romance

In a similar fashion, women seeking a foreign husband described turning to the Internet for romance as a shift from local economies of desire (where local men were identified as macho, poor, and as having dark skin) to global economies (where foreign men were imaged alongside the desire for tall, professional, light-eyed, and hard-working men). The fact that women place their faith in changing their lives via Internet exchanges with men from the United States speaks to a shift in understanding the self in relation to global nodes of power. These women replicate problematic views of modernism by associating the local as the degraded space of stagnation (or being left behind) in comparison to the foreign space of cyber-interactions, where U.S. men represented future possibilities, the professional class, global capitalism, and scientific and technological progress.

Many women identify Internet romance as similar to the process of discovering one's true self, unobstructed by the disciplining eye of their families and social norms of behavior. During interviews with mostly middle-class women at the Guadalajara Tours about their use of e-mail correspondence to find a foreign husband, many older women described the streets and bars as unsafe places to find romance in contrast to the safe and controlled environment of the Internet. The Internet also proved to be an ideal place for less restrictive forms of courtship. While women's bodies are guarded and watched closely, the Internet affords them the opportunity to communicate with or date multiple people and to develop sexual intimacy in a society that heavily moralizes women's sexual activities outside of marriage. Because masculinity depends on expressions of independence and fraternity with other males, men are afforded much more liberty to frequent bars, clubs, and other social spaces. Blanca, a fifty-two-year-old divorced homemaker, explains how restricted she feels her movements are:

> Right now I am very confined, I almost never go out. I go out once in a while into the street and they follow me, people speak to me, but I don't like to get to know people off the street because I think, I *think* that they think that I am easy, and I'm not easy, I'm not an easy kind of woman.

I asked her whether it was also difficult for women to meet people at bars. She says:

> Well, look . . . another time I went out with some friends, only one time, we went out at night. It's not difficult, they had come up to me, but in reality they are people that are drinking, that think that if a woman goes to a bar . . . the men think that if one goes to a bar alone, she is looking for a sexual encounter.[26]

Meeting people in bars or in the streets is associated with those who drink excessively, are uneducated, and are from the working class. These places, more broadly, are considered unsafe, as spaces for the potential rape or spoiling of women's moral standing. I interviewed Blanca in her luxurious home set within a gated community within the tourist section of Guadalajara. As a single mother, she was able to raise her son (without any support from her ex-husband), send him to college, and pay for his education by working as a nanny in the United States and, later, housing foreign students in her home in Mexico. She described this as very hard work and continues her search for a husband who can provide her with the luxuries she is used to without the backbreaking labor.

Thus women's search for foreign men online or at the tours (held at five-star hotels away from the purview of their peers), speaks to women's desire to secure both their local and transnational movement through moral and middle- to upper-class lifestyles, interaction, and travel. Virtual travel and dating in cyberspace sets up the desire for mediated relationships with men from the business class. Furthermore, U.S. men afford Latinas safe travel across borders in comparison to illegal migrants (or low-paid laborers), who slowly pass across the desert under the dangerous gaze of border surveillance. Through the connective tissue of Internet communication and technologies of travel, women bypass the gaze of the state and are safely transported to the middle class of the United States. Women's mobility alongside the icons of modernity—technology, Western men, and marriage—also ensure that women cross the border and maintain or even augment their class and social standing.

Many Latin American women also describe their search for love on the Internet as an enterprising activity that encourages their personal development (otherwise stunted by work or attention to family). They discover aspects of themselves—such as their fidelity and commitment to the family—to be highly valued and respected by U.S. men. Through Internet interactions with others outside the nation, they come to understand themselves and the world from a different perspective. U.S. films may provide an *apertura*,[27] or opening into different perspectives and worlds, but through the Internet, women envision themselves as actively narrating and altering their lives. The interactive technological capabilities of the Internet involve some level of reciprocity, which Anthony Giddens argues is central to modern relationships and more equitable sexual practices.[28] Women's association with the United States as a meritocracy, a more open, fair, and democratic society, bleeds into their interaction with U.S. men via e-mail, in which open communication is the central component of the romantic interlude.

Both men and women share the desire for safe travel across borders. For men, however, this travel is imagined as moving through anachronistic space, or a past permeated with the longing for familial unity, while women see their movement across virtual spaces and borders as travel into the modern future. While these men gain currency for their association with the mobile class, Latinas gain currency in electronic currents for their association with roots, their supposed grounding in a pure state of family cohesion before industrialization and capitalism. By importing a superior breed of women who will rescue the patriarchal family foundation of the nation, women embody the frontier that rejuvenates new forms of global and national masculinity.

The discursive construction of the Internet as a new frontier reflects the West's (both the frontier of the United States and the Western Hemisphere) continued desire to mine alterity for profit and pleasure.[29] The spirit of adventure embedded in the idea of the frontier foregrounds travel to unknown continents within progressive values of invention and discovery, of turning nature into culture. Yet, the ways users incorporate this metaphor in practice is less documented. In cyberspace, the limitless opportunities of capitalism meet the practice of neocolonial mapping in which colonists (and Internet users) are invited to imagine themselves as traveling back in time, exploring virgin territories, and reaping the benefits of raw resources.

Bride-hunting men encounter Web sites that remind them they are embarking on a journey into unknown territories, imagined alongside women's bodies.[30] National geographies are cleansed of danger (from Latin American men as well as other tourists who spoil the purity of women and their "culture") so that national landscapes such as Mexico and Colombia reemerge as safe virtual and tourist sites plentiful of fantasies of virgin lands waiting to be conquered. As stated by Wendy Hui Kyong Çhun, "cyberspace both remaps the world and makes it ripe for exploration once more . . . Cyberspace proffers direction and orientation in a world disoriented by technological and political change, disoriented by increasing surveillance and mediation, through high-tech orientalism."[31] Çhun argues that cyberspace is not outside time and space, but is a fictional practice that continues to be "oriented" by the sexual and racial tropes of the Orient, an otherness that is in continual threat of breaking the totalizing myth of Western culture.

This frontier myth not only charts men's adventures of cyber-travel, but also informs the interpretation of oneself in relation to geographic mappings. Various men I interviewed at the "Vacation Romance Tour" in Cali, Colombia, described Cali's geography of lawlessness as akin to the wild, wild west, a fluid and anachronistic space where legible grids of state power and order were unintelligible.[32] The experience of cyberspace as devoid of history and time or perhaps as a medium of simultaneous time and space is evident in the collapsing of globalization into colonization, the "wild west" of the United States at the turn of the nineteenth century and the contemporary effects of globalization in Cali, Colombia. Internet travel, e-mail correspondence, and Romance Tours ensure men they will not have to interact with the geographies of danger that lurk beyond their travel via the Internet and across tourist zones. Another participant explained his outlaw notion of love as one that departed from the hegemonic idea of romantic love in the United States (and its association with the rise of the women's movement, political movements centered on equality, and the rise of consumerism):[33]

> The eastern and Latin notion of "love" as basis for engagement and marriage is different. The closest analogy I can give you in our "system" of thinking is "the old wild west" of the U.S. where men took wives (and vice versa) quickly, so long as a man and woman were roughly compatible, e.g. age, height, etc. Another analogy is Australia of the old days, when men would take a wife quickly (and vice versa) and then adventure out to colonize the "outback" frontier.
>
> The Western notion of "love" is that two people "fall into love," and know each other well enough beforehand. The Eastern and Latin notion of "love" is that two people, roughly finding in each other what they are looking for, "grow into love" with each other over time as they are married.
>
> Which is "right," which is "wrong"? Which is "better"? No one knows, I think.[34]

In this description of the frontier as a place for romantic conquest, the idea of a partnership harkens back to a time when there was a clear division of labor between women who are the domestic caretakers and the men who weather the dangers of the world outside the home, or the "wild" west (populated by dangerous forces of nature, including native peoples). The uncharted space of the frontier coincides with the participant's sense that he is out of step with contemporary norms of relationships, family, and gender roles in the United States. Perceptions of the Internet as charting new territory partially structure participants' sense of themselves as rogue citizens. Cyber-commerce follows a slippery slope as an avenue for lawless behavior (such as sex tourism/prostitution) and as an alternative route for those who seek the unconventional.

Technology as Pliable Subjectivity

The promise of Internet marketing that you will become someone new, or even become the more "authentic you" unencumbered by the body, resembles the ways women described turning to cosmetic surgery at the Cali Tours. Colombian women's use of surgery demonstrates new avenues for understanding the body in relation to scientific configurations of pliability, democracy, and mobility. At the Cali Tour, which takes place in a region with the highest black or *moreno/a* population (due to the region's history with African slave labor), there were many more dark-skinned and working-class women than there were at the Guadalajara Tours in Mexico. And unlike in Mexico, where tradition carried more symbolic cultural weight, many female participants at the Cali Tour were young and had undergone cosmetic surgery. Because of the different political and economic climates and the fact that these women had less frequent access to Internet communication than did women in Guadalajara, snatching a man at the tour was a more urgent project. Some proudly flaunted extreme breast implants, others had a combination of breast and butt implants, and still others had undergone liposuction procedures. Celia, for example, was planning to have the fat sucked out of her stomach and transferred to her buttocks. I followed various couples on dates during and after the tour and spent the evening with them as the translator. I accompanied an African American man, Seth (who is a taxicab driver in his midforties) and his Afro-Colombian date, Celia, to a restaurant that catered to tourists.[35] During our meal together, Celia excitedly told us that she was going to have a liposuction procedure done, even though, as Seth reminded her, this was a dangerous surgery. Celia explained that she had saved up money for several months to afford the $1,300 for the liposuction (she would also use money that had been given to her by a previous U.S boyfriend). She told us that her sister, who is now married to a man in the United States and has had a successful liposuction procedure, contributed to her decision. While Seth and I both told her that her body was perfect, she explained that while she liked her body, she wanted to improve it by thinning her stomach and fortifying her behind. She said, "If I have the opportunity to change something about my body, to improve myself, then I will. Just like Seth made the choice to come to Cali [to the Tour], I also decided to improve my life by choosing to do the surgery. It's an investment in myself."[36]

Celia's desire, like that of many other women I interviewed, to invest in her body with the hopes of improving her chances of finding someone, indicates how women use beauty for their own aspirations and to willingly transform themselves into marketable products of exchange. Women described having to submit photos and physical descriptions with job applications. And women thirty or older complained they had a much harder time getting jobs and attracting foreign men who went after the young girls. These women thus participate in a well-worn development narrative, similar to the one for which many men congratulate themselves regarding their journey to Latin America: if you can change your life, you should, if you don't, you deserve the consequences that await you. Modern states encourage a citizenry organized around self-help, or what Foucault labels self-cultivation, because it shifts the gaze from larger social critique (revolution) and the state, to the self. Working on the body is equated with the embrace of an entrepreneurial spirit in which women's bodies become the object of self-improvement, self-help, and the promise of a democratic future. Hard work on oneself is equated with one's entrance into modern subjectivity and citizenship, where one can become transformed into the desired image and lifestyle. The shift in Cali from any reliance on the state—as a corrupt and violent force—has shifted to the arena of science, new technologies, the marketplace, consumption (rather than production), and individuals as the arena for change and mobility.

For Celia, her soft voice, graceful gestures (reminiscent of women groomed for beauty pageants), and desire for an enlarged behind articulate a complex desire for a more pliable construction of identity, ethnicity, and sexuality that has both local and global currency, that situates her subjectivity as both embodied and translatable in a broader context. Depending on the spaces and social situations Celia moves through, she feels that she can rework the meanings of her race and class. Because the tours are held at expensive hotels with men searching for "high quality" women (and the class implications that accompany their desire for quality), women must upgrade their bodies to blend into these tourist zones so they are not mistaken to be prostitutes or "green card sharks." Yet at the same time, these women garner currency in their local context and with foreign men for the embodied signs of authentic difference and hypersexuality. Perhaps influenced by beauty pageant culture, where women are groomed for respectable femininity, Celia alters her body in ways that confound the boundaries between sex work and marriage, increasing the visible markers of class, ethnic capital, and sexual appeal toward the goal of local mobility and/or movement across borders.

Women's sense of their bodies as pliable commodities is also a response to the foreign male's, and especially African American men's, erotic desire for large buttocks and a curvy body. Unlike the tours in Mexico, where the majority of men were white and Latino, 20 percent of the men at the Cali Tour were African Americans (including two from Europe). For Afro-Colombian women, emphasizing their physique may be more complex than merely conformity to the demands of the tourist market and patriarchal desire; it is also a chance to pleasurably accentuate characteristics (such as large butts) that have been degraded by mainstream notions of ideal beauty in Colombia, characterized by whiteness, straight hair, and large breasts.[37] Accustomed to being scrutinized on the grounds of physical beauty, these women capitalize on new cosmetic technologies and foreign configurations of desire to display their motivation and self-enterprising spirit in the face of limited opportunities for jobs, travel, and mobility at home.

Women's ideology of uplift resembles popular narratives of cosmetic surgery as seen in the onslaught of television reality programming. Cosmetic surgery proves to be an apt practice and popular metaphor for individuals who envision poverty as a personal flaw that can be remedied through willpower rather than a social problem based in unequal structures of power, opportunity, and continued repercussions of racial inequalities. In the popular reality show in the United States called *The Swan*, the power of science is embedded in the array of heroic surgeons (and one psychologist) who turn visible physical characteristics, such as crooked and yellow teeth, an overweight body, small breasts, and droopy facial features, away from a potential discussion of an economic system that fails to provide health care and education to those most impoverished and into a feat of science in which those who choose to improve themselves witness stunning possibilities.[38] This narrative is repeated in Colombian pageants in which television respondents focus on the success or failure of women's cosmetically altered additions, emphasizing an understanding of the body not as whole or natural, but as alienated pliable parts as well as a democratic surface available to all through scientific and artistic uplift. Beauty pageants and telenovelas also individualize beauty and romance as qualities attained through the entrepreneurial motivation of the woman, rather than as larger questions about how beauty is related to a patriarchal society, questions of access to expensive (and dangerous) surgery procedures, or examinations of the lucrative industry of beauty and fashion that accompany these popular events. In these contexts, it is evident that women's success, whether romantic or otherwise, is the responsibility of their own enterprising spirit and aesthetic alteration, rather than a question that is queried in relation to capitalism or the state.

What we witness in these notions of self is the sleight of hand from the alteration of the outer signs of the body to its meaning for the interiority of the self. For those transgendered subjects Allucquére Stone argues are the natural occupants of cyberspace, electronic and textual morphing of the body belies the harsh reality that prosthetic alterations by transgendered prostitutes speak to.[39] The association of flexibility with transgression violently erases bodies that cannot escape the hegemonic readings visible on their bodies. Such is the case for transgendered subjects whose labor resides squarely in their bodies, evident in the self-narratives by Brazilian prostitutes who inject silicone into erotic zones of the body, such as the hips and butt, as a means of increasing the return from their sexual labor.[40] The use of technologies for many subjects of the global South are relegated to the increased marketability of the body, which then reinscribes race, class, and sexual differences, rather than enabling their transcendence from markers of difference.

The prevalence of "choice" as a central tenet of neoliberal market values and individualism corresponds with recent constructions of the body as pliable via popular discourses of technology, science, and genetics. These new discourses and uses of technology and pliability continue the propaganda of neocolonialism, but also offer the possibility of providing Latin American women a decolonial strategy for performing themselves as ideal global citizens with superior reproductive and moral characteristics. The dangers of pliable subjectivity via technology and science have to do with the conflation of a consumer ethic and gene mapping, marketed as socially beneficial not only for the ridding of diseases, but for offering people the flexibility not only to *be* themselves, but to *produce* themselves, to alter the inner and outer reproduction of the self. It is worth recalling here that early cosmetic surgery was critical in erasing the markers of disease such as leprosy, enabling subjects to pass as able-bodied and free of disease. In this contemporary case, cosmetic surgery disassociates the violence and harsh economic conditions of everyday life in Cali from women's bodies. Similarly, the body as a malleable resource can serve as the raw material for making oneself into a more appropriate citizen in which the skin stretches its meaning across global space.[41]

For many women from Mexico and Colombia, technology offers a model of living that empowers the individual to act out against tradition and fate and to alter one's future. Thus these women's discourses of technology and foreign culture articulated their position as modern subjects in charge of, rather than at the mercy of, their fate. For some women from Cali, cosmetic surgery enables them to augment their body capital and chances of advancement in foreign romance, or to enter into a variety of labor markets. For women and men who do not choose to improve their chances of romance or personal uplift, this refusal to help oneself translates into a character flaw, or as a lack of motivation, initiative, and an unwillingness to change.

It is also apparent that a new language of privilege and modernity is being articulated through cosmetic alterations. It is through one's adherence to the norms of heterosexual femininity and male desire inscribed onto the body, rather than merely skin color, that one acquires a new language of global ascendancy. Cosmetic advances rearticulate long-enduring colonial legacies of racial difference as biologically based to create identities that are pliable, "democratically" available, and reliant on innovative technologies. Being civilized is equated with technologies of self-cultivation, with working hard, taking a risk, and jumping at opportunities that come your way. This is directly related to the role of U.S. state immigration in discouraging citizenship to those who may become a public charge, in other words, accepting only those whose enterprising spirit will contribute to the surplus labor of the nation.

Conclusion

Internet dating and marriage across the Americas reveal tensions in the debate over how subjects use technology to become part of the visible machinery of global mobility. Chicana/Latina scholarship is often left out of the debate on globalization that has tended to focus on the cosmopolitan subjects of flows rather than on those whose bodies disrupt the gears of technologies that promise flexibility and mobility. The understanding that a flexible sense of self is expressed in the departure from the body and geography is entangled in neoliberalism and a post-body utopia of liberal individualism.

Technological representations celebrating flexible identities elide the uneven access bodies have to flows, yet also generate new gendered tactics. Women turn to foreign men and circuits for a better life, but must confront the realities that their movement as spouses across borders does not assure equal rights as citizens without the fear of discrimination or estrangement from the benefits of belonging. I have demonstrated the ways technological embodiments of the self render insidious the narrative of development. Instead, values of rugged individualism take on new forms as the unsuccessful bodies of late capitalism are defined by their need for technological uplift, subjects who lack initiative. In "choosing" to alter themselves through the pliable modes afforded by technological practices, these women are encouraged to imagine themselves gaining membership into global citizenship. For women who choose to genetically enhance their bodies, improving the self through contact with modern technology and foreign culture offers a model for the ideal citizen whose worth can be exchanged for a higher value in the foreign marketplace. If we are to follow the flows of power across borders, then we are in dire need of a critical "postnational" American studies paradigm that interrogates the continued influence of Western individualism and hides the uneven consequences of flexible citizenship for diasporic Latina/o populations on both sides of the border. Corporate marketing of flexible strategies cautions us not to be too easily seduced by mobile identities as a means of achieving democracy, freedom from the constraints of the body, and liberal ideals of choice.

Notes

1. Michèle Ohayon, director, *Cowboy del Amor*, documentary by Emerging Pictures (86 minutes), February 10, 2006.

2. Jennifer González reminds us that online "passing" of racial identity has concrete consequences as much of the activity of "passing" is linked to racial fantasies of otherness. See Jennifer González, "The Appended Subject: Race and Identity as Digital Assemblage," in *Race in Cyberspace*, ed. Beth E. Kolko, Lisa Nakamura, and Gilbert B. Rodman (New York: Routledge, 2000).
3. These tours are usually three-day social events held at five-star hotels where couples actually meet one another. Men usually pay from $500 to $1,000 for the event (not including their travel expenses), while women are often invited free of charge.
4. Sherry Turkle, *Life on the Screen: Identity in the Age of the Internet* (New York: Simon & Schuster, 1995), 12.
5. Shannon McRae, "Coming Apart at the Seams: Sex, Text, and the Virtual Body," in *Wired Women: Gender and New Realities in Cyberspace*, ed. Lynn Cherny and Elizabeth Reba Weise (Seattle: Seal Press, 1996), 262.
6. Allucquére Rosanne Stone, *The War of Desire and Technology at the Close of the Mechanical Age* (Cambridge, Mass.: MIT Press, 1996), 180–81.
7. Mimi Nguyen, "Queer Cyborgs and New Mutants: Race, Sex, and Technology in Asian American Cultural Productions," in *AsianAmerica.Net: Ethnicity, Nationalism, and Cyberspace*, ed. Rachael C. Lee and Sau-Ling Cynthia Wong (New York: Routledge, 2003), 288. Nguyen's critique here also draws from Cynthia Fuchs, "'Death Is Irrelevant': Cyborgs, Reproduction, and the Future of Male Hysteria," in *The Cyborg Handbook*, ed. Chris Hables Gray (New York: Routledge, 1995).
8. This interview appears in more detail in Felicity Schaeffer-Grabiel, "Cyberbrides and Global Imaginaries: Mexican Women's Turn from the National to the Foreign," *Space and Culture: International Journal of Social Spaces* 7.1 (February 2004): 44.
9. From Arthur W. Frank, "Bringing Bodies Back In: A Decade Review," *Theory, Culture and Society* 7.1 (February 1990): 133.
10. Judith Butler, *Bodies that Matter: On the Discursive Limits of "Sex"* (New York: Routledge, 1993).
11. Felicity Schaeffer-Grabiel, "Planet-Love.com: Cyberbrides in the Americas and the Transnational Routes of U.S. Masculinity," *Signs: Journal of Women in Culture and Society* 31.2 (Winter 2006): 331–56.
12. María Josefina Saldaña-Portillo, *The Revolutionary Imagination in the Americas and the Age of Development* (Durham, N.C.: Duke University Press, 2003).
13. David Bell and Gill Valentine, eds., *Mapping Desire: Geographies of Sexualities* (New York: Routledge, 1995).
14. See also Fredrick Pike, *The United States and Latin America: Myths and Stereotypes of Civilization and Nature* (Austin: University of Texas Press, 1992).
15. Arjun Appadurai, *Modernity at Large: Cultural Dimensions of Globalization* (Minneapolis: University of Minnesota Press, 1996); and Manuel Castells, *The Rise of the Network Society* (Oxford: Blackwell, 2000).
16. Castells, *The Rise of the Network Society*, 442.
17. Ibid., 71.
18. Nestor García Canclini argues that consumption is the defining practice of citizenship and class relations between Latin Americans. See *Consumers and Citizens: Globalization and Multicultural Conflicts*, trans. George Yúdice (Minneapolis: University of Minnesota Press, 2001).
19. See Schaeffer-Grabiel, "Cyberbrides and Global Imaginaries," 33–48.
20. A democratic vision of Internet use is countered by the fact that cyber-segregation is alive and well. In fact, 80 percent of the Internet user population (globally and in the United States) is white. See C. Beckles, "Black Struggles in Cyberspace: Cyber-Segregation and Cyber-Nazis," *The Western Journal of Black Studies* 21.1 (Spring 1997): 12–19.
21. See Coco Fusco, "The bodies that were not ours," in *The Bodies That Were Not Ours and Other Writings* (New York: Routledge, 2001).
22. Recent media depictions of the Katrina disaster similarly saturated the popular imaginary with the conflicting images of the black poor Katrina victims uprooted by nature, and the image of Bush flying high above the natural disaster from his plane.
23. Charlotte Hooper analyzes the fabrication of corporate configurations of global manhood in *The Economist* magazine. Hooper, *Manly States: Masculinities, International Relations, and Gender Politics* (New York: Columbia University Press, 2001).
24. Donna Haraway, *Modest_Witness@Second_Millennium. FemaleMan_Meets_OncoMouse: Feminism and Technoscience* (New York: Routledge, 1997), 102.

25. See Castells, *The Rise of the Network Society*, and David Harvey, *The Condition of Postmodernity: An Enquiry into the Origins of Cultural Change* (Oxford: Blackwell, 1989).
26. Schaeffer-Grabiel, "Cyberbrides and Global Imaginaries."
27. The Spanish term *apertura* usually refers to a historical moment when Mexico's economy opened up to foreign trade and business. I use the term to bridge the economic, social, and cultural opening of Mexico through globalization.
28. Anthony Giddens, *The Transformation of Intimacy: Sexuality, Love, and Eroticism in Modern Societies* (Cambridge: Polity, 1992).
29. For a discussion of how the metaphor of the frontier plays out in cyber-interactions see Lisa Nakamura, "Race in/for Cyberspace: Identity Tourism and Racial Passing on the Internet," and Ziauddin Sardar, "Alt.Civilizations.Faq: Cyberspace as the Darker Side of the West," in *The Cybercultures Reader*, ed. David Bell and Barbara Kennedy (New York: Routledge, 2000).
30. Schaeffer-Grabiel, "Planet-love.com."
31. Wendy Hui Kyong Çhun, "Orienting Orientalism, or How to Map Cyberspace," in *AsianAmerica.Net*, ed. Lee and Wong, 7.
32. One particular male participant at the Cali Tour exclaimed that Cali was like the wild west, as he explained his decision to take the law in his own hands and hire a private detective to follow his Colombian fiancée, who was cheating on him.
33. For a discussion of romance as an expression of capitalism see Eva Illouz, *Consuming the Romantic Utopia: Love and the Cultural Contradictions of Capitalism* (Berkeley: University of California Press, 1997).
34. Anonymous e-mail interview, July 26, 2005.
35. They both knew I was a researcher writing a book on the matchmaking marriage industry and as such I did not charge for my translating services, although Seth paid for my meal and offered me a "tip" at the end of a long night.
36. Other women I spoke to similarly described their multiple surgery procedures as an investment in their self, careers, and futures.
37. Frances Negrón-Mutaner makes this argument for the Puerto Rican diaspora through a theorization of Jennifer Lopez's butt. See "Jennifer's Butt," in *Perspectives on Las Américas: A Reader in Culture, History, and Representation*, ed. Matthew C. Gutmann et al., 291–98 (Oxford: Blackwell, 2003).
38. For a discussion about cosmetic surgery in the barrage of television shows such as *Extreme Makeover*, *The Swan*, and others, see Brenda R. Weber, "The Economy of Sameness in ABC's Extreme Makeover," *Genders On-Line Journal* 41, http://www.genders.org/recent.html (accessed November 5, 2005).
39. Allucquére Rosanne Stone, "Will the Real Body Please Stand Up? Boundary Stories about Virtual Culture," in *The Cybercultures Reader*, ed. Bell and Kennedy, 439–59.
40. Don Kulick, *Travesti: Sex, Gender, and Culture among Brazilian Transgendered Prostitutes* (Chicago: University of Chicago Press, 1998).
41. Ibid., 68.

"Slow and Low Progress," or Why American Studies Should Do Technology

Carolyn de la Peña

Langdon Winner ends his *Autonomous Technology: Technics-out-of-Control as a Theme in Political Thought* with a corrective to what he believes is an inaccurate popular understanding of the message in Mary Shelley's *Frankenstein*. It is not, he argues, a monster story of the inevitable dangers of technological wizardry. Rather, it is a story of "the plight of things that have been created but not in the context of sufficient care" (313). It is easy to focus on the monster as the problem in the novel: certainly the physical threat he poses illustrates the danger of technological hubris. Looking more closely, however, we discover that the "horror" lies not in the monster's creation but rather in the inventor's failure to take responsibility for his machine. Frankenstein seeks recognition first. Only after his inventor refuses to acknowledge him does he begin to enact the horrors for which the novel is better known.

This essay suggests that scholars in American studies have something to learn from Mary Shelley. We in the United States frequently tell stories of technological redemption and technological damnation. We do not, however, spend much time considering stories of technological stewardship. A legacy of positivism has embedded our political, social, and cultural systems with a disturbing patina of technological "neutrality." And, in many ways, we as scholars have contributed to this legacy of positivism by failing to critique technology as both substance and ideology in American cultural life. The field of American studies has largely left questions of technology to others, in spite of our early leadership in innovative methods of technological analysis and cultural critique. And while discipline-based inquiries into technology have been immensely useful at revealing particular histories and consequences of American technology, they have not been primarily focused on issues of diversity, equity, and justice that are fundamental to our field. Nor have they been written with a particular focus on interdisciplinary connections that embed everyday actions within their larger political and cultural systems.

It is time to revitalize a "technology studies" core within our field. I have three goals for this piece within the broader context of our issue of *Rewiring the*

Nation: to demonstrate, using books that approach technology from within the field, the importance of an American studies methodological and contextual approach to understanding the place of technology in American life; to provide a primer by which scholars interested in incorporating technology studies into their particular areas of interest can begin to do so using books currently available across disciplines on technology; and to demonstrate that concerns fundamental to our field are inextricably linked to technology.

Defined here as the material or systemic result of human attempts to extend the limits of power over the body and its surroundings, technology has been an essential tool in our ability to feed, clothe, house, and protect our bodies. It has unified individuals across expanses of time and space in a manner that has extended pleasure and prevented pain. Few of us would give up our computers; many of us are lost without our cell phones and Blackberrys (in spite of the incessant work both enable). Airplanes allow us to breakfast in New York and lunch in Los Angeles. Levees allow us, when they hold, to live in spaces deemed uninhabitable by nature. Vaccinations relegate afflictions like polio (in this country) to the pages of history books. Technologies facilitate the very construction of "normal" life as we know it. (Consider, regardless of where you live and what cultures you identify with, how your breakfast this morning would have been changed without this nation's vast, mechanized systems of food production and distribution.) Undoubtedly, technology brings with it an enhancement of our personal pleasures and an elongation of our social networks. Yet beyond our specific points of entry (and often within them), technology also enables the creation, aggregation, and enactment of particular systems of power that imperil our personal pleasures and individual expressions.

Technology is slippery; to understand it one must simultaneously consider the "body" and the "system." Often its most important cultural meanings lie in technological ideologies—ways of looking at the world and ourselves that are enabled by innovations but not apparent in close studies of their invention or adoption alone. For example, in *American Technological Sublime* David Nye argues that Americans may actually want to live on the edge between awe and terror when it comes to technology. Nye looks to natural and manufactured "technological" sites in the United States since the nineteenth century, theorizing that people enjoyed following the progress of undertakings such as the Eric Canal because they offered an opportunity for transcendence. With few common rituals and no state religion, individuals found in technological achievements a common ground. At the same time, the intractability of nature, frequently experienced in natural disasters, regularly demonstrated that ground's instability. Building on Kant's "sublime," Nye argues that, for

Americans, technology was primarily a means through which individuals could unite through common experiences; technology enabled not merely inventions but also the awe necessary to validate "a belief in national greatness" (43). Nye's insights are gained by traversing historical periods and drawing together disparate "technological" objects. By doing so he discerns that early infrastructural projects such as bridges, canals, and skyscrapers originally "actively involv[ed] citizens, linking world and product, technology and human agency." However, later technological events, such as space launches, merely displayed "organizational and technical power" (279). Nye's work, especially when extended to an analysis of "sublime" displays such as the mediated experience of war, suggests that we may as individuals support technological systems because of how they make us feel rather than what they actually produce. Once visible signs of individual and communal material benefits, today's technological displays frequently offer little beyond the "awe" itself. It is a sensation that effectively masks the discrepancy between those who celebrate and those who actually benefit from technological events.

The assumption that culture is technology's most important product is unique to our field. American studies scholars who work on technology are not bound to tell the story of particular objects; nor are they limited by particular historical periods. They begin with questions about the present, questions about how we imagine ourselves and how that imagination is facilitated by technological forms. In so doing they have arrived at counterintuitive insights that enable individuals to better evaluate the actual impact of technologies on their daily lives.

It is not enough to refamiliarize ourselves with this body of work. New work is needed. The essential emphasis today in our field on work in transnational, ethnic, and sexualities studies has heightened our awareness of and our ability to analyze the relationship between individual subjectivities and complex systems. This approach, taken by several emerging scholars in what might be termed a "third wave" of technology studies, has already demonstrated that people are frequently most passionate about technological "progress" when it functions in ways directly opposed to systems of mechanized efficiency. I write in a moment heavy with evidence that technological systems have failed: it is ten degrees warmer in my town of Davis, California, than it was last summer; my students report that their younger siblings are being introduced to the "thrill" of soldiering through downloadable video games provided by the U.S. military; the news regularly reveals our government's technologically enabled spying on its private citizens. It is important to critique these systems and to work toward more equitable, sustainable distributions of technological power.

But it is also important—perhaps more so—to know the real source of our technological pleasures.

The Place of American Studies in Technology Studies

American studies scholars have long written about technology. There are two periods, however, in which the number of scholars and the strength of their work have been particularly important. The first period's central text is Leo Marx's 1964 *Machine in the Garden: Technology and the Pastoral Idea in America*, for which its author was named by Elaine Tyler May in her 1995 presidential address as a founder of one of our three schools of Marxism. Marx's was not the first book to ask critical questions about the role of technology in American culture—it was preceded by the pioneering work of nonacademics Lewis Mumford (*Technics and Civilization*, 1934) and John Kouwenhoven (*Made in America*, 1948). It was, however, the first work to take an interdisciplinary approach to the study of technology *as a material and a concept* in American life. Marx looked to the history of politics, literature, and art to understand the fundamental tension between furthering technology and celebrating nature embodied in evocations of an ideal American "middle landscape." His work demonstrated that artists and politicians both regarded mechanized production as essential in achieving the "happy balance of art and nature" (226). Yet, as Marx reveals, the pastoral ideal is chiefly important as a cultural concept not because it was achieved but because it became a mythology perpetuated by those who pursued technological "progress" (production, wealth, and power) in direct conflict with pastoral living.

Marx's ability to understand that technologies succeed when they produce both products and national (as well as personal) mythologies makes his work American studies. It is, of course, impossible to draw clear methodological lines between historians, sociologists, anthropologists, literary critics, and American studies practitioners when it comes to technology studies. There are important differences, however, between the questions asked by early American studies scholars of technology and the work of their more discipline-bound colleagues. This becomes clearer when we look at the books produced in the mid-1980s and 1990s by the second generation of American studies technology scholars. David Nye's *Electrifying America* and *The American Technological Sublime*, Cecelia Tichi's *Shifting Gears*, and Jeffrey Meikle's *American Plastic*—treated more fully in the sections below—should be singled out here as a starting point for those seeking the early edges of our field. Each took the technologies of everyday life as a point of entry (fairs, factories, lights, Bakelite). Each worked across disciplinary boundaries and disparate contexts to make cultural connections. And each insisted that the meanings individuals give to technologies are often more important—and in direct conflict with—those "applied" by their producers. In so doing, these works distinguished the approach American studies practitioners brought to studies of technology. We are entering a third period of scholarship, as signaled by the work of scholars such as Joel Dinerstein and Ben Chappell (discussed at the end of the essay) as well as that of the other scholars whose essays are included in this special issue.

Technology studies, since the mid-1990s, has had its most innovative and prolific expressions outside of the context of the American Studies Association. Science and technology studies (STS) has emerged as a field of study of vibrant contributions, primarily by sociologists and anthropologists, to the construction of scientific and technological knowledge and its impact on cultural life. Sociologists Steve Woolgar and Bruno Latour's 1979 *Laboratory Life: The Social Construction of Scientific Facts* argues that a network of concerns (among them prestige, funding, and personality) determine the process and result of scientific investigation. Latour's work has helped create a growing body of ethnographies of the individuals who produce scientific and medical knowledge. Among these are Sharon Traweek's *Beamtimes and Lifetimes*, an account of a high-energy physicist subculture, and Joseph Dumit's *Picturing Personhood* (discussed more fully below). There is much that we in American studies could gain from a greater collaboration with STS. First, we also have a rich tradition of ethnographic methods in cultural analysis; STS suggests ways we might use this specifically to ask questions of technology and science (STS also illuminates the difficulty of distinguishing between these fields, especially after World War II). Second, STS, particularly through the work of Latour, has used its understanding of the politics and subjectivity of scientific study to bridge the gap between academic and public discourse, a priority among the organization today. Third, STS tends to engage science and technology by studying its experts. American studies scholars could extend this to "technologists" outside of the institutional mainstream; we could also fruitfully contextualize expert ethnographies within broader historical and cultural contexts.

Currently, it is historians who produce the majority of the important books on technology in American life. The primary intellectual forum for scholars of technology in the United States continues to be the Society for the History of Technology (SHOT). The key journal is its *Technology and Culture*. As a result, historians' books are the majority in this bibliographic essay. Evidence suggests that some of these books are finding a receptive audience within American studies—*Soundscapes of Modernity*, Emily Thompson's analysis of the technological facilitation of sound in the early twentieth century, won the John Hope Franklin prize for the best book in American studies in 2003. Yet while many scholars in American studies operate in historical frameworks, the fact that we are not history bound as scholars, but rather are interdisciplinarily inclined, enables us to make the temporal and evidentiary leaps that constitute "connective" work in our field. And these broad leaps are particularly important in discerning Americans' complex relationships with technology.

This essay, then, suggests that technology studies as a field does not need to be resurrected as much as relocated. American studies as an organization was once the home for individuals to ask and find answers to critical questions concerning Americans' engagements with technologies. Yet, at least in the past decade, individuals trained in American studies have found STS and SHOT to be more friendly forums for their work. Prospective graduate students committed to studying the impact of technology and science on everyday life are more likely to enter sociology and anthropology PhD programs than American studies. Such an observation begs the question: does this matter? Certainly, American studies cannot, as a field, take up every issue at every moment. And there are, in fact, logical reasons why technology has remained a minor player in the field, among them the antipathy of the new left to technology and the propensity, particularly among early technology scholars, to construct uncritically white narratives of nation and progress. Technology, however, is not "every issue." It is one that deeply enables and restricts the ways we live, the ways we define ourselves and identify others, and the ways we act in and on the world. It is at the heart of questions that American studies is heavily invested in as a field.

Critical Issues in Technology Studies

It is impossible to include every important work on technology across period and disciplinary boundaries in an essay such as this. Here I offer those that engage in three critical areas of inquiry of interest to scholars in American studies: the production and aggregation of power; the resistance and collusion of consumers; and the process of subcultural meaning making.

A significant body of scholarship, in what I here define as technology studies, offers useful ways for envisioning the complex modes by which American power, past and present, aggregates and expands. Technology is not created or used by a monolithic group of individuals called "Americans," a point frequently missed by scholars in the field through errors in the precision of their claims. Racial, ethnic, gender, and class differences play a dramatic role in who gets to create technologies and to what ends those technologies are used. Recent work on the role of minorities in technological systems and as technology users and inventors helps us see how narratives of technological progress rely upon rendering invisible those excluded. It also allows us to see how national mythologies of technological heroism can obscure the real dynamics of race, gender, and imperialism in the inventive process.

A second way in which technology offers an important point of entry into cultural practices is through the complex interactions it facilitates between consumer "choice" and systems of power in market exchanges. Americans use technologies to define themselves as individuals: cosmetic surgeries, online dating services and MySpace, iPod life "soundtracks," and branded cars are only the most obvious consumer technologies that many rely upon to enhance personal pleasure and facilitate community formation. Technology studies scholars take seriously the complex reasons that consumers adopt or reject particular innovations that transform daily life. They reject models of technological determinism, instead revealing that frequently people adopt technologies that end up dramatically changing their lives only because they originally wanted things to stay the same. Scholars are opening up fruitful discussions about the limits of consumer choice, and the problematic relationships specifically created between that choice and systems of production within technological systems. These texts should be of interest to scholars working on consumer culture, teasing out the challenges posed by subcultural processes and products, and those seeking to embed the process of production more thoroughly in discussions of "choice" within the American marketplace.

A final set of books demonstrates the value of scholarship on technological subcultures and alternative definitions of technological "progress." While many scholars outside of our field suggest that technologies are inherently resistant to control, and others illuminate the fundamental difficulties in stepping outside of technological epistemologies, an emerging body of work within and around the field (including work in this volume) suggests that there is much to be gained by the attempt. American studies scholars have revealed the irrational, subjective, and often subversive relationships individuals have long pursued within technological systems. They might go further, identifying and illuminating alternative ideologies of technological "progress" outside of and in direct conflict with the systems themselves.

Technology and Systems of Power

In 1989 Thomas Hughes urged historians of technology to pay greater attention to systems. As he argues in *American Genesis*, technology includes both the objects that result from particular innovative processes and the way of thinking engendered by participating in those processes. "Technology," he asserts, "is the effort to organize the world for problem solving so that goods and services can be invented, developed, produced, and used" (5). Hughes's systems approach is useful for practitioners of American studies for several reasons. First, when we approach technology from a systems perspective we see clearly the enormous impact of the American military on our current technological landscape. Hughes juxtaposes the small-shop invention systems of Edison and Ford with the larger university and private laboratories that replaced small-scale inventors. These were the sites of development, frequently with military funding, for the massive systems of energy, production, communication, and transportation that came to characterize twentieth-century American life. And while Hughes does not raise the concern, it may be worth asking if our field's early fascination with Corliss engines and Fordist assembly lines has not served to obscure the actual systems that are arguably more essential and certainly more troubling today. Second, Hughes's systems approach allows us to debunk the myth of the heroic individual inventor too often perpetuated post Edison and Ford. By looking beyond the individual products of technological success to the particular climates that produced marketable inventions, Hughes allows us to see the way that narratives of technological heroism can mask larger systems of national control and internal segregation. American technological history is fraught with systems of privilege that have allowed particular subjects to emerge as inventors while others have, at best, enjoyed marginal success.

Technological systems have been used within and beyond the United States as a means for individuals to achieve professional, political, and social power. Such a statement does not dismiss the fundamental achievements of technological systems and their designers. Nor does it deny that many engineers, inventors, and scientists have worked selflessly for the greater good of humanity. It is important, however, in the context of this issue to demonstrate that technology should be of fundamental concern to any American studies scholar committed to revealing and addressing social inequity. Particularly relevant for scholars interested in the processes by which individual technologies are used to establish systems of political and social dominance is Michael Adas's recent *Dominance by Design*. His study brings together U.S. Indian land policies, Philippine engineering projects, the Panama Canal, cold war kitchen debates, and the Iraq Wars in order to demonstrate that technologies have been more powerful as instruments of ideology than as direct tools for military force in furthering American dominance. Adas's work should be considered alongside David Nye's America as *Second Creation.* Both enable us to understand why some Americans came to believe in the righteous superiority of their technology and how they used that belief to remove resources from those less technologically skilled in the name of "progress" (Adas quotes one congressman who in 1830 argued for the removal of Native Americans from western lands by stating that failing to do so would "obstruct the march of science" [96]). Adas's case studies illuminate the importance of technology in imperialism: it was the industrial revolution that created a class of individuals eager to carry out the "civilizing mission" (missionaries, administrators, and teachers), produced a stock of finished goods in need of "open markets," and engineered connections between distant infrastructures. Machines and tools often constituted the "front line" in imperial efforts; the first step in "opening" Japan was the delivery of gadgets designed to suggest (if not reveal) an implied superior American military force.

For Adas, the "vast engineering projects" of engineers and scientists have served as de facto foreign policy. They have enabled believers to repeatedly disregard cultural knowledge, certain that demonstrations of American technological superiority would in and of themselves win the hearts and minds of non-Western peoples (147). It is a framework within which 9/11 appears predestined. By fusing our technological goods with our imperial designs—and continually underestimating those we deem "technologically inferior"—we have created a "paradoxical vulnerability" whereby our technological symbols themselves are appropriate targets for those who would attack our ideologies.

American studies scholars may also find useful two recent books on how technological systems facilitate imperialism at home. In *Making Technology Masculine* historian Ruth Oldenziel argues that "there is nothing inherently or naturally masculine about technology" (10). Her engineering history reveals that the very category of technology, upon which such systems of knowledge were based, was profoundly affected by the professionalization of systems designers in the late nineteenth and early twentieth centuries. Chief among them were the engineers who professionalized their field between 1890 and 1920 by excluding draftsmen, skilled mechanics, and women. Oldenziel uses historical records, individual biographies, and fiction to reveal how technology functions "as part of a narrative production . . . in which men are the protagonists and women have been denied their part" (14). Male authors eager to combat "mobs of scribbling women" (123–25) produced numerous novels dedicated to the exploits of heroic engineers, tales that circulated among the broader public, solidifying the ideal technological mind as the white male systems and nation builder. (Cecelia Tichi's *Shifting Gears* has a chapter on engineers in fiction that augments this analysis.) Professional associations drew barriers of professional training and certification around engineering as a field. The very word "technology" reveals a narrowing of how we define important innovations and their makers: whereas previously the idea was expressed by the phrase "the useful arts," a category that included industrial and domestic innovations, by the late nineteenth century, the term referred only to the industrial, effectively limiting "inventor" status to men. Oldenziel's work can be contextualized more broadly by the essays in Jennifer Terry and Melodie Calvert's *Processed Lives*.

Historian Rayvon Fouché's *Black Inventors in the Age of Segregation* suggests that technological systems created both barriers to entry and opportunities for self-definition among inventors of color in the United States prior to World War II. Fouché argues against the notion that inventors operate as "heroic" individuals, and instead focuses on the networks of financial support and knowledge of legal and business institutions necessary for particular inventors and their products to enjoy market success. Within his analysis we see would-be inventors, such as Granville T. Woods, "navigat[ing] a sea of problems" (79). For Woods and others, success came only after they abandoned the goal of building their own systems and began instead to build parts for systems developed by white inventors.

Fouché's work also suggests that technology operated as a powerful system of identity formation among some inventors of color. And it was frequently an identity that enabled inventors to work against justice and inclusion for people of color *because* it enabled them to imagine themselves as members of a (white) technological elite. Of Lewis H. Latimer, Fouché writes, "he wanted to fade into this technical world . . . where he was no longer seen as a black man, but as a raceless member of this environment" (84). Moving away from stories of individual inventors, focusing instead on the systems within which they were embroiled, Fouché explains why minority inventors have often ended up creating "tools of white uplift" that have "reinforced racial inequality and the hegemony of white American culture" (23). Fouché's book might fruitfully be considered along with journalist Teresa Riordan's *Inventing Beauty* and Nowalie Rooks's *Hair Raising* as they explore the relative success of female inventors of color during a similar period.

A number of scholars have explained how technological systems—as objects and epistemologies—have changed our definition of "normal" bodies in the United States. Joseph Dumit and Keith Wailoo have contributed important work on how our technologically enabled systems of body knowledge produce problematic conclusions. In *Picturing Personhood* Dumit looks at biomedical knowledge as generated by brain scans. He argues that such technological images have "a persuasive power that is out of proportion to the data they are presenting" (17). Dumit's work here should be considered alongside Lisa Cartwright's *Screening the Body* in that both demonstrate how technologically generated visions of "normal" can present as facts distinctions that are actually generated by cultural constructions of difference. As Dumit reveals, there is a great deal of variation among types of brain scans deemed "accurate," in spite of the fact that different scans reveal very different information. The variation itself is a sign of the competing cultures of expertise and prestige among experts. And the acceptance of brain scans as accurate gauges of brain health—in popular culture and courtrooms—reveals more about our propensity to seek biological explanations for difference than it does about the arrival of an accurate "brain science."

Keith Wailoo's *Drawing Blood* complements Dumit's analysis. According to Wailoo, "questions of gender, race, lifestyle, and professional politics [have been] inevitably linked to the construction of medical knowledge" in the United States (ix). Technological systems can render disease visible and invisible. Choleria, a common nineteenth-century disease of the blood in women disappeared in the twentieth century when hospital technology for blood screening could not locate a specific ailment. Pernicious anemia, on the other hand, was diagnosed first only when hospitals became sites for university research, staffed by physicians trained to find blood illnesses. The diagnosis served the interests of physicians and the Eli Lily company (makers of a liver extract remedy) who were seeking to establish their authority; as a result, iron remedies produced for the newly diagnosed "disease" cured 80 percent of sufferers while leaving the other 20 percent (who had additional symptoms) unassisted. According to Wailoo, hematologists and pharmaceutical companies sought to develop systems of disease diagnosis and treatment that complemented their own social and economic networks. They were systems then in which "professional identity and disease identity were mutually constitutive" (3).

Technological systems have determined how we diagnose and interpret disease, often with powerful cultural impacts. Prior to World War II, sickle cell anemia was believed by hematologists to be a blood disease. The condition, which effects primarily African Americans, allowed medical professionals to function as "race police" who used diagnostic technology to "pierce through duplicity and false pride" to identify those with black ancestry who were passing for white (147). Yet in the 1950s when new technologies diagnosed the disease as molecular and found that the risk lessened for individuals of racially mixed heritage, disease knowledge instead suggested the possible benefits of interracial unions. Wailoo, like Dumit, finds that systems of medical technology are not neutral. They can, on occasion, challenge the cultural biases from which they emerge; nevertheless, careful attention must be paid to those system creators who have the power to develop and interpret the technologies firsthand.

Resisting or Colluding Consumers?

Technology studies has long pioneered inquiries into the meanings consumers give to the objects they consume. Frequently, technological innovation produces objects for which there is no clear consumer demand (cellophane's creators, for example, initially had no idea what to do with it). And when "experts" do have an idea of what to do with an innovation, consumers frequently reject those ideas in favor of their own. Technological inventions often dramatically change individuals' definitions of "normal life" in the space of a generation (consider air-conditioning or the green lawn). As a result, books in technology studies are crucial to illuminating the complex process by which products take on cultural meaning.

Early work on technological consumption came from the scholars working within a social construction of technology paradigm. The central text for this approach was *The Social Construction of Technological Systems*, an edited volume from Wiebe Bijker (who also produced *Of Bicycles, Bakelites, and Bulbs* some years later using a similar approach), Thomas Hughes, and Trevor Pinch. Their argument has become a standard assumption in the field today: social needs and desires determine which technological options will be taken up by consumers and which abandoned. Bijker found that the "ordinary" bicycle (one with a large front wheel and a smaller back wheel) was actively endorsed by young men in search of thrill rides; it was rejected, however, and ultimately replaced by the familiar "standard" because women and older men preferred safety over speed. A single technological object can, in fact, be many individual "artifacts" to different users. This "interpretative flexibility," as Bijker calls it, enables technological products to acquire multiple meanings through negotiations, any one of which could reveal entirely different "proper" uses for particular objects. (Gail Cooper, in *Air-Conditioning America*, uses this approach to reveal the myriad forms that air-conditioning technology took before becoming adopted for mass residential use.) As a group, these scholars dismissed "technological determinism," or the belief that technologies themselves possess internal logics that lead to particular social outcomes. In so doing, they elevated the importance of "lay" knowledge in technological meaning making, and moved the conversation from sites of production (Edison's laboratory, the Corliss demonstration at the World's Fair) to those of consumption. This pioneering work of historians, sociologists, and engineers provided a foundation for the 'second wave' of American studies scholarship.

Three texts should be considered here as foundational studies of technological consumption in the United States. Cecelia Tichi's *Shifting Gears: Technology, Literature, and Culture in Modern America*, David Nye's *Electrifying America: The Social Meanings of a New Technology*, and Jeffrey Meikle's *American Plastic: A Cultural History* reveal the peculiar mixture of the miraculous and mundane that have characterized individuals' technological experiences. They also suggest that one of the most important affects of technological consumption has been the ability *to think through machines*. At the same time that actual technological artifacts and processes allow us to know and act on our environments, our experience of knowing and using those artifacts and processes changes the way we know ourselves.

David Nye's *Electrifying America* works on both the production and consumption sides to understand the popular appeal of electricity between 1880 and 1940. The book remains important for two reasons. First it considers the contrasting ideologies of producers (inventors and promoters, mostly male) and consumers (those without expert knowledge, male and female) as equal contributors to the meaning of electricity. Like Wailoo, Nye argues that electrical engineers framed electricity and its products in a way that solidified their own expertise. Yet Nye looks to popular electrical spectacles, including world's fairs, the illumination of urban centers, and traveling electrical performers—as well as electric novelties such as belts and tie lights—to understand the importance of electricity as a consumer spectacle in this early period. Second, *Electrifying America* provides ample evidence that the success of "productive" electrical systems depended largely upon the popularity of "unproductive" displays developed expressly for consumer pleasure. As such, technological gadgets and performances emerge as "visible correlative(s) for the ideology of progress," correlatives that stressed the pleasure and grandeur of electric systems even while those systems, in reality, frequently electrocuted people through their imperfect applications (35). (Here Carolyn Marvin's *When Old Technologies Were New* is also instructive.)

In *Shifting Gears*, Cecelia Tichi augments this study of technological dramas with a study of technological language. She finds, in the same period, popular literature increasingly relying upon mechanized metaphors (Robert Herrick's *Together* from 1908 has a character wonder "if we are just machines with the need to be oiled now and then"). Popular medical descriptions made disability analogous to having a "flaw in [one's] casting," while books on health described hearts as "motors" and food as "fuel" (30–35). Tichi's book is important because it looks outside of the world of technological production and consumption to locate a new technological good: the human body. It is usefully positioned next to Nye's as both reveal that technological innovations are not merely experienced as physical human aids; they are also embedded in new epistemologies of who we are and how we relate to the world. (*Shifting Gears* is usefully augmented with Anson Rabinbach's *The Human Motor* and Martha Banta's *Taylored Lives*, books that contextualize the human-machine discourse within actual labor practices. One might also consult Terry Smith's *Making the Modern* to incorporate the importance of visual renderings of machines during the same period.)

In the introduction to *American Plastic* Jeffrey Meikle acknowledges the difficulty scholars face in determining the meaning of technological products. One must understand the material—the complex processes and expertise that enable an invention to take form. At the same time, one must decode the myriad motivations for individual consumers to actually use such an object, paying particular attention to how consumers push back on producers, as well as to the differences in consumer meaning across time, space, and gender. Postwar consumers cared little about advances in molecules and polymers; they bought into plastic for convenience (such as the "damp-cloth cleaning" enabled by plastic furniture), not chemistry (173). Meikle's work situates consumption within production; it also echoes back to the careful attention scholars such as Kowenhowen placed on materiality. Where it is particularly useful to American studies scholars is its ability to cross the line between technology and popular culture. Meikle considers "plastic" as both a technological product and a cultural concept. Plastic was meant to create objects for consumption. It also created, on the part of those who celebrated and critiqued it, visible evidence of cultural change. Plastic transcended previous material boundaries, suggesting the limitlessness of the technological future as viewed from the 1950s. It also embodied cheapness, waste, and insatiability for novelty. Meikle's work demonstrates the importance of looking to cultural concerns to understand technological success and failures. At the same time, he suggests the difficulty in critiquing—or challenging technological consumption—when its materials become deeply embedded in the practice of "normal" life.

Two additional books offer insights into consumers' motivations for accepting technologies that dramatically alter the experience of everyday life. Emily

Thompson's *Soundscapes of Modernity* suggests that at times major technological innovations succeed because they are perceived as enhancing consumer control and directly combating the undesirable effects of previous technologies. Along with David Sterne's *The Audible Past*, Thompson's work marks an emerging body of scholarship in sound studies. She reveals that acoustic technology offered a way for individuals to limit their exposure to sound, to block out the technologically enabled cacophony of streetcars, radios, and automobiles that filled urban spaces in the early twentieth century. In Thompson's work we see how the very act of selective listening is itself a technological invention—one that depends on an elaborate network of acoustic materials, systems of sound measurement, and networks of sound control. She allows us to understand that individual acts of technological consumption during the period were often driven by the desire to create a private soundscape (thereby showing us that the iPod is not a new phenomenon). *Soundscapes* also demonstrates the importance of considering technology as a producer of material and immaterial "goods." It is intuitive, perhaps, to study the radio to understand how technological sound modifies our environment. Thompson's work suggests, however, that "negating" technologies also play a key role in acclimating individuals to life within technological systems. (For another perspective on the "engineering" of the urban soundscape during this period, see Joel Dinerstein's analysis of big-band swing music in *Swinging the Machine*.)

Claude Fischer's *America Calling* adds two important arguments to considerations of consumer choice in technological systems. First, people frequently choose technologies in order to avoid change. Second, "technology can be both a tool for an individual user and, aggregated, become a structure that constrains the individual" (19). Fischer calls into question assumptions that new technologies create consumer "anxiety," finding instead in consumer records and through interviews that many original adopters found easy comfort in the telephone. The historical record may reveal the dramatic changes caused by the technology; yet, individual users primarily used phones "to widen and deepen existing social patterns rather than to alter them" (262). (One could fruitfully juxtapose this with Kenneth Haltman's material analysis of the candlestick telephone in *American Artifacts*.)

Fischer's analysis allows us to see the ways in which technological objects enable consumers to experience large systems as intimate realms of individual choice. Women, especially those in rural areas, frequently rejected the notion that telephones should be reserved for business or emergency use, choosing instead to make social calls with those too distant to see in person. Telephones were then not foreign objects but sounds of intimacy. At the same time, the mass adoption of telephones also limited intimacy: it eliminated the face-to-face contact citizens had with police in neighborhoods in times of distress; it diminished the frequency of men and children's shopping trips as women increasingly relied upon phone-facilitated home delivery.

It is important in studies of technological consumption that we do not overlook the place of technological labor. As Venus Green suggests in *Race on the Line*, innovations that create consumer ease often diminish the experiences of technological workers. And what may appear "technologically neutral" from our perspective of consumer use may be the result of discriminatory practices within the system. *Race on the Line* is primarily a historical analysis of the work of female telephone operators within Bell Systems. It is particularly intriguing, however, for what it reveals about the way that consumer technologies have expanded and restricted the intellectual and physical power of female operators. Because customers wanted to use telephones within familiar social networks, female operators were hired to work the lines. (This role of women as "feminizers" of threatening technologies is also explored in John Kasson's *Civilizing the Machine*.) The consumer preference expanded opportunities for some women, particularly white operators in isolated regions, who by necessity acquired the skill required to work the lines. The result was an early hybrid: the female operator/engineer. Yet as technology enabled calls to be made without operators, thereby improving consumers' experiences, the skill required to operate declined. While still female, operators were no longer given the freedom to operate the technical aspects of phone service or move about freely. Such "improvements" were in fact frequently initiated to defeat workers' demands for wage increases.

Green demonstrates that technologies consumers view as racially "neutral" in fact have profound effects on racial practices within their larger systems. Many defenders of segregation within the industry have used arguments about the appropriateness of black bodies: white operators refused to sit in the close proximity to blacks that was required by the particular configuration of operating work; white telephone users resisted the physical intimacy required in exchanges with operators of color. Operating work did ultimately desegregate, but it did so only after white flight (itself enabled by a combination of infrastructural technology and racial discrimination) diminished the available white labor force in cities. And perhaps not coincidentally, desegregation was accompanied by additional consumer "improvements" in phone service, improvements that enhanced user experiences while lowering yet further operators' wages and autonomy. Green's work ultimately challenges consumer scholars to consider the labor that facilitates material goods and technological services.

While it may be easy to focus primarily on the reception work facilitated by advertising, marketing, and consumer records, innovations always affect both sides "of the line."

Over the past ten years a number of scholars have begun to look closely at the importance of the body in technology studies. Several of their texts, taken together, allow us to see body technologies as ground zero in the conflict between systems of control and objects of freedom in technology studies. Carl Elliott's recent *Better than Well*, considered alongside David Serlin's *Replaceable You*, suggests that individuals often choose technologies for physical enhancement to achieve what they deem an "authentic self." Elliott examines the rise in pharmaceutical and surgical body enhancements, combining philosophical, psychological, and cultural studies modes of analysis. His work considers technologies that facilitate visible physical change as well as those that facilitate psychological changes, paying particular attention to the motivations provided by individuals who choose what he loosely terms "enhancing technologies." Among them are professionals who use beta blockers to overcome social anxieties (the introduction to the book is provided by *Listening to Prozac*'s Peter Kramer), parents who give their boys growth hormones, individuals who undergo sex reassignment surgery, and voluntary amputees who aim to become the "person they've always been."

Particularly useful for American studies scholars is Elliott's attempt to contextualize the specific use of technology within larger structures of American consumer identity and mediated social networks. Purchasing the self is not an expressly technological phenomenon; self-help books have made profits since the eighteenth century. What is technological is the closed circuit Elliott presents between technologically generated ideas of "normal" behavior available through public discourse and the technologies that enable private consumers to achieve it. Elliott suggests that television viewing itself, a solitary activity that exposes couch sitters to "normal" worlds of highly extroverted, attractive individuals, can help explain not just the increasing numbers of people who seek enhancements, but the numbers of people who justify their reason for doing so as a search for a more "authentic self."

David Serlin's *Replaceable You* demonstrates the importance of considering the "choice" individuals make to partake of enhancement technologies within larger social and political structures. His case studies of media coverage of postwar "Hiroshima Maidens" programs to outfit amputated servicemen with prosthetic limbs, the first publicly acclaimed transsexual, and a lesbian who underwent hormone treatments to achieve heterosexuality suggest that these postwar choices were frequently made by individuals seeking social visibility.

As Serlin explains, such physical "rehabilitation" was frequently a way for the socially excluded to become "more tangibly and visibly American than ever before" (14). Serlin does not suggest that these people were victims of false consciousness or pawns of the medical establishment. Rather he demonstrates that individuals can dramatically alter their physical selves, often in ways that deny difference and embrace homogeneity, in search of a *social authenticity* that is possible only by altering their actual physical bodies.

Particularly in his account of the "Hiroshima Maidens," a group of Japanese women whose trip to the United States to undergo plastic surgery was highly publicized by the American press, Serlin reveals the paradox of technological choice within technological systems. The success of American technology to conceal their disfigurement (itself a result of radiation burns caused by American technology), put a friendly face on American ingenuity. Similarly, prosthetic legs served as technological showpieces for restoring war amputees to "manhood," and drew attention from the technological source of their original disfigurements. Serlin aptly demonstrates the paradox in such engineering: individuals turned to technological modifications to be visible, thereby effectively strengthening the normalizing forces that rendered them invisible to begin with. (Scholars particularly interested in additional work on cosmetic surgery should consult Elizabeth Haiken's *Venus Envy* and Kathy Peiss's *Hope in a Jar*.)

Other scholars have considered technologies that allow the body to transcend physical limitations. Emerging with Donna Haraway's *Primate Visions*, this scholarship has been augmented by recent texts that explore the limits of transcendent physicalities in cyberspace. These, however, are fruitfully contextualized by Gaby Wood's *Edison's Eve*, a text that considers Western attempts since the Enlightenment to blur the boundaries between the human and nonhuman through technological innovations. The uneasy truce between technological enhancement and obsolescence is as old as mechanics itself. Wood reveals that Enlightenment audiences willingly participated in the optical illusions of defecating ducks and "automatic" card players; nineteenth-century fiction writers found eager consumers for stories of women made over by men desiring to have particular traits removed; consumers hoped that Thomas Edison's talking doll might record their own children's voices so that early death would enable continued "life" in ceramic form.

Such quests revealed active desires to use technology to transcend the limits of human imperfection and stave off the inevitability of death. At the same time, they suggest that technologies developed by individuals to transcend human limits inevitability reproduce those limits. Edison's dolls failed because he knew little about children (he spent little time with his own), and Vancanson's duck never really digested properly (neither did Vancanson). Wood's work provides a useful historical backdrop against which to consider *How We Became Posthuman*, a close study of the assumptions and actions of scientists in cybernetics and informatics who, according to N. Katherine Hayles, have already developed technologies that pose fundamental challenges to the sovereignty of human intelligence and may render anachronistic the binary splits we pursue between "men" and machines.

Historical frameworks are useful foundations for scholarship on cyberspace; one must be careful not to assume that a medium will enable individuals to transcend structures of embodied power, even if it possesses the capacity to enable them to do so. As the authors of the anthology *Race in Cyberspace* argue, the Internet has frequently been constructed as something apart from the real world when in fact "when it comes to questions of power, politics, and structural relations, cyberspace is as real as it gets" (4). As a constructed environment, the Internet cannot work outside of social and cultural systems of power. Within the environment, individuals can confuse, question, and subvert such systems; they can also reproduce them (see, for example, the use of Internet blogs in Kathleen LeBesco's *Revolting Bodies*). Lisa Nakamura finds in *Cybertypes* that people often create avatars very different from their own physical bodies; the opportunity could enable people to experience the world from a different perspective, thereby increasing cultural knowledge. It can also enable participants to reproduce and embody racial and gender stereotypes.

Two final books suggest fruitful directions American studies scholars might take to interpret the effects of Internet technologies on individuals and the systems they inhabit. Sherry Turkle's *Life on Screen: Identity in the Age of the Internet* uses an ethnographic approach to look at the experience of diverse Internet users. She takes seriously the meanings they make from the networks they participate in and the games they play. And while it is important to ask the critical question of when identity "play" becomes dangerous, her ethnographic research demonstrates that Internet technologies may also enable individuals to develop models of "psychological well-being" that "acknowledge the constructed nature of reality, self, and other" (263).

The risk of these intimate consumptive activities arises when the simulation begins to obscure the real. Vincent Mosco suggests that cyber-technology all too easily lends itself to cyber-mythologies; these then become stories we tell ourselves (much like Marx's "middle landscape") to obscure the actual perils posed by technological systems. In *The Digital Sublime* Mosco uses discourse analysis to reveal the ways in which technological mythologies—including but not limited to cyberspace—obscure what are actually "irreconcilable" conflicts between the ideals of virtue through consumable technologies and the realities of power within the technological structures that produce them. Mosco analyzes mythologies surrounding cyberspace, looking specifically at why people have frequently claimed that they signify the end of history, geography, and politics and how these mythologies affect individual engagements with Internet technology.

The Digital Sublime ultimately suggests that while the "digital revolution" has, on a personal level, improved access to information and communication networks, it has also, on a systemic level collapsed jobs and increased individual labor (television reporters are now also editors), enabled the proliferation of precision marketing (our purchases are digitally tracked), and the facilitated state-supported surveillance of private citizens (157–59). This book reminds us of the challenge to American studies scholars who work on the consumer side of technological meanings: history suggests that rhetorics of consumer pleasure may obscure more than they illuminate.

Technological Futures?

Scholarship on systems and the body helps us better understand where we have been with technology. But where do we go from here? Or, as Marita Sturken and her coeditors in *Technological Visions* have framed the question: "Is it possible to think about technologies outside of the frameworks they create?"

Several books in the field suggest that the answer is yes. These can be divided into three categories: those that question the ideology of technological progress by showing us that its effects are often anything but; those that demonstrate how individuals can take a "master's tools" approach to technology, using it to subvert the systems it was invented to support; and those that truly move entirely outside traditional frameworks of "technological progress" to hint at the full range of (yet unexplored) meanings that individuals and subcultural groups give to innovation in their everyday lives.

Edward Tenner's *Why Things Bite Back* and Ruth Schwartz Cowan's *More Work for Mother* look to history to demonstrate that technologies, on their own, frequently fail to produce the benefits that their creators and promoters anticipate. Cowan's work is the better known within American studies; her research into the history of housework and technological innovation argues that "labor saving" innovations such as stoves and vacuum cleaners frequently end up increasing the total labor spent on housework, precisely the opposite effect of their stated intent. Tenner calls these unintended consequences "revenge effects," and provides useful examples in objects as diverse as turnstiles (designed to facilitate orderly entrance and exits but also responsible for stampede deaths by preventing quick exists in emergencies), carp (intended to increase fish supplies in overharvested waters but also responsible for killing off their competitors), and climbing shoes (designed to improve climbing safety but also accountable for putting insufficiently trained climbers on the rock, thereby increasing injuries).

Both texts should be considered in conversation with Winner's *Autonomous Technology* (which opened this essay). Seen together, these texts reveal the historical cycle whereby technologies emerge as immediate, dramatic enablers of consumer ease only, over time, to create pernicious new problems. Written in the late 1970s, Winner's book was influential in raising the question of whether elaborate technological systems can, in fact, be controlled. By juxtaposing repeated claims by "technocratic" rulers that technologies could be developed to create rational economic and military systems with the repeated failures of those systems, Winner successfully characterizes American technologies as "tools without handles" (29).

That these three books could make such similar arguments over thirty years suggests that Americans are not easily deterred from pursuing technological solutions to complex social and political problems. The pursuit creates a cycle of "technological fixes," a term defined in Lisa Rosner's edited anthology *The Technological Fix* as a "cheap fix using an inappropriate technology that creates more problems than it solves" (1). And while it is beyond the scope of this essay to define what is and what is not an "appropriate technology," the concept does allow us to see a pattern. Much like the perpetuators of Mosco's *Digital Sublime*, we come to realize the limitations of particular technological innovations only as we generate new mythologies of "limitlessness" for the next system to emerge. *The Technological Fix* analyzes studies of proposed technological solutions, concluding that only under very specific conditions (when the problem is carefully defined, when individuals working on it understand its technical and cultural context, and when they maintain a willingness to find nontechnical solutions) can technologies actually solve problems created by their predecessors.

Ron Eglash et al.'s edited volume *Appropriating Technology* is one of several texts to look to subcultures for alternative definitions of technological progress and visions for technological futures. In his introduction Eglash argues that individuals are rarely "passive recipients of technological products and scientific knowledge." This analysis, while not new, is in this context important: Eglash and his contributors look for ways in which individuals use technological systems to facilitate "critique, resistance, and outright revolt" (vii). Particularly useful is their attention to technological life cycles. Technologies that originally appear to benefit corporate and political forces can, over time, enable individuals to challenge those very institutions. Environmental activists, for example, have successfully used scientific tools to measure and control atmospheric toxins, demonstrating that problem with binaries wherein technologies either "cause" environmental pollution or offer quick fixes to ameliorate environmental problems. Additionally, body builders frequently use their own self-generated medical knowledge to challenge conventional and legal definitions for the appropriate use of muscle enhancement technologies such as synthetic steroids. Together the essays reveal that while technologies are not neutral, they are malleable. Individuals frequently use technological tools to produce effects and epistemologies directly in conflict with the systems that created the tools to begin with.

The Network Society: A Cross-Cultural Perspective is Manuel Castells's edited volume that follows his 1996 multivolume series, *The Rise of the Network Society*. It extends and diversifies Castells's line of inquiry into what he terms a "fundamental dilemma of our world": the fact that we live in a system wherein we are dominated by networks of global power that are frequently used for capital aggregation and social control, and yet these same networks can enable the formation of alternative networks that facilitate cultural interaction and are unified by what he terms "a common belief in the use value of sharing" (43). The authors in the volume are not American studies scholars. Yet, their work demonstrates the usefulness of looking to transnational and comparative national studies, approaches currently prioritized in the field, to discern technology's cultures and effects. Castells and Pekka Himanen contrast the American model of technological networks, as modeled by Silicon Valley, with state-funded democratic networks of funding used in Finland, revealing that both produce profit and encourage creative freedom among technological entrepreneurs. Jeffrey Juris studies the use of Internet networks among global justice activists, revealing that technologies of control contain within them tools of cohesion and engagement across distance that can themselves force dramatic remakings of the systems themselves. Many authors envision technology networks ultimately moving power away from centers and to the edges, to "nodes of coordination" that cross borders of nation and power. Ultimately these authors demonstrate the possibility that emerging, flexible technological *networks* can, if put in the hands of subversives, successfully challenge the power structures put in place through previous technological *systems*.

Two last texts suggest that American studies scholars may have a unique contribution to make in redefining technological "progress." We can, and should, consider the myriad ways in which individuals use "systems" technologies to enhance self-expression and diversity. Music downloads, file sharing, cell-phone-enabled protests, and local cable productions are all evidence of the vital ways in which individuals can work within technological systems to question their very legitimacy. But technological systems are not the only, or perhaps even most important, source for technological meaning making in American life. As the ample work on body scholarship reveals, people desire innovations in order to change themselves—frequently in pursuit of pleasure. Technology may then function "on the ground" in ways quite different—and even antithetical—to the systems and products scholars have tended to focus on in technology studies.

Ben Chappell's essay in *Technicolor* finds Latino members of Texas low-rider clubs honing advanced technological skills to build distinctive cars. Technology, within this subculture, works best when it enables self-expression and diversity. At the same time, it achieves through its "low and slow" goals (to modify cars so that they ride as low as possible to the ground and to drive them slowly to avoid damage and to showcase their looks and sounds), a definition of technological progress fundamentally different from that of automotive engineers. Joel Dinerstein's *Swinging the Machine* argues that Americans participated in technological systems only because they found ways to "swing the human machine." By taking as a starting point a position that could be cultivated only within American studies (that African drumming practices via big-band swing mediated the assembly line through the Lindy Hop), Dinerstein is able to argue for the importance of subcultural practices within technological studies. His is an analysis in which technological systems succeed because they produce goods: they enable the delivery of products more rapidly and efficiently from one place to another. But they also succeed because of a parallel world of meaning making in which those same technologies are raw materials to produce people. By stylizing and reproducing machine rhythms in the 1930s, big-band swing enabled listeners to feel the human body challenge—and transcend—the sensation of industrial production. The exchange, Dinerstein argues, enabled machines to be viewed as agents of progress in spite of the other more apparent perils they posed to human agency and creativity. Much like Nye's examination of electrification, but with an emphasis on nonwhite, "lay" contributions to technological knowledge, Dinerstein's book ultimately demonstrates that the success of technological systems depends on our ability to alter those systems to increase our "somebodyness." Together Dinerstein and Chappell suggest that by asking questions on our terms as scholars of culture we discover ways in which technology enables diverse expression and enhances human experiences—for bodies *within* their systems. A more accurate understanding of our culture of technology consists not merely of posing challenges to the notion of "progress," but in illuminating the fact that individuals have long found the best products of technological systems the experiences and expressions that exist apart from, and often in opposition to, the "work" of the production floor.

Coda

> I love light . . . Light confirms my reality, gives birth to my form.
>
> —Ralph Ellison, *Invisible Man*

At the beginning—or what is actually the end—of Ralph Ellison's *Invisible Man*, the narrator discloses that he possesses 1,369 filament light bulbs. Into these bulbs flows electric power he has stolen from Monopolated Light & Power. Out of it flows the energy that warms the underground space and the sound that plays its mood: Louis Armstrong's "Black and Blue." Ellison describes the space as one of "vital aliveness" for the narrator. Yet it is a site of odd contradictions. On the one hand, the narrator uses his light to set himself apart and resist white society, making visible the blackness it denies and nurturing fantasies of revenge against its inhabitants. On the other hand, the narrator uses the light—specifically his mastery of the technical knowledge needed to create it—to demonstrate his connection to white society. "Though invisible," he explains. "I am in the great American tradition of tinkers. That makes me kin to Ford, Edison, and Franklin" (7).

Ellison's "invisible man," who uses technology to both sever and solidify his relationship with society, demonstrates why technology issues are at the heart of our field. Technologies as individual objects and personal engagements facilitate individual expression and social cohesion. And they direct the flows of labor, capital, and politics that channel and constrict those expressions and engagements. They are simultaneously sites of production and consumption whose meanings are multilayered and often contradictory. When Ellison's protagonist flips on the light and turns on the record player, he manipulates technology for personal pleasure. He steals from the power company, generating a warmth denied to him aboveground and a soundscape of his own design. At the same time, that consumption renders him a unit on the grid, a participant in systems of power that hinder his experience of pleasure. And ultimately, he interprets his participation in this exchange as enhancing his own power through the cultural myth of heroic "tinkering." In so doing, he fails to see his reliance on the grid as an act that may ultimately reinforce his own invisibility.

For more than fifty years scholars have created a body of work in "technology studies" that suggests we are not going to locate ourselves on that grid by

GPS alone. It requires the "connecting minds" that Gene Wise once saw as unique to American studies scholars: individuals trained to see the historical particularities of "normal" and to find the reinforcement of and the resistance to power in the actions and objects of everyday life. Considering these books, and placing them within the canon of the field, is a good start if we hope to better understand how technological assumptions guide the production of all knowledge; enable assessments of desirable bodies, peoples, and nations; and cultivate particular manifestations of power. Changing those assumptions, however, will require new books—those that can transcend the linear narratives of technological history and master the steps of our long-running dance with the machine. As the best scholarship here reveals, American studies scholars excel at making technology relevant to the practice of everyday life. The challenge to create a "context of care" is ours.

I'd like to thank those scholars who generously shared their technology studies reading lists with me. I am also particularly grateful to Joel Dinerstein, Christina Cogdell, and Jeffrey Meikle for the critical readings they gave to the various versions of this essay.

Bibliography

Adas, Michael. *Dominance by Design: Technological Imperatives and America's Civilizing Mission.* Cambridge, Mass.: Harvard University Press, 2006.

———. *Machines as the Measure of Men: Science, Technology, and Ideologies of Western Dominance.* Cornell Studies in Comparative History. Ithaca: Cornell University Press, 1989.

Banta, Martha. *Taylored Lives: Narrative Productions in the Age of Taylor, Veblen, and Ford.* Chicago: University of Chicago Press, 1993.

Bijker, Wiebe. *Of Bicycles, Bakelites, and Bulbs: Toward a Theory of Sociotechnical Change.* Cambridge, Mass: MIT Press, 1995.

Cartwright, Lisa. *Screening the Body: Tracing Medicine's Visual Culture.* Minneapolis: University of Minnesota Press, 1995.

Castells, Manuel. *The Rise of the Network Society.* Information Age, vol. 1. 2nd ed. Malden, Mass.: Blackwell, 2000.

Castells, Manuel, ed. *The Network Society: A Cross-Cultural Perspective.* Northampton, Mass.: Edward Elgar, 2004.

Chinn, Sarah E. *Technology and the Logic of American Racism: A Cultural History of the Body as Evidence.* Critical Research in Material Culture. New York: Continuum, 2000.

Cogdell, Christina. *Eugenic Design: Streamlining America in the 1930s.* Philadelphia: University of Pennsylvania Press, 2004.

Cooper, Gail. *Air-Conditioning America: Engineers and the Controlled Environment, 1900–1960.* Baltimore: Johns Hopkins University Press, 1998.

Corn, Joseph, ed. *Imagining Tomorrow: History, Technology, and the American Future.* Cambridge, Mass.: MIT Press, 1986.

Cowan, Ruth Schwartz. *More Work for Mother: The Ironies of Household Technology from the Open Hearth to the Microwave.* New York: Basic Books, 1983.

Dinerstein, Joel. *Swinging the Machine: Modernity, Technology, and African American Culture between the World Wars*. Amherst: University of Massachusetts Press, 2003.
Dumit, Joseph. *Picturing Personhood: Brain Scans and Biomedical Identity*. In-Formation Series. Princeton, N.J.: Princeton University Press, 2004.
Eglash, Ron, Jennifer L. Croissant, Giovanna Di Chiro, and Rayvon Fouché, eds. *Appropriating Technology: Vernacular Science and Social Power*. Minneapolis: University of Minnesota Press, 2004.
Elliott, Carl. *Better than Well: American Medicine Meets the American Dream*. New York: W. W. Norton, 2003.
Ellison, Ralph. *Invisible Man*. 1952 New York: Vintage Books, 1990.
Fischer, Claude S. *America Calling: A Social History of the Telephone to 1940*. Berkeley: University of California Press, 1992.
Fouché, Rayvon. *Black Inventors in the Age of Segregation: Granville T. Woods, Lewis H. Latimer, and Shelby J. Davidson*. Johns Hopkins Studies in the History of Technology. Baltimore: Johns Hopkins University Press, 2003.
Giedion, Sigfried. *Mechanization Takes Command: A Contribution to Anonymous History*. New York: Norton, 1969.
Green, Venus. *Race on the Line: Gender, Labor, and Technology in the Bell System, 1880–1980*. Durham, N.C.: Duke University Press, 2001.
Haiken, Elizabeth. *Venus Envy: A History of Cosmetic Surgery*. Baltimore: Johns Hopkins University Press, 1997.
Haraway, Donna. *Primate Visions: Gender, Race, and Nature in the World of Modern Science*. New York: Routledge, 1989.
Hayles, N. Katherine. *How We Became Posthuman: Virtual Bodies in Cybernetics, Literature, and Informatics*. Chicago: University of Chicago Press, 1999.
Hughes, Thomas Parke. *American Genesis: A Century of Invention and Technological Enthusiasm, 1870–1970*. 1989; Chicago: University of Chicago Press, 2004.
Hughes, Thomas, Wiebe Bijker, and Trevor Pinch, eds. *The Social Construction of Technological Systems: New Directions in the Sociology and History of Technology*. Cambridge, Mass: MIT Press, 1987.
Jenkins, Virginia Scott. *The Lawn: A History of an American Obsession*. Washington, D.C.: Smithsonian Institution Press, 1994.
Kasson, John. *Civilizing the Machine: Technology and Republican Values in America, 1776–1900*. New York: Viking, 1976.
Kolko, Beth E., Lisa Nakamura, and Gilbert B. Rodman, eds. *Race in Cyberspace*. New York: Routledge, 2000.
Kouwenhoven, John Atlee. *Made in America: The Arts in Modern Civilization*. 1948; New York: Octagon Books, 1975.
Kramer, Peter D. *Listening to Prozac*. New York: Penguin Books, 1997.
Latour, Bruno, and Steve Woolgar. *Laboratory Life: The Social Construction of Scientific Facts*. Beverly Hills: Sage Publications, 1979.
LeBesco, Kathleen. *Revolting Bodies? The Struggle to Redefine Fat Identity*. Amherst, Mass.: University of Massachusetts Press, 2004.
Lehrman, Nina, Ruth Oldenziel, and Arwen Mohun, eds. *Gender and Technology: A Reader*. Baltimore: Johns Hopkins University Press, 2003.
Marvin, Carolyn. *When Old Technologies Were New: Thinking about Electric Communication in the Late Nineteenth Century*. New York: Oxford University Press, 1988.
Marx, Leo. *The Machine in the Garden: Technology and the Pastoral Ideal in America*. New York: Oxford University Press, 1964.
Meikle, Jeffrey L. *American Plastic: A Cultural History*. New Brunswick, N.J.: Rutgers University Press, 1995.
Mumford, Lewis. *Technics and Civilization*. New York: Harcourt, Brace, 1934.
Nakamura, Lisa. *Cybertypes: Race, Ethnicity, and Identity on the Internet*. New York: Routledge, 2002.
Nelson, Alondra, and Thuy Lin N. Tu, eds., with Alicia Headlam Hines. *Technicolor: Race, Technology, and Everyday Life*. New York: New York University Press, 2001.
Nye, David E. *America as Second Creation: Technology and Narratives of New Beginnings*. Cambridge: MIT Press, 2003.

———. *American Technological Sublime.* Cambridge, Mass.: MIT Press, 1994.
———. *Electrifying America : Social Meanings of a New Technology, 1880–1940.* Cambridge, Mass.: MIT Press, 1990.
Oldenziel, Ruth. *Making Technology Masculine: Men, Women, and Modern Machines in America, 1870–1945.* Amsterdam: Amsterdam University Press, 1999.
Peiss, Kathy. *Hope in a Jar: The Making of America's Beauty Culture.* New York: Metropolitan Books, 1998.
Peña, Carolyn Thomas de la. *The Body Electric : How Strange Machines Built the Modern American.* American History and Culture. New York: New York University Press, 2003.
Prown, Jules David, and Kenneth Haltman, eds. *American Artifacts : Essays in Material Culture.* East Lansing: Michigan State University Press, 2000.
Rabinbach, Anson. *The Human Motor: Energy, Fatigue, and the Origins of Modernity.* New York: Basic Books, 1990.
Riordan, Teresa. *Inventing Beauty: A History of the Innovations That Have Made Us Beautiful.* New York: Broadway Books, 2004.
Rooks, Noliwe M. *Hair Raising Beauty, Culture, and African American Women.* New Brunswick, N.J.: Rutgers University Press, 1996.
Rosner, Lisa. *The Technological Fix: How People Use Technology to Create and Solve Problems.* Hagley Perspectives on Business and Culture. New York: Routledge, 2004.
Serlin, David Harley. *Replaceable You: Engineering the Body in Postwar America.* Chicago: University of Chicago Press, 2004.
Sinclair, Bruce, ed. *Technology and the African-American Experience: Needs and Opportunities for Study.* Cambridge, Mass.: MIT Press, 2004.
Smith, Terry. *Making the Modern: Industry, Art, and Design in America.* Chicago: University of Chicago Press, 1993.
Sturken, Marita, Douglas Thomas, and Sandra J. Ball-Rokeach, eds. *Technological Visions: The Hopes and Fears That Shape New Technologies.* Philadelphia: Temple University Press, 2004.
Tenner, Edward. *Our Own Device: The Past and Future of Body Technology.* New York: Alfred A. Knopf, 2003.
———. *Why Things Bite Back: Technology and the Revenge of Unintended Consequences.* New York: Knopf, 1996.
Terry, Jennifer, and Melodie Calverts. *Processed Lives: Gender and Technology in Everyday Life.* New York: Routledge, 1997.
Thompson, Emily. *The Soundscape of Modernity: Architecture and the Culture of Listening in America, 1900–1933.* Cambridge, Mass.: MIT, 2002.
Tichi, Cecelia. *Shifting Gears: Technology, Literature, Culture in Modernist America.* Chapel Hill: University of North Carolina Press, 1987.
Traweek, Sharon. *Beamtimes and Lifetimes: The World of High Energy Physicists.* Cambridge, Mass.: Harvard University Press, 1988.
Turkle, Sherry. *Life on the Screen: Identity in the Age of the Internet.* New York: Touchstone, 1997.
Vaidhyanathan, Siva. *Copyrights and Copywrongs: The Rise of Intellectual Property and How It Threatens Creativity.* New York: New York University Press, 2001.
Wailoo, Keith. *Drawing Blood: Technology and Disease Identity in Twentieth-Century America.* The Henry E. Sigerist Series in the History of Medicine. Baltimore: Johns Hopkins University Press, 1997.
Winner, Langdon. *Autonomous Technology: Technics-out-of-Control as a Theme in Political Thought.* Cambridge, Mass.: MIT Press, 1977.
———. *The Whale and the Reactor: A Search for Limits in an Age of High Technology.* Chicago: University of Chicago Press, 1986.
Wood, Gaby. *Edison's Eve: A Magical History of the Quest for Mechanical Life.* New York: Anchor Books, 2003.

Digital Junction

Debra DeRuyver and Jennifer Evans

Introduction

Ten years ago, Randy Bass, project director for the American Studies Crossroads Project (*www.georgetown.edu/crossroads/*) asked, "What will we be looking at when the World Wide Web is invisible?"—invisibility referring to the way that a technology has of disappearing once it becomes a ubiquitous part of our lives.[1] He predicted in his essay that by 2006 the Web would, in fact, be invisible and that "'primary materials' will comprise a significant answer to the question." While the digitization of primary sources has largely been under the purview of commercial enterprises or our neighbors in the information sciences, American studies professionals and our digital humanities compatriots have been actively involved in placing some primary sources online. Carl Smith, for example, professor of English and American studies at Northwestern University teamed up with the Chicago Historical Society and Northwestern's Academic Technologies to digitize sources and create (in 1996) The Great Chicago Fire and The Web of Memory (*www.chicagohs.org/fire/index.html*), one of the best of the early online exhibitions/essays/archives. Graduate students, perhaps more familiar with the new technology, have also been involved in memorable projects. See, for example, Kelly Quinn's Learning from Langston Terrace (*www.wam.umd.edu/~kaq/langston.html*), which contains digitized primary sources and some historical background on this 1930s racially segregated New Deal housing project. But, as both of the aforementioned projects illustrate, putting primary sources online is not enough. As Bass continues in his answer, "the real power of these materials will not come from sheer access to primary resources, but the connections that can be made across them and the visibility of the process of the work being done on them."

While much has been written on teaching with technology and the need to instruct students in the use of online archives and primary source materials,[2] the "connections that can be made across" primary sources and the "visibility of the process of the work being done on them" in our own scholarship has not kept up with the pace at which primary sources are being digitized and made available. One of the reasons for this, as Jerome McGann points out, is that "we're illiterate."[3] As scholars, most of us do not know how to use "any of the languages we need to understand how to operate with our proliferating digital technologies—not even elementary markup languages." Furthermore, he points out, our illiteracy places us on the margins of discussions and decisions that are being made every day with regard to the digital transformation of our cultural heritage—the primary sources that are of vital interest to us as scholars and educators.

Some Web sites we traversed for this article were designed, or utilized particular technologies, to help foster connections. With great difficulty, for example, one can overlay a printed 1907 Sanborn Fire Insurance map and current street map to see the changes, but in digital form this is easily accomplished. Without expensive audiovisual equipment and prohibitive technical skills, one cannot create one's own "folk songs" by remixing and overlaying a range of sounds taken from New York's Lower East Side, but on Folk Songs for the Five Points (*www.tenement.org/folksongs/*), one can easily experiment and, in so doing, not only create something new but also develop a greater understanding, appreciation, and connection to the material through the process.

But, we were disappointed to realize that for the vast majority of online primary sources, the information professionals involved in the project have not made the simplest of connections—a hyperlink between two Web pages—available to their users. For example, we ran across a photographer whose photographs were held in a collection in Louisiana and at the Library of Congress (LOC). The Louisiana site mentioned that the bulk of the photographer's work was at LOC, but did not provide a link to the scanned images LOC had online, while LOC did not mention at all that this photographer's work was also held by other institutions. This kind of hypertextual cross-connecting is supposed to be one of the great features of the Web, but without a concerted effort and a focus on something other than simply making one's own materials accessible, even the simplest of connections will not be made.

While sites that promote internal connections via technology or the mixing of primary sources with exhibition and interpretation are steps in the right direction, connections on the order imagined by Bass have, in large part, remained elusive over the past ten years. As the Web becomes as invisible as our television sets, it becomes increasingly difficult to imagine it presenting different configurations of primary sources and scholarship. Overcoming our technical illiteracy would help us make the Web visible again and give us the vocabulary to develop or participate in new, connected, scholarly environments, such as the Networked Interface for Nineteenth-Century Electronic Scholarship (*www.nines.org/*) currently under development by McGann and others. Bass, himself, is hoping to move the American Studies Crossroads Project in this direction, pursuing research and development along the lines of online exhibitions, knowledge sharing, and other special projects of interest to the American studies community.[4]

But perhaps the first step toward this new connectivity is to raise our awareness of what the digital landscape of online primary sources looks like today. In 1996, the same year that Bass posed his question, Patricia Limerick, in her presidential address to the American Studies Association, humorously related her feelings of being overwhelmed by the amount of reading she faced: "At any given moment, the scoreboard is several thousand to one: thousands of things I should be reading, and would *profit* from reading, and only one of me." She went on to ask those in American studies involved in designing electronic projects to keep her dilemma in mind and to remember that "while easier access to information is fine, we are in much greater need of methods and strategies for filtering, sorting, managing, synthesizing."[5] While there have been some notable efforts within American studies, ten years later her plea still carries weight.[6] The amount and range of information available online for teaching and researching American studies topics can be overwhelming. It is practically impossible to keep up with the multitude of notices about new archives, let alone existing collections that are being expanded regularly. So, where does one start?

This essay addresses that question, discussing strategies for locating online primary sources, specifying a selection of Web sites as example starting places, and pointing out some of the particulars to consider when utilizing online resources. The greater part of the article delves into six types of primary sources available online: government documents, born digital materials, maps, oral histories, graffiti, and historic newspapers. For some of the materials that we chose not to look at in this article, including advertisements, literature, built environment, ephemera, personal papers, and photographs, we have included a list of helpful links at the conclusion of the article.

There are similarities between researching and teaching with print materials and doing so with digital collections. Regardless of format, it takes time to locate primary sources, and while search engines, like card catalogs, are useful, they cannot totally replace a well-informed librarian or the human thought behind a portal site. It is therefore useful to begin one's search for online primary sources by paying a visit to one's friendly and knowledgeable university librarian. As professionals in the information science field, librarians and archivists are ahead of the curve when it comes to knowing where to find appropriate digital resources. They can point one directly to materials of interest or to topical Webliographies, multitopic online "portals," and print resources that contain lists of links (sometimes annotated), reviews of Web sites, and digital collections.[7]

Some multitopic portals we like to use include Librarian's Internet Index (*lii.org*), a searchable, annotated list of links selected, described, and organized by a team of librarians; History Matters (*www.historymatters.gmu.edu/*), a joint project of the Center for History and New Media at George Mason University and CUNY's American Social History Project, which provides a bounty of information for teaching the U.S. History Survey course, including, among others items, primary sources, how-to guides, reprinted Web reviews from the *Journal of American History*, and searchable annotated links to more than 850 history Web sites; Best of History Web Sites (*www.besthistorysites.net/*), a searchable list of more than 1,000 annotated history sites and sites containing lesson plans, particularly plans aimed at grades K–12; Voice of the Shuttle (*vos.ucsb.edu/index.asp*), a searchable list of humanities and humanities related resources; Digital History (*www.digitalhistory.uh.edu/*), a multipurpose site that includes a hypertext online American history textbook, multimedia primary sources, exhibits, and lists of links to external history resources; and the WWW Virtual Library (*www.vlib.org/*), a searchable list of links that is the oldest Web site catalog in existence, started in 1991 by Tim Berners-Lee, creator of the World Wide Web. H-Net: Humanities and Social Sciences Online *(www.h-net.org/*), a well-known site of topical Listservs, can also be used to locate sources mentioned in discussion logs, online reviews, and resource lists maintained by Listserv moderators.

After portals, the next place to visit in one's online pursuit of primary sources might be a favorite search engine. But, while we use them to find just about everything else we need online, search engines are not very useful in directly locating primary sources. Most digitized primary sources are stored in large databases accessible only through user input into a site-specific search form that then initiates a specific online query into the database in which it resides. Most digitized primary sources are thus part of what's termed the Deep Web. While some engines search some databases (phone directories are a popular example) in the Deep Web, it is estimated that the unsearchable portion of the Deep Web contains more than 500 times the information of the searchable Web. Search engines can be useful, but typically only to find pages that then allow one to query a database for primary sources—not to find primary sources themselves. Search engines, therefore, work best for locating sites with narrow collections; sites with broad or multiple collections would need to have extensive descriptive information of their complete contents online in order for their pages to come up in a generic search of a topic by a search engine. One strategy for overcoming this problem on sites with online collections is to create links to index pages that automatically query and pull descriptive information out of their databases. By doing this, search engine spiders can crawl the information and include a link back to the site within search engine results. Unfortunately, not many sites include such index pages.

Finally, multicollection sites are important stops on one's search. The best-known multicollection site is American Memory at the Library of Congress (LOC) site (*www.memory.loc.gov/ammem/index.html*), which contains more than 9 million digitized multimedia items across more than a hundred collections, which are searchable across collections. The site also has numerous interpretive essays and lesson plans. It is worth noting that twenty-three of these collections do not physically reside at LOC. Their collecting institutions won grants from LOC to digitize and to share the collections online. The National Archives and Records Administration (NARA) site (*www.archives.gov/research/tools/checklist.html*) is another well-known site containing online access to 50 million electronic records through their Archival Access Database as well as 124 thousand multimedia items through their Archival Research Catalog.

While not nearly as large as LOC or NARA, additional multicollection sites of interest to those in American studies include large public libraries such as New York Public Library Digital (*www.nypl.org/digital/index.htm*), which contains a searchable digital gallery of 450,000 items from the library's collections as well as other searchable collections and online exhibits; consortiums of institutions such as the Online Archive of California (*www.oac.cdlib.org/*), which contains 170,000 items from several collections and 8,000 finding aids from more than 150 institutions in California; and university research libraries such as Documenting the American South (*www.docsouth.unc.edu/index.html*), from the University of North Carolina Library, which contains materials (primarily text, but also oral histories and photographs) from ten thematic collections and Duke University Libraries Digitized Collections (*www.library.duke.edu/specialcollections/collections/digitized/index.html*), which contains twelve unique collections, several of which specialize in American advertising. It is worth noting that there are also subscription-based multicollection sites such as RLG's Cultural Materials site (www.*culturalmaterials.rlg.org*), which contains more than 246,000 digitized items.

This may seem like a lot of items, but these sites represent only a fraction of the physical materials the institutions hold, as they themselves usually point out. For that reason, unless your research is oriented around a particular online collection or your finds are incredibly serendipitous, a bricks-and-mortar archive may still be the best option. Still, even if one eventually ends up in an archive, much legwork can be accomplished online, and while armchair research may be millions of dollars away, the plethora of online materials is a bounty for those who wish to integrate more primary sources into their teaching or utilize modern online exhibits and hypertextual interpretation instead of traditional textbooks. While there are some notable fee-based services that provide these possibilities to teachers and students, comparable free sources can usually be found with a little looking.[8]

Once one has located some promising online resources for either teaching or research, one needs to consider several important variables. First and foremost is the authority and authenticity of the collection and the provider. How to employ specific critical thinking skills and questions to assess the value of online material is a topic that has been much discussed. At the Public History Resource Center (*www.publichistory.org/reviews/rating_system.html*), we evaluate online history sites from two perspectives. First we employ general criteria, examining a site, for example, for its timeliness and permanence, authority and bias. Second, we utilize history-specific criteria to evaluate a site from the perspective of its interpretation of materials, primary source documents, educational items, and promotion of a community of interest. To the best of our knowledge, we created the first (and still the only) history-specific rating system for evaluating Web resources. While we have not rated the Web resources discussed in this article, we have indicated in the list of links at the end whether or not they contain a substantial amount of (ⓘ) interpretation, (Ⓟ) primary sources, (📚) educational items, and (🌎) promotion of a community of interest.

While the Public History Resource Center (and most portals for that matter) parses Web sites by topic or theme, this article takes a different approach. By arranging the sites below by format of materials, we intend to draw attention to materials that may not be used as frequently as others in teaching and research and to the unique history governing the online availability of a particular type of material. After culling hundreds of sites, we chose these to illustrate the spectrum of sites available within each type. Some are well known, while others are not. In some cases, a site was chosen because it employed a noteworthy piece of Internet technology, or it may have illustrated a particular point.

Although both the quality and quantity of Web resources have greatly increased since 1999, when we first started writing Web reviews at the Public History Resource Center, there is still much to be accomplished, particularly in terms of making connections both within and between sites, sources, and scholarship. In the past ten years there has been a great deal of pressure on institutions to simply get their materials up and online without regard to placing their collections or exhibits in the context of the rest of the digiverse. Will the time come when we'll see the creation of one search page (archive.google anyone?) that will search across collections in the Deep Web, allowing users to find, in one search, all of the available digitized resources on a topic? Will we see the development of interactive digital scholarly environments (American studies Wiki?) conducive to the kinds of associations and interpretive connections we've come to expect from practitioners of American studies? Let's see what the next ten years has in store.

Maps

Due, in part, to their often-fragile physical condition and/or unwieldy size, maps are regularly overlooked as important primary sources and teaching tools. Institutions, however, have not only heightened the visibility of map collections by providing online access to catalogs, bibliographies, and collection samples but also have increased accessibility by posting high-resolution scans that can be used without having to make a trip to the repository. While, admittedly, navigating a large map on a small computer screen can be challenging, technology, as the following examples show, can mitigate these difficulties. We were surprised at how elegant some of the online presentations are and how fascinating much of the online map material is. The teaching and research possibilities are endless, and what is most exciting are the applications that exploit the strength of the digital to create opportunities that are impossible with a printed map, such as overlaying current street maps with historical maps, viewing maps from different institutions side by side, and creating custom maps on demand.

There are thousands of map sites on the Web containing maps of all sizes, quality, subjects, and time periods. To help people navigate the myriad options, Tony Campbell, a retired map librarian at the British Library, compiled and regularly updates one of the most comprehensive portal sites for finding map-related information. Map History/History of Cartography: The Gateway to the Subject (*www.maphistory.info*) provides, in one of its many sections, annotated links to digital map collections. Also, to aid educators and students, in particular, David Stephens, professor of geography at Youngstown State University, wrote "Making Sense of Maps" (*historymatters.gmu.edu/mse/maps/*) for History Matters' Making Sense of Documents series. Stephens "offers an overview of the history of maps and how historians use them, a breakdown of the elements of a map, tips on what questions to ask when analyzing maps, an annotated bibliography, and a guide to finding and using maps online."

With 4.5 million items, the Library of Congress has one of the largest collections of maps in the world. There are ten map-specific collections in LOC's American Memory (*memory.loc.gov/ammem/browse/ListSome.php?category=Maps*), gathered under such headings as Civil War, National Parks, Railroads, and World War II, some of which are accompanied by contextual essays. In addition to critical thinking exercises in the Collection Connections (*memory.loc.gov/ammem/ndlpedu/collections/map/*) section, we stumbled onto a special presentation, Zoom into Maps (*lcweb2.loc.gov/ammem/ndlpedu/features/maps/introduction.html*), which is designed specifically as an introduction to the library's map holdings and to the possibilities cartography brings to the study of history, geography, genealogy, and social studies. Zoom into Maps is divided into nine sections: Hometown, USA: Local Geography; Exploration and Discovery; Migration and Settlement; Travel and Transportation; Environmental History; Military Maps; Pictorial Maps; Maps of Today; and Unusual Maps. Each section highlights one feature map, poses questions about the map, and links to related maps, other special presentations, collections, or other repositories.

Structured similarly to Zoom into Maps, the Newberry Library's Hermon Dunlap Smith Center for the History of Cartography has brought together high-quality images of historic maps, lesson plans, and supplemental materials in its site Historic Maps for the K–12 Classroom (*www3.newberry.org/k12maps/*). The eighteen core maps are arranged under six themes, such as "geography of American communities" and "historical geography of transportation." For each of the themes, there are lesson plans for four grade levels: K–2, 3–5, 6–8, and 9–12. These plans may be too simplistic for undergraduate or graduate courses, but with the inclusion of supplementary materials—additional maps, documents, background information, and bibliographies—the site not only provides a model for other sites but also encourages multiple uses for people at all educational levels.

In most cases, digital maps are posted by either the collecting institution as in the examples above or by the creating organization, such as the more than 21,000 nautical charts; hydrographic, topographic, and geodetic surveys; city plans; Civil War battle maps; and landscape perspective sketches of the Office of Coast Survey, National Oceanic and Atmospheric Administration (*nauticalcharts.noaa.gov/csdl/ctp/abstract.htm*). MapTech's Historical Maps (*historical.maptech.com/*) and Cartography Associates' The David Rumsey Collection (*www.davidrumsey.com/index.html*) are two noteworthy exceptions.

In a unique partnership between a for-profit company and a group of volunteers, historic United States Geological Survey (USGS) topographical maps of fourteen New England and Mid-Atlantic states are available on the Web. MapTech provides the Web space, and volunteers scan and process the maps. There are approximately 2,000 maps from the 1890s to the 1950s on the site, showing buildings, roads, railroads, waterways, and elevations. The site is easy to navigate with a quad index, town index, and image map. We did not have any difficulties viewing the one-megabyte files with a DSL connection. We could see the smallest detail by clicking once on the image but could not zoom or pan, and it was somewhat difficult to navigate within the map without a navigator view as some of the other sites use. (Texas Bird's-Eye View from the Amon Carter Museum (*www.birdseyeviews.org/index.php*) offers an excellent example of a Flash-based zoom, pan, and navigation feature, enabling Web visitors to see the tiniest of details from the carriages in the streets to the advertisements on the buildings' sides.)

Because the site includes maps of many of the same areas from different time periods, the maps can be used to explore the development of these cities and towns, as well as urban growth, population shifts, and environmental changes. Similarly, Sanborn Fire Insurance Maps, particularly when used in conjunction with other primary sources such as the census, can convey tremendous amounts of information about, among other things, architecture, socioeconomics, and urban planning. The University of Virginia Library has made the 1907 and 1920 Sanborn maps of Charlottesville available online in a user-friendly format. One can browse original maps, peruse the street and building index, or search by modern streets. Most interestingly, the historic maps have been georeferenced together and overlaid with modern streets and railroad data. The image files are available for downloading in zipped files.

Pushing technology even further, David Rumsey has posted more than 13,600 maps from his private collection, which focuses on rare eighteenth- and nineteenth-century North and South America maps, ranging from antique atlases to maritime charts to manuscript maps, online (*www.davidrumsey.com/index.html*). The site is known not only for its maps but also for the innovative technology designed specifically for the online presentation and study of maps. With the Insight Browser, which does not require any program to be downloaded onto your computer, maps from different time periods can be viewed side by side. With the GIS browser, users can overlay current geospatial data with historical maps. This feature is not to be used on a slow computer or with a dial-up connection. Each change we made to the map took from thirty to ninety seconds to load. The Collections Ticker is a Flash-based thumbnail viewer, similar to a rolling filmstrip that provides yet another way of accessing the collection.

The following three sites challenge the way we think about maps and how they are created. Produced by A9.com, a subsidiary of Amazon.com, BlockView Images (*maps.a9.com/*) allows the user to virtually walk down the streets of twenty-four major cities. Using trucks equipped with digital cameras and GPS receivers, the creators of this site captured scenes along public streets.

A cartogram resizes each country according to some other variable than land area. Some of the other variables available include: population, imports/exports, and transportation methods. In collaboration with the University of Michigan and the University of Sheffield, over a hundred cartograms and their underlying data are currently available for viewing and downloading, and there are plans for adding more in late September 2006.

Produced by the U.S. Department of the Interior, the National Atlas' Map Maker allows users to customize, view, and print out their own maps. The researcher can choose from hundreds of layers of demographic and geographic layers of information in the following categories: agriculture (farmland); biology (invasive species); environment (hazardous waste); geology; cities, towns, and counties; crime; and energy consumption. One can look at, for example, the fetal and infant mortality rate for 1995–1999 or the average wage per job. The site also includes articles that provide additional information about the National Atlas maps, the intersection of geographic information and public policy decisions, government programs, and U.S. history. The raw date from which the map layers are derived is available for download on the site.

Archives of Websites, E-mails, and Other Born Digital Materials

A self-referential issue created by the explosion of e-mails, Web sites, instant messaging, chat rooms, MUDS (multi-user domains), MOOS (MUD object oriented), MUSHes (said to stand for multi-user shared hack, habitat, or hallucination), newsgroups, products of collaboration software, e-literature, banner advertisements, and other "born digital" materials is how and whether these materials will, themselves, be preserved and archived.[9] Electronic Records Management (ERM) is a growing field with implications for business (see *www.thesedonaconference.org*, for example) and government agencies alike (see NARA's Electronic Records Archive, *www.archives.gov/era/*, for example). Looming concerns include legal issues regarding records and evidence preservation, volume of materials, format obsolescence, capturing nonstatic records such as Web sites, and cataloging. For example, due to policy concerns regarding the "capture and transfer of Federal public Web sites at the end of a term of a Presidential Administration" NARA conducted its "2004 Presidential Term Web Harvest" (*www.webharvest.gov/collections/peth04/*), capturing .gov and .mil sites prior to January 20, 2005.

While businesses and government agencies work on creating and implementing standards and guidelines for the management of their materials, a whole generation of personal electronic expression and self-creation is either slipping away to digital dustbins or being haphazardly electronically copied and warehoused. These materials are important not only to those in the future who, for example, wish to better understand a particular figure by analyzing her first blog on MySpace (*www.myspace.com/*), but also to those interested in the creation of cultural memory, explorations of everyday life, and so on.

A giant in the field of warehousing digital material is the Internet Archive (*www.archive.org/index.php*), which, as of June 17, 2006, boasts digital copies of 36,409 movies, 83,052 audio recordings (including 36,325 concerts), and 29,564 texts. Typically these materials have been uploaded by individuals or in partnership with other organizations. For example, in the audio section you'll find a collection of more than 2,000 items from Democracy Now (a daily independent news program) alongside 300 lectures from the Tse Chen Ling Buddhist Center, alongside a collection of 792 78-rpm recordings contributed by individual users. While the selection is impressive, it is the Internet Archive's born digital collection that is its unique contribution to the world of online archives. The Internet Archive boasts a collection of 55 billion Web pages. Type in any URL into its "Way Back Machine" and you'll likely retrieve a list of dates linking to archived versions of the Web site collected on those days. Asterisks designate when the archived site is different from that collected on the previous date. Material is largely contributed by Alexa Internet (alexa.com). One of us was pleased(?) and chagrined(?) to find that her personal home page was archived beginning on April 28, 1999, available for future researchers—to determine what, we don't know.

Most large online collections of archived Web pages in the United States have been collected in collaboration with the Internet Archive. They offer a tool, "Archive-It," on a subscription basis to institutions wishing to preserve Web sites as online collections (*www.archive-it.org*). A list of forty-three available collections through Archive-It is available at *www.archive-it.org/all_collections.php*. While some collections may be of little or no interest to those in American studies (e.g., Tennessee School Report Cards 2005), others, such as the Quaker and Peace Web Archives, Canadian Labour Unions, a series of three collections on Latin American politics including the Latin American

Government Documents Archive (*lanic.utexas.edu/project/archives/lagda/*), and a series of collections on Virginia politics including Archiving the Web: Virginia's Political Landscape, Fall 2005 (*www.lva.lib.va.us/whatwehave/webarchive/index.htm*) are lookworthy.

The Internet Archive has also worked in collaboration with the Library of Congress (LOC) on several of its Web Archive projects. National libraries, like LOC, have traditionally been responsible for gathering and preserving the cultural, intellectual, and creative materials deemed of interest to future generations. To this end, the national libraries of several Western European countries, Australia, and Canada, along with the LOC and the Internet Archive, have formed the International Internet Preservation Consortium (*www.netpreserve.org*). The Minerva Web preservation project (Mapping the INternet Electronic Resources Virtual Archive; *www.loc.gov/minerva/*) is LOC's initiative in this realm.

Minerva currently has commissioned or hosts four Web archives, all of potential interest to American studies scholars and students: The 107th Congress Web Archive contains 588 official Web sites of members of Congress or congressional committees archived once on December 12, 2002; the Election 2002 Web Archive contains 4,000 Web sites that were archived multiple times between July 1 and November 30, 2002. These sites were published by candidates, citizens, advocacy groups, political parties, the press, and so on. Sites were captured on different schedules with candidate sites being captured daily. On election day approximately 1,800 of the sites were captured hourly. The Election 2000 Web Archive contains a collection of nearly 800 sites archived daily between August 1, 2000, and January 21, 2000; and the September 11th Web Archive contains more than 30,000 Web sites archived daily from September 11, 2001, to December 18, 2001 (the original collection dates were through December 1).

The September 11th Web Archive illustrates many of the technical and cataloging issues facing this type of collection. LOC offers two ways into the archive, one cataloged, one not. Uncataloged access involves first clicking to an alphabetical index page of Web site addresses harvested in the archive. From there, click on any address and you are transported to a "Way Back Machine" interface (as described above) embedded in a Minerva wrapper. The archive is physically hosted by LOC. However, when one begins clicking into the actual pages of the archive, the trouble begins. Clicking into archived pages from the first link in the index, *1-800-vermont.com/*, the Web site of the Vermont Department of Tourism and Marketing, one wonders why, exactly, this site was included. It is a government site, but the main point it seems to make about September 11 is that it was business as usual for the promotion of the fall colors tour in Vermont. We randomly viewed site after site and found nothing of direct interest; it was as if September 11 hadn't even happened in this digisphere. When we finally chose a site whose address seemed to promise more directly related material, *11sep2001wherewereyou.org/*, we were pleased to see a small list of links to stories contributed by users. However, pleasure quickly turned to disappointment when each link returned a page saying it was "not in archive." While the technological issues at play in the incorrect display of or missing pages of archived Web sites are the same as those in the Internet Archive as a whole (e.g., the auto-harvesting of Web sites is disrupted by the use of frames, dynamic content, forms, log-ins, executable programs, low priority assigned to gathering images, time-outs, Deep Web content, etc.), to discover them associated with archived pages sponsored by the Library of Congress is a more disappointing experience. Browsing the much smaller collection of 2,300 Web sites cataloged by *webarchivist.org* was more satisfying. The catalog allows one to enter the collection by producer name, type (e.g., government, religious, etc.), country, and language. Randomly poking through the cataloged collection produced far fewer "huh?" moments, and far fewer technological bloopers.

Perhaps not surprisingly, a more captivating 9/11 digital archive is the one created by historians—not librarians or archivists. The September 11th Digital Archive: Saving the Histories of September 11, 2001 (*911digitalarchive.org/*) is a site with which many may already be familiar. Organized by the American Social History Project (*www.ashp.cuny.edu/*) and the Center for History and New Media (CHNM) (*chnm.gmu.edu/*), the site is home to digitized collections of photographs, video recordings, audio recordings, and flyers, among others. Its born digital materials largely consist of a collection of some 2,000 e-mails that people received or sent on or shortly after September 11 and a smaller collection of archived Web Logs from that time period. Although the collection is not as massive as LOC's, this site speaks more clearly to the collective creation of cultural memory(ies) and could easily be used in the classroom, for undergraduate and perhaps some graduate research papers, as well as by the casual surfer, while the LOC collection may provide a more useful sample for conducting pointed research-oriented inquiries into the specific public responses of a particular organizational Web site over time.

In the rush to digitize existing records, let us not neglect the task of collecting and preserving the born digital materials being created now. American studies professionals today have an opportunity to get involved through policy creation, recommendations, and even the creation of their own digital collection, like

CHNM's *911.digitalarchive.org*. Toward that end, CHNM's director of research projects, Daniel Cohen, and center director and founder, Roy Rosenzweig, have recently put out a guidebook, *Digital History: A Guide to Gathering, Preserving, and Presenting the Past on the Web* (University of Pennsylvania Press, 2005). In addition to detailed instructions for creating history Web sites, the book includes a chapter, "Exploring the History Web," which contains a history of the history Web, and a chapter on "Preserving Digital History." The text is freely available online at *chnm.gmu.edu/digitalhistory/*.

Oral Histories

Critics have often seen oral histories as a lesser form of historical evidence with more inherent problems and potential pitfalls than the written record. (See Step-by-Step Guide to Oral History, *www.dohistory.org/on_your_own/toolkit/oralHistory.html*, for a summary of some of these issues.) As with historical records of any type, however, questions of authenticity, motivation, and accuracy must be raised; oral history is no different, and in many cases, may be the only account available of a particular event, experience, or person's life. History Matters' Making Sense of Oral History (*historymatters.gmu.edu/mse/oral/online.html*) provides an excellent essay on interpreting oral history, which also includes an annotated bibliography and list of exemplary oral history sites.

Ranging from well-defined, short-term projects, such as Remembering the 20th Century: An Oral History of Monmouth County (*www.visitmonmouth.com/oralhistory/*) to the ongoing collecting projects of universities and libraries, such as the Black History Project at Virginia Tech (*spec.lib.vt.edu/archives/blackhistory/oralhistory/*), the breadth, depth, and quality of oral history programs varies widely. What these institutions place online also ranges from transcripts in PDF or HTML to audio excerpts to streaming video or some combination of all of the above.

Library of Congress's Veterans History Project (*www.loc.gov/vets/*) is one of the largest oral history projects in the United States and includes not only personal histories (audio- and videotaped interviews, written memoirs) but also related material, such as letters, postcards, v-mail, personal diaries, photographs, drawings, and scrapbooks. The project's primary focus is on firsthand accounts of U.S. veterans from the following twentieth-century wars: World War I, World War II, Korean War, Vietnam War, Persian Gulf War, and Afghanistan and Iraq conflicts. In addition, those U.S. citizen civilians who were actively involved in supporting war efforts (such as war industry workers, USO workers, flight instructors, medical volunteers, etc.) are also invited to share their stories. As of 2004, nearly 25,000 personal recollections had been collected, with more than 2,400 collections digitized. Some of the interviews have transcripts; others do not.

Users can either browse the collection by last name, war and military branch, state of residence, or race/ethnicity, or they can use the straightforward search screen to narrow results. By searching for women in the Army Nurse Corps during World War II, for example, one of us was pleasantly surprised to see that her grandmother's oral history had been donated to the project. The project has also created online exhibits *www.loc.gov/vets/stories/themes.html*, on themes such as courage, buddies, sweethearts, family ties, community, POWS, military medicine, and more. These self-contained exhibits are a good way to begin exploring the site or to use in the classroom. As an example, "African Americans at War: Fighting Two Battles," presents interview clips, newspaper articles, and photographs of eleven highlighted individuals.

The Veterans History Project has many partners, such as the Rutgers Oral History Archive (*oralhistory.rutgers.edu/*), which has online more than 400 oral history transcripts and a collection of digitized and/or transcribed diaries, letters, memoirs, and photographs of the men and women who served on the home front and overseas during World War II, the Korean War, the Vietnam War, and the Cold War. It is unclear what the overlap is between the LOC site and the partner sites. We did not find any of the Rutgers oral histories in the LOC database, so researchers may want to check partner sites (a list can be found at *www.loc.gov/vets/vets-partners.html*), as well as the LOC site. The Veterans History Project also encourages students to conduct interviews and offers suggestions and samples of student work (*www.loc.gov/vets/youth-resources.html*).

An example of a site consisting solely of oral histories with students as interviewers is the *Oral History Project: A Culture and Heritage Exchange Initiative* (*oralhistory.minds.tv/*). According to the site, "the objective of the Oral History Project is to build understanding between generations, help students master and produce interactive multimedia content (including online streaming videos) and provide the results of their work to the public as a whole." Since the last posting was in summer 2005 and only one of the four sections, Veteran Stories, is populated with interviews, it was not clear to us if the site is still being actively maintained. The approximately 175 videos, however, illustrate quite clearly the potential (and problems) with novice oral historians.

While the videos in the example above opened in Windows Media Player without difficulty, that was not the case with every site visited while doing research for this article. Created by the University of Southern Mississippi

(USM) Libraries and drawing from the holdings of the USM Center for Oral History, the Civil Rights in Mississippi Digital Archive (*www.lib.usm.edu/~spcol/crda/index.html*) makes available transcripts of 145 oral histories of individuals involved in the civil rights movement. Civil rights leaders, such as Charles Cobb, Charles Evers, Aaron Henry, and Hollis Watkins, are included, as are "race-baiting" governor Ross Barnett, national White Citizens Council leader William J. Simmons, and State Sovereignty head Erle Johnston. Audio excerpts are available for six of the interviews. In one of the excerpts, for example, researchers hear Edward L. McDaniel, who was instrumental in organizing the Ku Klux Klan in 1962 in Mississippi, explain how he was brought into the Ku Klux Klan. The audio is delivered using an IBM HotMedia™ Java applet, and no player software is required on your computer. Unfortunately, however, we could not get the audio files to play on a Macintosh computer.

In some cases, oral histories and/or contextual materials may be incorporated into larger, institution-wide online catalogs. Photographs and manuscripts related to the Civil Rights oral histories, such as FBI file #157-333 regarding the arrest of freedom riders on July 14, 1961, and a transcription of the diary of Dean Hay, a Presbyterian minister from Nebraska, in which he details his trip to Hattiesburg, Mississippi, in February 1964, can be accessed through USM's Digital Archive (*aquila.lib.usm.edu/cdm4/browse.php*). It is important to keep in mind that histories may exist online in several locations, even within the same institution. The Web site of the Regional Oral History Office (ROHO) at the University of California, Berkeley (*bancroft.berkeley.edu/ROHO/index.html*) clearly states that histories from its collection can be found in the Online Archive of California (OAC) and additional ones can be found as PDFs on the office's main site. If we had searched OAC first, we are not sure we would have found the additional ROHO oral histories easily.

Of all of the other oral history sites we visited, the Hanashi Oral History Video Archive (*www.goforbroke.org/oral_histories/oral_histories_video.asp*) was the only one that required registration. Once registered though, one can access the complete collection of 275 full-length video interviews of Nisei WWII veterans. Unregistered users can view 39 sample video clips, read historical background materials, experience interactive maps, and explore the Virtual Veteran Experience, which chronicles the lives of five veterans each quarter, incorporating Hanashi oral histories, archival and personal photographs, and texts and transcripts.

For a more upbeat collection, turn to the jazz holdings of the Schomburg Center for Research in Black Culture, one of four major research centers of the New York Public Library. In the Video Oral History Gallery (*www.nypl.*

org/research/sc/scl/MULTIMED/JAZZHIST/jazzhist.html), twenty-one of the video interviews are available either in part or in full. Transcripts are available for the clips, and streaming full interviews are available for Natt Adderley, Chico Hamilton, Marian McPartland, and Arthur Taylor. The video quality was relatively poor, certainly much poorer than the clips on the Regional Oral History Office site (*bancroft.berkeley.edu/ROHO/collections/av_online.html*), but there are no substitutes for hearing a musician explain his craft with instrument in hand.

While we have focused on oral histories that offer online transcripts and/or multimedia presentations, we came across many project sites that list only the names of interviewees or in very general terms describe their institution's holdings. The New York Public Library's Oral History Project and Archive, for example, has conducted more than 500 interviews with people in the dance community. Researchers can view a list of participants at *www.nypl.org/research/lpa/dan/background.htm* and can search for oral histories in the online catalog, but no actual materials are available online. The same is true for the Oral History Interviews at the Niels Bohr Library at the American Institute of Physics (*www.aip.org/history/ohilist/*). As it becomes easier and less expensive to digitize audiovisual materials, in particular, and as access to high bandwidth continues to increase, it is likely that more programs will place materials online. The New York Public Library's Web site, for example, states there are plans to digitize the entire corpus of transcripts from its American Jewish Committee Oral History Collection (*www.nypl.org/research/chss/jws/oralhistories2.cfm*) in the near future.

Government

Where once using government records required a trip to the National Archives, the state capitol, or the county seat, a wide variety of materials is now available online. These documents range in date from the colonial records included in the Maryland Online Archives to the twentieth-century records in the federal agencies' Freedom of Information Act electronic reading rooms. While many of the government sites post only the documents, second-party sites, such as the National Security Archive, provide historical essays, and other sites, such as the Salem Witch Trials, incorporate government records with historical literature, photographs, and manuscripts into an online exhibit. Increasingly, multimedia offerings, particularly audio files, are available at sites such as the University of Virginia's Miller Center for Presidential Recordings and the Oyez Supreme Court multimedia site. Government information is also available in some of the newest electronic formats, including podcasting (*www.firstgov.gov/Topics/Reference_Shelf/Libraries/Podcasts.shtml*).

Under the Freedom of Information Act, all federal agencies are required to disclose records requested in writing by any person and must make four distinct categories of materials available electronically as well as in paper format for "public inspection and copying." FOIA applies only to federal agencies and does not create a right of access to records held by Congress, the courts, or by state or local government agencies. While agencies may withhold information pursuant to nine exemptions and three exclusions contained in the statute, these electronic reading rooms can be an important source for documents, covering a wide variety of topics. For a list of the Department of Justice's FOIA reading rooms, see *www.usdoj.gov/04foia/04_2.html*; a list of all other federal FOIA reading rooms can be found at *www.usdoj.gov/04foia/other_age.htm*.

Despite the perception by some that government records are dull, we were pleasantly surprised by the amount and variety of material available. The State Department's FOIA Reading Room (*www.foia.state.gov/SearchColls/colhelp.asp*), for example, includes a collection of telephone conversation transcripts from former secretary of state Henry Kissinger, documents related to the creation of the Central Intelligence Agency (CIA), and, perhaps unexpectedly, a collection of documents related to the 1974–1976 review of Amelia Earhart's disappearance in 1937. The CIA reading room (*www.foia.cia.gov*) contains collections of declassified documents about the Vietnam War (*www.foia.cia.gov/nic_vietnam_collection.asp*), relations with China (1948–1976) (*www.foia.cia.gov/nic_china_collection.asp*), and prisoners of war or soldiers missing in action (*www.foia.cia.gov/pow_mia.asp*), among other topics. The FBI's alphabetical listing of available digital files (*foia.fbi.gov/alpha.htm*) ranges from Louis Armstrong to Columbine High School to Ku Klux Klan.

The reading rooms are typically straightforward to navigate, most arranging the documents into subject areas and with search capabilities. Many of the documents in these reading rooms contain redactions and usually are image files or pdfs without transcriptions or much, if any, contextual information.

The National Security Archive (*www.gwu.edu/~nsarchiv/*) is an "independent non-governmental research institute and library located at The George Washington University in Washington, D.C. The Archive collects and publishes declassified documents acquired through the Freedom of Information Act (FOIA)." The National Security Archive provides free access to declassified records in its "electronic briefing books" (*www.gwu.edu/~nsarchiv/NSAEBB/index.html*), which are arranged into these subject areas: Europe, Latin America, nuclear history, China and East Asia, U.S. intelligence, Middle East and

South Asia, the September 11th sourcebooks, humanitarian interventions, and government secrecy. A typical briefing book, such as The Atomic Bomb and the End of World War II: A Collection of Primary Sources (*www.gwu.edu/~nsarchiv/NSAEBB/NSAEBB162/index.htm*), contains an introductory essay with footnotes followed by links to scanned documents or photographs and a short explanatory text about the primary sources. In addition to its Web site, the National Security Archive has more than twenty microfiche collections available for purchase (may be available in libraries), the majority of which are part of the Digital National Security Archive, a subscription-based database available through ProQuest (*nsarchive.chadwyck.com/*).

Outside of those documents opened via FOIA, there are many government documents available online. Researchers can locate records at the National Archives (*www.archives.gov*) through the Access to Archival Databases (AAD) (*aad.archives.gov/aad/*), including World War II army enlistment records and records for passengers who arrived at the Port of New York during the Irish famine and the Archival Research Catalog (*arcweb.archives.gov/arc/basic_search.jsp*), which is particularly strong for finding photographs and other audiovisual materials, as well as descriptions of textual materials.

Many of the eleven National Archives' Presidential Libraries (*www.archives.gov/presidential-libraries/*) also have primary sources available online. The Truman Presidential Library and Museum (*www.trumanlibrary.org/photos/av-photo.htm*), for example, offers photographs, audio material, political cartoons, and documents. Many of these items are tied to lesson plans or ideas for using them in the classroom. The Franklin D. Roosevelt Presidential Library and Museum (*www.fdrlibrary.marist.edu/online14.html*) also has an extensive collection of online materials, including audio files and transcripts of important speeches, such as "Fear Itself," "New Deal," and "Day of Infamy," as well as documents and photographs. If you can bear to sit through the opening commercials for consumer products, the History Channel (*www.historychannel.com/broadband/home/*) offers an interesting collection of presidential videos and audio files, much of which came from the National Archives. Visitors can watch historical clips of Franklin Roosevelt signing the bill that ended Prohibition, the first nationally televised presidential debate, between Nixon and Kennedy, or Reagan answering reporters' questions about the Strategic Defense Initiative, also known as "Star Wars."

To find audio files of meetings and conversations secretly recorded by Presidents Roosevelt, Truman, Eisenhower, Kennedy, Johnson, and Nixon, visit the flagship site of the University of Virginia's Miller Center's Presidential Recordings Program (*www.whitehousetapes.org/*). Given the nature of the original recordings and the lack of complete transcripts, finding conversations relevant to one's interest can be difficult. The center, however, has created multimedia clips, with transcripts synchronized to the audio, specifically for use in the classroom. The virtual exhibits on civil rights, Vietnam War, and the space race are in synchronized multimedia format as well. In addition to the recordings, access to oral histories, presidential speeches, documents, and secondary sources can be found through the Scripps Library and Multimedia Archive (*millercenter.virginia.edu/scripps/index.html*), also part of the Miller Center.

Executive records are not the only type of government materials available on the Web. With support from the National Science Foundation, the National Endowment for the Humanities, Northwestern University, the M. R. Bauer Foundation, FindLaw, and the law firm of Mayer, Brown Rowe & Maw, the Oyez Supreme Court Multimedia Project (*www.oyez.org/oyez/frontpage*) makes available audio files from the Supreme Court. (Oyez is a term used three times in succession to introduce the opening of a court of law.) According to the Web site, all audio in the Court recorded since 1995 is included, as are selected audio clips from earlier years for such cases as *Roe v. Wade*, *Miranda v. Arizona*, and *United States v. Nixon*. In addition to the audio, much of which is captioned, the site also includes transcripts of arguments and a link to the written opinion.

A very different example of court records is the Salem Witch Trials: Documentary Archive and Transcription Project site (*etext.virginia.edu/salem/witchcraft/*), a part of the University of Virginia's Electronic Text Center, consisting of seventeenth-century court records; contemporary books about witchcraft; literary works by Hawthorne, Longfellow, Whittier, and Freeman; record books; and historical maps of Salem Village, Salem, and Andover. To its credit, the site does include both transcriptions of the court records and scans of the original documents, but the two formats are not directly linked, making it difficult to compare them. The maps display the chronology of the accusations and show how they spread across the towns of Massachusetts Bay. The site also contains other information helpful in exploring the trials, including biographical profiles, a collection of images containing portraits of notable people involved in the trials, pictures of important historic sites, historical paintings and published illustrations taken from nineteenth- and early-twentieth-century literary and historical works. Also included is a syllabus for "Salem Witch Trials in History and Literature" (*cti.itc.virginia.edu/%7Ebcr/relg415_02/syllabus.html*).

Not to be outdone by the federal government, states are also making headway in the digital world. As one of the largest examples, the Archives of Maryland

Online (*aomol.net/html/index.html*) currently provides access to more than 400,000 documents "that form the constitutional, legal, legislative, judicial, and administrative basis of Maryland's government." The documents are both searchable and browseable, and for most of the documents, a scanned image and transcription are available. In addition to constitutional, executive, judicial, and legislative records, the site also contains land, probate, and military records as well as slave-related documents, such as *Slave Statistics of St. Mary's County Maryland* (1864) by Agnes Callum. Entries give date of registration, name of the owner, and, if applicable, name of person acting for the owner. For each enslaved person, the records show his or her name, sex, age, physical condition, term of servitude, date of emancipation, and, if applicable, information on and compensation for military service. Under the "Early State Records" category, are a number of newspapers, including the *Cecil Whig* (1870–1874), *American and Commercial Daily Advertiser* (1802–1807), and *Kent County News* (1965). The quality of the microfilm, published volumes, or original materials varies considerably; some of it is quite poor. By selecting either the jpg or tiff image and then zooming in, however, we were usually able to decipher the text. Unfortunately, some of the volumes require password access; the site states, "this is done because of the file size or format, to honor copyright restrictions or because the files are restricted to staff access only."

On June 9, 2006, the John F. Kennedy Presidential Library and Museum (*www.jfklibrary.org*) announced an initiative to digitize the entire collection of papers, documents, photographs, and audio recordings at the library and make them available on the Web. This and other projects, such as Ron Suskind's site The Price of Loyalty: The Bush Files (*thepriceofloyalty.ronsuskind.com/thebushfiles/*), indicate that the corpus of online historical government information, in all of its variety and complexity, will continue to grow.

Graffiti

Occasionally a professor or a graduate student will embark on an online project meant to be ongoing—added to by each subsequent class of students enrolled in a particular course. These may grow for a number of years, but then additions and updates to the online portion of the projects dwindle and may even stop altogether, leaving the sites to slowly decay as external links begin to fail or technology and design leave them behind. See Virtual Greenbelt (*www.otal.umd.edu/~vg/*) and The Centennial Exhibition of 1876 (*www.history.villanova.edu/centennial/*), for example. While we cannot say why these particular sites appear to be on hiatus, a number of factors typically come into play, including time, budget, the moving on of key individuals, or a change in course syllabus and design.

A project that we hope continues long into the future is the Urban Archives (*urbanarchives.org/index.html*) started by three graduate students at the University of Washington. In collaboration with their undergraduate students, they have taken a mutual interest in studying communication in public spaces and turned it into a project with a Web site (*urbanarchives.org/projects.html*) that presents selected student projects (both undergraduate and graduate). In a move that increases the likelihood of long-term accessibility, the archive has collaborated with the University of Washington Libraries Digital Collections (*content.lib.washington.edu/uaweb/index.html*) to host the collection of, currently, 571 photographs of items across six different categories: ghost signs, yard art, electric signs, found documents, Seattle views, and graffiti, which has its own subcategories. While some of the items in the collection may at first solicit a chuckle—after all, "found documents" is just our way of making garbage important—imagine what we could do today with a collection of photographs of garbage from a hundred years ago! And, although the archive's digital collection may not yet be comprehensive enough to support research projects external to these students, its value will undoubtedly increase as it continues to grow.

While online archives and blackbooks of graffiti are popular among style writers and fans and may be of use to those in American studies (for example, see *aerosoldreams.com/* and *www.graffiti.org*, which contains an extensive list of links to similar sites), another example of a professionally curated collection is the Vietnam Graffiti Project (*www.vietnam.ttu.edu/graffiti/index.htm*). Part of the Virtual Vietnam Archive at Texas Tech University, the project is composed largely of photographs of nearly 400 graffitied bunk canvases from two troop transport ships. Providing an intriguing glimpse into the minds of young men going off to, or returning from, war, the project is one of the more unusual ones on the Web.

Newspapers

Whereas once we blackened our fingertips turning pages of the daily newspaper, or sat hunched and bleary-eyed scrolling through microfilmed pages of historic papers, the end of such days may be in site. Most major newspapers in the United States have Web sites where one can read the daily news and access, sometimes for a fee, articles from the past ten years or so. Many national papers, such as the *Washington Post* (*pqasb.pqarchiver.com/washingtonpost/search.*

html) and the *New York Times* (*pqasb.pqarchiver.com/nytimes/advancedsearch.html*), have partnered with ProQuest to provide free online searching of articles, ads, and listings dating back to 1877 and 1851, respectively, with full-text files available for purchase. ProQuest also sells its Historical Newspaper product (*www.il.proquest.com/products_pq/hnp/*), which contains complete runs of nine newspapers, the earliest of which is the *Hartford Courant*, dating back to 1734 (the *Hartford Courant* is also available piecemeal through *pqasb.pqarchiver.com/courant/advancedsearch.html*). Other dailies, such as the *Oklahoman* (*www.newsok.com/theoklahoman/archives/*, covering 1901 and later), have created their own historic archives, accessible via subscription.

In addition to ProQuest, several commercial providers of primary sources have notable collections available online.[10] With the completion on June 21, 2006, of their digital collection *Early American Newspapers, Series I, 1690–1876,* Readex provides access to some of the earliest American historic newspapers (*www.readex.com/readex/index.cfm?content=96*). This three-part series encompasses more than 1,000 titles, or 4 million digitized pages, dating from 1690 to 1922. Accessible Archives (*www.accessible.com/about.htm*) contains a much smaller collection focused on slavery and the Civil War that includes an abolitionist paper, *The Liberator*; a selection of nineteenth-century African American newspapers; the *Pennsylvania Gazette*; and Civil War–era articles from three newspapers.[11] Paper of Record (*www.paperofrecord.com/*) emphasizes historic Canadian newspapers and contains more than 8 million digitized pages. The site's international collection includes twenty-eight U.S. titles and selections from Mexico and Latin America. Most titles go back to the mid-nineteenth or early twentieth century. *NewspaperArchive.com* has more than 39 million pages of digitized newspapers (more than 900 international titles) online, including three titles from as early as the eighteenth century. This site offers free access to its archive through 1977 for public libraries and K–12 schools, as well as twenty free subject-based archives (*www.newspaperarchive.com/DesktopModules/ViewHtml.aspx?htfile=FreeArchives.htm*) on popular history topics such as the *Titanic*, the Kennedy assassination, Martin Luther King Jr., and the Winter Olympics.

While commercial entities are, at the moment, the primary providers of this type of digitized content, there are significant noncommercial projects available online and more to come on the digital horizon. Over the past few years several institutions have begun digitizing and making historic newspaper collections available online free of charge. In part, this was a result of changes in OCR technology (optical character recognition, the means by which a machine recognizes words in digital pictures, which then renders them searchable) and the development of affordable software solutions. An interesting subset of recently digitized historic newspapers includes college newspapers. See, for example, issues from 1934 to1992 of the *Gettysburgian* (*208.42.237.18/Archive/client.asp?skin=Gettysburg*), the student newspaper of Gettysburg College; Penn State University's digital archive of the *Collegian* (*www.libraries.psu.edu/historicalcollegian/*), 1887–1955; Ithaca College's *Ithacan*, (*www.ithaca.edu/library/archives/ithacan.php*), 1931–2002; and the University of Richmond's *Collegian* (*oncampus.richmond.edu/academics/library/digital/collegian/index.htm*), 1914–2003.

While some public libraries have moved to digitize an individual paper in their collection (see, for example, Brooklyn Public Library's *Brooklyn Daily Eagle* [*www.brooklynpubliclibrary.org/eagle/*], 1841–1902, or Barrington Illinois Library's *Barrington Review* [*contentdm.barringtonarealibrary.org/index.html*], 1914–1930),[12] large-scale grant programs like those administered through the Institute of Museum and Library Services via the Library Services and Technology Act (LSTA)—which was reauthorized in 2003 and earmarks money for each state and territory—have funded several large statewide collaborative projects. Examples include the Historic Missouri Newspaper Project (*newspapers.umsystem.edu*), which currently has a small selection from fourteen newspapers available online; Utah Digital Newspapers (*www.lib.utah.edu/digital/unews/*), which has about 500,000 pages available online,[13] taken from approximately forty newspapers ranging in dates from the mid-nineteenth century to the mid-twentieth century; and Colorado's Historic Newspaper Collection (*www.coloradohistoricnewspapers.org*), which contains eighty-six newspapers published in Colorado from 1859 to 1928 totaling 291,000 digitized pages. In addition to the archive, CHNC also makes educational materials and a guide for searching historical newspapers—which includes a discussion of searching around potential OCR errors—available online.

Many of the online newspapers discussed above have been digitized from microfilm originally produced by the commercial entities now providing digital content, often for projects funded through the National Endowment for the Humanities' (NEH) United States Newspaper Program (USNP), begun in 1982. With the USNP scheduled for completion in 2007, the NEH has now partnered with the Library of Congress to create the National Digital Newspaper Program (*www.loc.gov/ndnp/*). Begun in 2004, the program developed an initial set of standards and guidelines for the digitization of historic newspapers. The first six grants were awarded in 2005 as part of a two-year pilot program to digitize, per grantee, 100,000 pages of papers limited to 1900 through 1910. In addition to permanently hosting the digitized collections, as part of this initial phase, LOC will also produce a selection of digitized papers published in Washington, D.C., during that time period.

The first of these test bed projects is now available online. See, for example, the University of Kentucky (*www.uky.edu/Libraries/NDNP/*) (to search the digitized *Bourbon News* from 1900 to1910, go to *kdl.kyvl.org/cgi/t/text/text-idx?;page=simpleext&c=kynews*) or the University of California Riverside's California Newspaper Project (*cdnc.ucr.edu/*), which currently makes available the first few years of the *San Francisco Call.* NEH envisions this project continuing for the next twenty years, with grantees in every state and territory digitizing papers from 1836 to 1922. Why those dates? While anyone following the copyright wars will recognize 1922 for its significance—1923 marks the boundary for material in the public domain—1836 might be a bit more elusive because it has to do with current OCR technology and its inability to recognize words in the fonts typically used by colonial newspapers published prior to that date.[14] With the NEH's and LOC's leadership in this area there is much to look forward to in the digiverse of freely available historic newspapers!

Notes

All websites mentioned in this article were accessed between January 20 and July 3, 2006.

1. Randy Bass, "Can American Studies Find a Whole in the Net?" www.georgetown.edu/crossroads/guide/asins96.html (accessed July 3, 2006). Reprinted from *American Studies in Scandinavia* (Fall 1996), Odense University Press.
2. There are a number of available texts devoted to teaching with technology. See, for example, the following resources: the March 2003 issue of the *Journal of American History* (vol. 89, no. 4), which contains a section of articles devoted to using technology to teach American history; The American Studies Crossroads Project, *www.georgetown.edu/crossroads/index.html*, which contains numerous resources under the heading "Technology and Learning"; History Matters: The U.S. Survey Course on the Web, *www.historymatters.gmu.edu*; and articles from the *Journal of the Association for History and Computing*, online at *mcel.pacificu.edu/jahc/JAHCindex.HTM.*
3. Jerome McGann, "Culture and Technology: The Way We Live Now, What Is to Be Done?" *New Literary History* 36.1 (Winter 2005): 71–82.
4. Randy Bass, "Re: Q re. Crossroads for Sept AQ," personal e-mail, July 3, 2006.
5. Patricia Nelson Limerick, "Insiders and Outsiders: The Borders of the USA and the Limits of the ASA: Presidential Address to the American Studies Association, October 31, 1996," *American Quarterly* 49.3 (Fall 1997): 449–69.
6. For example, David Phillips's American Studies Web, created in 1994 and maintained on Crossroads, or David Silver's Resource Center for Cyberculture Studies, created in 1996 and currently available through the University of Washington at *www.com.washington.edu/rccs/* (accessed July 3, 2006).
7. For an example of a topical Webliography, see "An Online Bibliography for the Study of Woman Suffrage," at *www.historycooperative.org/journals/ht/37.2/sparacino.html.* For an example of a print resource, see *The History Highway: A Twenty-First-Century Guide to Internet Resources* (Armonk, N.Y.: M. E. Sharpe, 2006), ed. Dennis Trinkle and Scott A. Merriman.
8. For a discussion of textbook e-supplements, see David Jaffee, "Scholars Will Soon Be Instructed through the Eye: E-Supplements and the Teaching of U.S. History," *Journal of American History* 89.4 (March 2003), online at *www.historycooperative.org/journals/jah/89.4/jaffee.html* (accessed July 3, 2006).

9. For example, see a small online collection of banner art at *http://www.bannerart.org/* (accessed July 3, 2006).
10. For a review of ProQuest, Readex, and Accessible Archives, please see Gail Golderman and Bruce Connolly, "Old News," *Library Journal* 130 (Fall 2005): 22–30.
11. Newspapers from the Civil War, a popular historical topic, are also available at *www.libraries.psu.edu/digital/newspapers/civilwar/* and the well-known Valley of the Shadow, *valley.vcdh.virginia.edu/* (accessed July 3, 2006).
12. Historic newspapers collections are popular choices for public libraries to digitize due, in part, to their appeal to family historians and genealogists.
13. John Herbert, personal e-mail: "FW: [mlib-digfeedback] Qre. pages available for dig newspapers," July 5, 2006.
14. For an interesting discussion of the limits of OCR technology in recognizing colonial era text, see Katherine Stebbins McCaffrey, "American Originals," *Common-Place: The Interactive Journal of Early American Life* 3.3 (April 2003), online at *common-place.dreamhost.com//vol-03/no-03/mccaffrey/mccaffrey-2.shtml* (accessed July 3, 2006).

Links to Digital Resources in American Studies

Ⓣ Ⓟ Ⓒ ⓘ—Teaching, Primary Source, Community, Interpretation

ADVERTISING (including print and TV)

- Ad Land Ⓟ Ⓒ
  http://ad-rag.com/
- Ad*Access Ⓟ
  http://scriptorium.lib.duke.edu/adaccess/
- Banner Art Collective Ⓟ Ⓒ
  http://www.bannerart.org/
- The Commercial Closet Ⓟ Ⓒ ⓘ
  http://www.commercialcloset.org/
- Emergence of Advertising in America Ⓟ
  http://scriptorium.lib.duke.edu/eaa/index.html
- The Living Room Candidate Ⓣ Ⓟ ⓘ
  http://livingroomcandidate.movingimage.us/index.php
- Outdoor Advertising Association of America Creative Library Ⓟ
  http://www.oaaa.org/creativelibrary/
- The Trade Card Place Ⓟ ⓘ
  http://www.tradecards.com/

AMERICAN LITERATURE

Many of the largest collections of American literature, including poetry, drama, and periodicals, are available through fee-based services such as Chadwyk-Healy, ProQuest, Readex, Alexander Street Press, Accessible Archives, and Thomson Gale.

- African American Women Writers of the 19th Century Ⓟ
  http://digital.nypl.org/schomburg/writers_aa19/
- Brogan, Martha, with assistance from Daphnée Rentfrow. *A Kaleidoscope of Digital American Literature.* Washington, D.C.: Digital Library Federation and Council on Library and Information Resources, 2005. (printonly)
- Dime Novels and Penny Dreadfuls Ⓟⓘ
  http://www-sul.stanford.edu/depts/dp/pennies/home.html
- Early American Fiction Ⓟ
  http://etext.virginia.edu/eaf/
- Electronic Literature Directory Ⓟ🌎
  http://directory.eliterature.org/index.php
- Electronic Text Center, University of Virginia Ⓟ
  http://etext.lib.virginia.edu/
- Making of America Ⓟ
  U of MI: *http://www.hti.umich.edu/m/moagrp/*
  Cornell: http://cdl.library.cornell.edu/moa/
- North Carolina History and Fiction Digital Library 📚Ⓟⓘ
  http://digital.lib.ecu.edu/historyfiction/
- UbuWeb Ⓟ🌎
  http://www.ubu.com/
- The Walt Whitman Archive 📚Ⓟ🌎
  http://www.whitmanarchive.org/
- Wright American Fiction Ⓟ
  http://www.letrs.indiana.edu/web/w/wright2/

BORN DIGITAL

- 2004 Presidential Term Web Harvest Ⓟ
  http://www.webharvest.gov/collections/peth04/
- Archive-It Ⓟ
  http://www.archive-it.org/
- Archiving the Web: Virginia's Political Landscape, Fall 2005 Ⓟ
  http://www.lva.lib.va.us/whatwehave/webarchive/index.htm
- Digital History: A Guide to Gathering, Preserving, and Presenting the Past on the Web
  http://chnm.gmu.edu/digitalhistory/
- International Internet Preservation Consortium
  http://www.netpreserve.org/about/index.php
- Internet Archive Ⓟ
  http://www.archive.org/index.php
- Minerva Web Preservation Project (Mapping the Internet Electronic Resources Virtual Archive) Ⓟ
  http://lcweb2.loc.gov/cocoon/minerva/html/minerva-home.html
- September 11th Digital Archive: Saving the Histories of September 11, 2001 Ⓟⓘ
  http://911digitalarchive.org/

BUILT ENVIRONMENT

- Architecture in the Parks: A National Historic Landmark Survey Ⓟⓘ
  http://www.cr.nps.gov/history/online_books/harrison/harrison0.htm
- Archive of American Architecture Ⓟⓘ
  http://www.bc.edu/bc_org/avp/cas/fnart/fa267/
- Cities/Buildings Database Ⓟ
  http://www.washington.edu/ark2/
- City Sites (New York and Chicago) Ⓟⓘ
  http://artsweb.bham.ac.uk/citysites/
- Code City Ⓟⓘ
  http://www.tenement.org/codecity/

- Curating the City: Wilshire Blvd.
  http://www.curatingthecity.org/
- Historic American Buildings Survey/Historic American Engineering Record (HABS/HAER) Collections
  http://memory.loc.gov/ammem/collections/habs_haer/
- Historic Pittsburgh
  http://digital.library.pitt.edu/pittsburgh/
- Monticello Explorer
  http://explorer.monticello.org/
- Triangle Factory Fire
  http://www.ilr.cornell.edu/trianglefire/
- Virtual Greenbelt
  http://www.otal.umd.edu/~vg/
- Within These Walls
  http://americanhistory.si.edu/house/

EPHEMERA

- Curt Teich Postcard Archives Ⓟ
  http://www.lcfpd.org/teich_archives/index.cfm?fuseaction=home.view
- Digital Dress: Historic Costume Collection Ⓟ
  http://dlxs.lib.wayne.edu/cgi/i/image/image-idx?xc=1;page=searchgroup g=costumegroupic
- Dr. Seuss Went to War: A Catalog of Political Cartoons by Dr. Seuss Ⓟ
  http://orpheus.ucsd.edu/speccoll/dspolitic/
- Feeding America: The Historic American Cookbook Project Ⓟ
  http://digital.lib.msu.edu/projects/cookbooks/index.html
- Nursery and Seed Trade Catalog Image Gallery Ⓟ
  http://www.nal.usda.gov/speccoll/seedcatalogimagegallery.shtml
- "A Summons to Comradeship": World War I and II Posters and Postcards Ⓟ
  http://digital.lib.umn.edu/warposters/warpost.html
- Texas Tides Ⓟ
  http://tides.sfasu.edu/home.html

- Visual Materials of American Immigration and Ethnic History Ⓟ
  http://www.ihrc.umn.edu/collage/index.htm

- World Trade Organization History Project ⓅⒾ
  http://depts.washington.edu/wtohist/

GOVERNMENT

- Archives of Maryland Online Ⓟ
  http://aomol.net/html/index.html

- FBI's alphabetical listing of available digital files Ⓟ
  http://foia.fbi.gov/alpha.htm

- Foreign Relations of the United States Online Volumes ⓅⒾ
  http://www.state.gov/r/pa/ho/frus/c1716.htm

- Freedom of Information Act sites
  - Department of Justice's FOIA Reading Rooms **Portal**
    http://www.usdoj.gov/04foia/04_2.html
  - All other federal FOIA Reading Rooms **Portal**
    http://www.usdoj.gov/04foia/other_age.htm
  - State Department's FOIA Reading Room Ⓟ
    http://www.foia.state.gov/SearchColls/colhelp.asp
  - CIA FOIA Reading Room Ⓟ
    http://www.foia.cia.gov

    Vietnam War *http://www.foia.cia.gov/nic_vietnam_collection.asp*
    Relations with China (1948–1976) *http://www.foia.cia.gov/nic_china_collection.asp*
    Prisoners of War or soldiers Missing in Action *http://www.foia.cia.gov/pow_mia.asp*

- History Channel's speeches and videos ⓅⒾ
  http://www.historychannel.com/broadband/home/

- National Security Archive ⓅⒾ
  http://www.gwu.edu/~nsarchiv/
  - National Security Archive Electronic Briefing Books
    http://www.gwu.edu/~nsarchiv/NSAEBB/index.html
  - The Atomic Bomb and the End of World War II: A Collection of Primary Sources
    http://www.gwu.edu/~nsarchiv/NSAEBB/NSAEBB162/index.htm

    - Digital National Security Archive, a subscription-based database through ProQuest.
      http://nsarchive.chadwyck.com/

- National Archives and Records Administration 📚Ⓟ
  http://www.archives.gov
    - Access to Archival Databases (AAD) *http://aad.archives.gov/aad/*
    - Archival Research Catalog *http://arcweb.archives.gov/arc/basic_search.jsp*
    - Presidential Libraries *http://www.archives.gov/presidential-libraries/*
    - Truman Presidential Library and Museum Online Materials
      http://www.trumanlibrary.org/photos/av-photo.htm
    - The Franklin D. Roosevelt Presidential Library and Museum
      http://www.fdrlibrary.marist.edu/online14.html

- Oyez Supreme Court Multimedia Ⓟⓘ
  http://www.oyez.org/oyez/frontpage

- Podcasts by Federal Agencies **Portal**
  http://www.firstgov.gov/Topics/Reference_Shelf/Libraries/Podcasts.shtml

- Presidential Recordings Program Ⓟ
  http://www.whitehousetapes.org/
    - The Scripps Library and Multimedia Archive *http://millercenter.virginia.edu/scripps/index.html*

- The Price of Loyalty: The Bush Files Ⓟⓘ
  http://thepriceofloyalty.ronsuskind.com/thebushfiles/

- Salem Witch Trials: Documentary Archive and Transcription Project 📚Ⓟ
  http://etext.virginia.edu/salem/witchcraft/
    - Syllabus for Salem Witch Trials in History and Literature
      http://cti.itc.virginia.edu/%7Ebcr/relg415_02/syllabus.html

GRAFFITI

- Aerosol Dreams Ⓟ🌎
  http://aerosoldreams.com/

- Graffiti.org Ⓟ🌎
  http://www.graffiti.org/

- Urban Archives 📚Ⓟⓘ
  http://urbanarchives.org/index.html

- Vietnam Graffiti Project Ⓟ ⓘ
  http://www.vietnam.ttu.edu/graffiti/index.htm

INDEXES

- ArchivesUSA (fee-based)
  http://archives.chadwyck.com/
- Best of History Web Sites
  http://www.besthistorysites.net/
- Digital History
  http://www.digitalhistory.uh.edu/
- Edsitement
  http://edsitement.neh.gov/
- History Matters
  http://historymatters.gmu.edu/
- IMLS Digital Collections Registry
  http://imlsdcc.grainger.uiuc.edu/
- Librarian's Internet Index
  http://lii.org/
- Repositories of Primary Sources
  http://www.uidaho.edu/special-collections/Other.Repositories.html
- UNESCO Archives Portal
  http://www.unesco.org/cgi-bin/webworld/portal_archives/cgi/page.cgi?d=1
- University of Iowa Online Communication Studies Resource
  http://www.uiowa.edu/~commstud/resources/
- Voice of the Shuttle
  http://vos.ucsb.edu/
- WWW-VL
  http://vlib.org/

LARGE MULTICOLLECTIONS

- American Memory, Library of Congress Ⓟ ⓘ
  http://memory.loc.gov/ammem/index.html

- Documenting the American South Ⓟⓘ
  http://docsouth.unc.edu/index.html
- Duke University Ⓟⓘ
  http://library.duke.edu/specialcollections/collections/digitized/index.html
- New York Public Library Digital Gallery Ⓟⓘ
  http://www.nypl.org/digital/index.htm
- Online Archive of California Ⓟ
  http://www.oac.cdlib.org/

MAPS

- BlockView Images Ⓟ
  http://maps.a9.com/
- Historic Maps in K–12 Classrooms Ⓟⓘ
  http://www3.newberry.org/k12maps/
- Library of Congress, American Memory Map Collections Ⓟⓘ
  http://memory.loc.gov/ammem/browse/ListSome.php?category=Maps
- Library of Congress, Learning Page for Map Collections Ⓟ
  http://memory.loc.gov/ammem/ndlpedu/collections/map/
- Library of Congress, Zoom into Maps, LOC Ⓟ
  http://lcweb2.loc.gov/ammem/ndlpedu/features/maps/introduction.html
- Map History/History of Cartography: The Gateway to the Subject **Portal**
  http://www.maphistory.info
- Maps of the French and Indian War by the Massachusetts Historical Society Ⓟⓘ
  http://www.masshist.org/maps/MapsHome/Home.htm
- MapTech Historical Maps Ⓟ
  http://historical.maptech.com/
- National Atlas Ⓟⓘ
  http://nationalatlas.gov/
- The Office of Coast Survey's Historical Map & Chart Collection, National Oceanic and Atmospheric Administration Ⓟ
  http://nauticalcharts.noaa.gov/csdl/ctp/abstract.htm

- The David Rumsey Collection, Cartography Associates Ⓟ
  http://www.davidrumsey.com/index.html
- Sanborn Maps of Charlottesville, 1907 and 1920 Ⓟⓘ
  http://fisher.lib.virginia.edu/collections/maps/sanborn/index.html
- Making Sense of Maps, History Matters
  http://historymatters.gmu.edu/mse/maps/
- Worldmapper Ⓟ
  http://www.sasi.group.shef.ac.uk/worldmapper/index.html

NEWSPAPERS

- Accessible Archives Ⓟ
  http://www.accessible.com/about.htm
- *Barrington Review* (Illinois) Ⓟ
  http://contentdm.barringtonarealibrary.org/index.html
- *Brooklyn Daily Eagle* Ⓟ
  http://www.brooklynpubliclibrary.org/eagle/
- *The Collegian* (Penn State University) Ⓟ
  http://www.libraries.psu.edu/digital/newspapers/historicalcollegian/
- *The Collegian* (University of Richmond) Ⓟ
  http://oncampus.richmond.edu/academics/library/digital/collegian/index.htm
- Colorado's Historic Newspaper Collection Ⓟ
  http://www.coloradohistoricnewspapers.org
- *Early American Newspapers, Series I, 1690–1876* Ⓟ
  http://www.readex.com/readex/index.cfm?content=96
- *The Gettysburgian* Ⓟ
  http://208.42.237.18/Archive/client.asp?skin=Gettysburg
- *The Hartford Courant* Ⓟ
  http://pqasb.pqarchiver.com/courant/advancedsearch.html
- Historic Missouri Newspaper Project Ⓟ
  http://newspapers.umsystem.edu

- Historical Newspaper (ProQuest) Ⓟ
  http://www.il.proquest.com/products_pq/hnp/
- *The Ithacan* (Ithaca College) Ⓟ
  http://www.ithaca.edu/library/archives/ithacan.php
- National Digital Newspaper Program Ⓟ🌎ⓘ
  http://www.loc.gov/ndnp/
  - University of Kentucky *http://www.uky.edu/Libraries/NDNP/*
  - California Newspaper Project *http://cdnc.ucr.edu/*
- *The New York Times* Ⓟ
  http://pqasb.pqarchiver.com/nytimes/advancedsearch.html
- Newspaper Archive Ⓟ
  http://www.newspaperarchive.com
  - 20 free subject-based archives are listed at *http://www.newspaperarchive.com/DesktopModules/ViewHtml.aspx?htfile=FreeArchives.htm*
- *The Oklahoman* Ⓟ
  http://www.newsok.com/theoklahoman/archives/
- Paper of Record Ⓟ
  http://www.paperofrecord.com/
- Utah Digital Newspapers Ⓟ
  http://www.lib.utah.edu/digital/unews/
- *The Washington Post* Ⓟ
  http://pqasb.pqarchiver.com/washingtonpost/search.html

ORAL HISTORY

- The 20th Century: An Oral History of Monmouth County Ⓟ
  http://www.visitmonmouth.com/oralhistory/
- American Jewish Committee Oral History Collection
  http://www.nypl.org/research/chss/jws/oralhistories2.cfm
- Black History Project at Virginia Tech Ⓟ
  http://spec.lib.vt.edu/archives/blackhistory/oralhistory/
- Civil Rights in Mississippi Digital Archive Ⓟ
  http://www.lib.usm.edu/~spcol/crda/index.html
  - Digital Archive *http://aquila.lib.usm.edu/cdm4/browse.php*

- Hanashi Oral History Video Archive Ⓟ ⓘ
  http://www.goforbroke.org/oral_histories/oral_histories_video.asp

- Making Sense of Oral History, History Matters
  http://historymatters.gmu.edu/mse/oral/online.html

- New York Public Library's Oral History Project and Archive
  http://www.nypl.org/research/lpa/dan/background.htm

- Oral History Interviews at the Niels Bohr Library
  http://www.aip.org/history/ohilist/

- Oral History Project: A Culture and Heritage Exchange Initiative Ⓟ
  http://oralhistory.minds.tv/

- Regional Oral History Office Ⓟ
  (http://bancroft.berkeley.edu/ROHO/index.html)

- Rutgers Oral History Archive Ⓟ
  http://oralhistory.rutgers.edu/

- Step-by-Step Guide to Oral History
  http://www.dohistory.org/on_your_own/toolkit/oralHistory.html

- Veterans History Project Ⓟ ⓘ
  http://www.loc.gov/vets/
  - themes *http://www.loc.gov/vets/stories/themes.html*
  - suggestions for students *http://www.loc.gov/vets/youth-resources.html*

- Video Oral History Gallery Ⓟ
  http://www.nypl.org/research/sc/scl/MULTIMED/JAZZHIST/jazzhist.html

- Voices of the Colorado Plateau Ⓟ
  http://archive.li.suu.edu/voices/

PERSONAL PAPERS

- American Journeys Ⓟ ⓘ
  http://www.americanjourneys.org

- Battle Lines: Letters from America's Wars Ⓟ ⓘ
  http://www.gilderlehrman.org/collection/battlelines/index_good.html

- Camping with the Sioux: Fieldwork Diary of Alice Cunningham Fletcher Ⓟ ⓘ
  http://www.nmnh.si.edu/naa/fletcher/

- Dear Miss Breed: Letters from Camp . . . ⓟⓘ
  http://www.janm.org/exhibits/breed/title.htm
- Thomas A. Edison ⓟⓘ
  http://edison.rutgers.edu/index.htm
- Thomas Jefferson ⓟⓘ
  http://memory.loc.gov/ammem/collections/jefferson_papers/
- In the First Person: An index to letters, diaries, oral histories, and personal narratives **Portal**
  http://www.inthefirstperson.com/firp/index.aspx
- Letters of Delegates to Congress ⓟ
  http://memory.loc.gov/ammem/amlaw/lwdg.html
- Letters to Doct. James Carmichael & Son ⓟⓘ
  http://carmichael.lib.virginia.edu
- Making Sense of Letters and Diaries, History Matters
  http://historymatters.gmu.edu/mse/letters/
- Sterling Family Papers, University of Maryland ⓟⓘ
  http://www.lib.umd.edu/dcr/collections/sterling/index.html
- Territorial Kansas Online, 1854–1861 ⓟⓘ
  http://www.territorialkansasonline.org/cgiwrap/imlskto/index.php

PHOTOGRAPHS

- Analyzing Photographs, History Matters
  http://historymatters.gmu.edu/mse/sia/photo.htm
- Charles W. Cushman Photograph Collection ⓟⓘ
  http://webapp1.dlib.indiana.edu/cushman/index.jsp
- Early Visual Media ⓟⓘ
  http://www.visual-media.be/visualmedia.html
- Teenie Harris ⓟ
  http://www.cmoa.org/teenie/info.asp
- Image Archive of the American Eugenics Movement ⓟⓘ
  http://www.eugenicsarchive.org/eugenics/

- Frances B. Johnston Photograph Collection Ⓟ
  http://louisdl.louislibraries.org/FJC/Pages/home.html
- Library of Congress Prints and Photographs Online Catalog Ⓟ [includes additional material not in American Memory]
  http://www.loc.gov/rr/print/catalog.html
- Library of Virginia, Digitized Photo Collections Ⓟ
  http://www.lva.lib.va.us/whatwehave/photo/index.htm
- Making Sense of Documentary Photography, History Matters
  http://historymatters.gmu.edu/mse/Photos/
- National Library of Medicine Ⓟ ⓘ
  - Images from the History of Medicine
    http://wwwihm.nlm.nih.gov/
  - History of Medicine Online Syllabus Archive
    http://www.nlm.nih.gov/hmd/collections/digital/syllabi/index.html
  - Visual Culture and Public Health Posters
    http://www.nlm.nih.gov/exhibition/visualculture/vchome.html
- Picturing Modern America Ⓟ ⓘ
  http://www.edc.org/CCT/PMA
- Photography Collection of the Denver Public Library Ⓟ
  http://www.photoswest.org
- San Francisco Historical Photograph Collection, San Francisco Public Library Ⓟ
  http://sflib1.sfpl.org:82/screens/browse.htm
- U.S. Gary Steel Works Photograph Collection Ⓟ ⓘ
  http://www.dlib.indiana.edu/collections/steel/

Contributors

Carolyn de la Peña

Carolyn de la Peña is an associate professor of American studies at the University of California, Davis. Her research and teaching interests focus on popular technology and science, material and business cultures, and food. De la Peña's current project is a cultural history of artificial sweeteners in the United States. She has written previously on the machines of Krispy Kreme, the video immersion environments of Prada Soho, and the I-ON-A-CO electric belt of Henry Gaylord Wilshire. Her book *The Body Electric: How Strange Machines Built the Modern American* was published in 2003.

Debra DeRuyver

Debra DeRuyver is director of membership for the International Leadership Association. She is ABD in American studies from the University of Maryland, where she taught courses on electronic publications and virtual exhibitions, online activism, and dance in American culture. She holds a master's degree in American studies from California State University, Fullerton, and a bachelor's degree in English literature from the University of Michigan. She previously worked at the National Archives and Records Administration as a reference archivist. In 1999, DeRuyver was one of four founding editors of the Public History Resource Center (PHRC) (http://www.publichistory.org). She has continued to serve as a managing editor since that time, publishing more than seventy-five reviews written by information scientists or subject specialists. In 2000, PHRC won the National Council on Public History's Student Project Award. She can be reached at editors@publichistory.org.

Joel Dinerstein

Joel Dinerstein is an assistant professor of English at Tulane University, where he also teaches in the American studies program. He is the author of *Swinging the Machine: Modernity, Technology, and African American Culture between the World Wars* (2003), which was awarded the 2004 Eugene M. Kayden Book Prize, given annually to the best book in the humanities published by an American university press. He is presently at work on *The Cool Mask: Jazz, Noir, and Existentialism, 1940–1960*, a book that will analyze the emergence of the concept of cool in American society.

Susan J. Douglas

Susan J. Douglas is the Catherine Neafie Kellogg Professor of Communication Studies at the University of Michigan and chair of the department. She is author of *The Mommy Myth: The Idealization of Motherhood and How It Undermines Women* (with Meredith Michaels, 2004); *Listening In: Radio and the American Imagination* (1999), which won the Hacker Prize in 2000 for the best popular book about technology and culture; *Where the Girls Are: Growing Up Female with the Mass Media* (1994 and 1995); and *Inventing American Broadcasting, 1899–1922* (1987).

Jennifer Evans

Jennifer Evans is an archivist at the National Archives and Records Administration in the Office of Presidential Libraries, where she is responsible for Web site content and development. She holds master's degrees in history and library science from the University of Maryland and earned her bachelor's degree in English literature from Randolph-Macon Woman's College. She previously worked as an archivist in the University of Maryland Libraries' Special Collections Division. In 1999, Evans was one of four founding editors of the Public History Resource Center (PHRC) (http://www.publichistory.org). She has continued to serve as a managing editor since that time, publishing more than seventy-five reviews written by information scientists or subject specialists. In 2000, PHRC won the National Council on Public History's Student Project Award. She can be reached at editors@publichistory.org.

Nicole R. Fleetwood

Nicole R. Fleetwood is assistant professor of American studies at Rutgers University, New Brunswick. She writes and teaches in the fields of media studies, technology studies, gender studies, and theories of race and ethnicity.

Rayvon Fouché

Rayvon Fouché is an associate professor of history and African American studies at the University of Illinois at Urbana-Champaign. As a historian of technology, he explores the multiple intersections and relationships between cultural representation, racial identification, scientific practice, and technological design. He is the author of *Black Inventors in the Age of Segregation* (2005) and coeditor of *Appropriating Technology: Vernacular Science and Social Power* (2004).

Carma R. Gorman

Carma R. Gorman is an associate professor in the School of Art and Design at Southern Illinois University, Carbondale, where she teaches courses on the history of graphic design and industrial design. She has published reviews and articles on early-twentieth-century design in *Design Issues* and *Winterthur Portfolio*; has edited a primary-source anthology, *The Industrial Design Reader* (2003); and, with David Raizman, is coediting a book titled *Literature, Audiences, and Objects: Alternative Narratives in the History of Design*. She is a past president of the Design Studies Forum, an affiliated society of the College Art Association.

Caren Kaplan

Caren Kaplan is director of the Cultural Studies Graduate Group and associate professor in women and gender studies at the University of California, Davis. She is the author of *Questions of Travel: Postmodern Discourses of Displacement* (1996) and the coeditor of *Scattered Hegemonies: Postmodernity and Transnational Feminist Practices* (1994), *Between Woman and Nation: Transnational Feminisms and the State* (1999), and *Introduction to Women's Studies: Gender in a Transnational World* (2001, 2005). Her current research focuses on militarization, transnational consumer culture, and location technologies.

Robert MacDougall

Robert MacDougall is an assistant professor of history and an associate director of the Centre for American Studies at the University of Western Ontario. In 2005, he was a visiting scholar at the American Academy of Arts and Sciences in Cambridge, Massachusetts. He is completing a book on the political and cultural history of the telephone in the United States and Canada.

David E. Nye

David E. Nye is the author or editor of fourteen books that concern technology and American culture, most recently *Technology Matters: Questions to Live With* (2006). He has been a visiting professor at Cambridge, Leeds, Oviedo, Harvard, MIT, Notre Dame, and the Netherlands Institute of Advanced Study. He has held major grants from the NEH, ACLS, Fulbright, Carlsberg Fond, Levehulme Trust, and the Danish Humanities Council. The recipient of a number of prizes and awards, in 2005 he received the Leonard da Vinci Medal, the highest award of the Society for the History of Technology. He is writing a comparative cultural history of electricity in American culture since 1940.

Andrew Ross

Andrew Ross is professor of American studies and director of the Metropolitan Studies Program at New York University. He is the author of several books, including *Fast Boat to China: Corporate Flight and the Consequences of Free Trade–Lessons from Shanghai*; *Low Pay, High Profile: The Global Push for Fair Labor*; *No-Collar: The Humane Workplace and Its Hidden Costs*; and *The Celebration Chronicles: Life, Liberty, and the Pursuit of Property Value in Disney's New Town*. He has also edited several books, including *No Sweat: Fashion, Free Trade, and the Rights of Garment Workers* and, most recently, *Anti-Americanism.*

Ricardo D. Salvatore

Ricardo D. Salvatore is professor of history at Universidad Torcuato Di Tella, Buenos Aires, Argentina. He is the author of *Wandering Paysanos. State Order and Subaltern Experience: Buenos Aires Province during the Rosas Era* (2003) and *Imágenes de un Imperio: Estados Unidos y las formas de reprsentación en América Latina* (2006). He has recently edited *Culturas Imperiales: Experiencia y representación en América, Asia y Africa* (2005). He has coedited four books: *The Birth of the Penitentiary in Latin America* (1996), with Carlos Aguirre; *Close Encounters of Empire: Writing the Cultural History of U.S.–Latin American Relations* (1998), with Gilbert Joseph and Catherine LeGrand; *Caudillismos Rioplatenses. Nuevas miradas a un viejo problema* (1998), with Noemí Goldman; and *Crime and Punishment in Latin America: Law and Society since Colonial Times* (2001). Professor Salvatore has published widely about social control, criminology, peasant-state relationships, market culture, and economic welfare. He is working on a manuscript titled "Imperial Knowledge: U.S. Strategies for Knowing Latin America, 1890–1945," an attempt to understand the cultural representations of the United States' informal empire in Latin America during the era of Pan-Americanism.

Felicity Schaeffer-Grabiel

Felicity Schaeffer-Grabiel is an assistant professor in the Department of Feminist Studies at the University of California, Santa Cruz. She is working on a manuscript titled, "Cyberbrides across the Americas: Transnational Imaginaries, Marriage, and Migration." Her published articles include "Planet-Love.com: Cyberbrides in the Americas and the Transnational Routes of U.S. Masculinity" (2006) and "Cyber-brides and Global Imaginaries: Mexican Women's Turn from the National to the Foreign"(2004).

Sunny Stalter

Sunny Stalter is a PhD candidate in English at Rutgers University, New Brunswick. Her dissertation examines the role of the subway in New York modernist literature and culture.

Julie Sze

Julie Sze is an assistant professor of American studies at the University of California, Davis. Her research focuses on race, class, gender, and the environment; the culture and politics of the environmental justice movement; and urban environmentalism and environmental health. Sze has worked on environmental justice issues and with community-based organizations nationally for the past decade. Her forthcoming book, *Noxious New York: The Racial Politics of Urban Health and Environmental Justice* (MIT Press, 2006) examines environmental justice activism in New York City, asthma politics, and garbage and energy policy in the age of globalization and privatization.

Siva Vaidhyanathan

Siva Vaidhyanathan, a cultural historian and media scholar, is the author of *Copyrights and Copywrongs: The Rise of Intellectual Property and How it Threatens Creativity* (2001) and *The Anarchist in the Library: How the Clash between Freedom and Control is Hacking the Real World and Crashing the* System (2004). Vaidhyanathan is currently an associate professor of Culture and Communication at New York University and a fellow at the New York Institute for the Humanities. He writes and edits a Weblog called sivacracy.net.

Caitlin Zaloom

Caitlin Zaloom is assistant professor of American studies and metropolitan studies in the Department of Social and Cultural Analysis at New York University. She is the author of *Out of the Pits: Traders and Technology from Chicago to London* (2006). Her current projects explore new dimensions of American capitalism in the emerging science of neuroeconomics and in the politics and culture of personal debt.

Index

Printed in the United States
74227LV00004B/15